# Biological Control of Invasive Plants
# in the United States

Publication of this book was made possible in part by a grant from the Western Society of Weed Science in its continuing efforts to promote weed science educational activities and foster cooperation among state, federal, provincial, and private agencies in matters of integrated weed management.  The Oregon State University Press and the editors gratefully acknowledge this support.

# Biological Control of Invasive Plants in the United States

*editors:*

    *Eric M. Coombs*

    *Janet K. Clark*

    *Gary L. Piper*

    *Alfred F. Cofrancesco, Jr.*

Oregon State University Press
Corvallis

This material was made possible, in part, by a Cooperative Agreement from the United States Department of Agriculture Animal and Plant Health Inspection Service Plant Protection and Quarantine (USDA-APHIS-PPQ), and National Biological Control Institute (NBCI). It may not necessarily express APHIS' views. Additional funding was provided by the Center for Invasive Plant Management (CIPM) at Montana State University, Bozeman.

The paper in this book meets the guidelines for permanence and durability of the Committee on Production Guidelines for Book Longevity of the Council on Library Resources and the minimum requirements of the American National Standard for Permanence of Paper for Printed Library Materials Z39.48-1984.

**Library of Congress Cataloging-in-Publication Data**
Biological control of invasive plants in the United States / editors, Eric M. Coombs ... [et al.].
　　p. cm.
　Includes bibliographical references and index.
　ISBN 0-87071-029-X (alk. paper)
　1. Invasive plants—Biological control—United States. 2. Weeds—Biological control—United States. I. Coombs, Eric M.
　SB612.A2B56 2004
　632'.5—dc22

　　　　　　　　　　2004002695

**Oregon State University Press**
102 Adams Hall
Corvallis OR 97331-6407
541-737-3166 • fax 541-737-3170
http://oregonstate.edu/dept/press

**OREGON STATE
UNIVERSITY**

# Dedication

We wish to dedicate this book to the memory of our dear friend and colleague, Barbra Mullin (1950-2002), who was the motivating force behind this project. Through the Montana Department of Agriculture and numerous professional organizations, Barbra spent many years in the service of improving our public and private lands by promoting awareness and implementation of integrated weed control projects.

# Contents

## III. New and Ongoing Biological Control Projects in the United States

# Foreword

Thank you for your interest in biological control of weeds. We have sought advice and input from some of the nation's top experts to provide you with a compilation of information about understanding and safely implementing biological control as part of a long-term, integrated weed management program.

This new publication represents a collaboration of 63 authors representing 10 universities, five federal agencies, four private organizations, and four overseas agencies. As such, this project reflects the multidisciplinary effort required to improve the practice of classical biological control of invasive plants, particularly as we focus more on ecological systems than on discrete organisms.

Section I of this book provides a discussion of the classical biological control process and concludes with an International Code of Best Practices for Classical Biological Control of Weeds. We highly endorse and recommend the Code's guidelines to increase safety and effectiveness and to reduce costs. We do not recommend the use of accidentally introduced species for weed management until they are proven safe and effective under current testing protocols and are sanctioned by the U.S. Department of Agriculture.

Invasive weeds and their approved biological control agents are described in Section II. We have focused this compilation on weeds and control agents of the continental United States. The common names given for agents are often unofficial and ascribed according to the agents' roles in relation to biological control; they have proven to be helpful field descriptions for practitioners. The book concludes with short synopses of other biological control projects in the research stage. Readers can find out more about the progress of these projects by contacting the partners involved or searching the Web.

We express our deep admiration for and acknowledgment of all those pioneers of biological control of weeds who provided the foundation and historical data for this project. This book is based on a previous publication, *Biological Control of Weeds in the West*, published by the Western Society of Weed Science in 1996 and edited by Norman E. Rees, P. C. Quimby, Jr., Gary L. Piper, Eric M. Coombs, Charles E. Turner, Neal R. Spencer, and Lloyd V. Knutson.

Our great appreciation goes to the numerous authors and reviewers who contributed generously to this project, and to the patient staff at Oregon State University Press. We are honored to work with them all. Biological control of weeds is necessarily a global discipline, so we also thank our colleagues and collaborators throughout the world, particularly in Australia, Canada, South Africa, Switzerland, and the United Kingdom. And finally, we thank our families and employers who gave us time and encouragement to work on this collaborative project over the past two years.

Biological control of weeds is a dynamic discipline and will continue to grow and evolve. We welcome readers' updates and suggestions for future editions.

*Eric M. Coombs,* Oregon Department of Agriculture
*Janet K. Clark,* Center for Invasive Plant Management
*Gary L. Piper,* Washington State University
*Alfred F. Cofrancesco, Jr.,* U.S. Army Corps of Engineers

# Introduction

## Ernest S. Delfosse

Biological control of weeds is stronger globally now than ever before. For example, at one of the workshops held during the XI International Symposium on Biological Control of Weeds (Canberra, Australia, 27 April – 2 May 2003), participants were asked about the support for their programs. All participants (representing several countries) indicated that there was more support for biological control of weeds than four years ago, when the X International Symposium on Biological Control of Weeds was held. There are several reasons for this increased support. One of the most important is that land and natural area managers have realized that they have few effective tools other than biological control for managing invasive weeds. Citizens are requesting more programs. Also, customer-led coalitions have been established for specific weeds, such as leafy spurge (*Euphorbia esula*), saltcedar (*Tamarix* spp.), and melaleuca (*Melaleuca quinquenervia*), to help support research that delivers biological control agents.

This book contains three sections that cover many of the important issues in biological control. Section I, *The Theory and Practice of Biological Control*, contains 16 chapters on scientific subjects (ecology; foreign exploration; host specificity testing; plant pathology; release, handling, establishment and monitoring of agents; nontarget impacts; integrated weed management; economics of biological control; and the role of private industry) and administrative procedures (reviewing applications for release of biological control agents, permitting, documentation, and the International Code of Best Practices for Classical Biological Control of Weeds). Section II, *Target Plants and the Biological Control Agents*, discusses 39 target weeds involving release of biological control agents. Section III, *New and Ongoing Biological Control Projects in the United States*, discusses projects for 15 target weeds that are in progress, but for which releases of biological control agents have not been made. This section reflects the strong support for biological control of weeds mentioned above.

For the purposes of this discussion, biological control is defined as the use of live natural enemies of pests to reduce pest population levels below that which would occur in the absence of the natural enemies. Excellent discussions of the definitions and types of biological control are given in standard references such as Barbosa (1998), Bellows and Fisher (1999), DeBach (1964), Hoy and Herzog (1984), and Huffaker and Messenger (1976). Most authors agree on three types of biological control (classical, augmentative, and conservation). Wapshere et al. (1989) added a fourth type, "broad-spectrum biological control," to take account of perhaps the most common type of biological control, the use of grazing animals (Table 1).

It is important to note that biological control is defined by characteristics of how natural enemies are used, and not by the type of target pest. Thus, for classical biological control, normally, small numbers of natural enemies (e.g., insects, mites, nematodes, plant pathogens) are inoculated into field plots for long-term control; for augmentative biological control, large numbers of agents are usually applied for shorter-term control; field (environmental) conditions are modified to favor natural enemy survival in

Table 1. Types of biological control (modified from Wapshere et al. 1989)

| Type | Characteristics |
| --- | --- |
| Classical or inoculative | Exotic, host-specific natural enemies introduced against exotic or native pests |
| | Relatively small numbers of individuals inoculated into field plots |
| | Ecological |
| | Long-term |
| Augmentative or inundative | Native or exotic natural enemies released against native or exotic pests |
| | Large numbers of agents mass-reared and released |
| | Technological |
| | Short-term |
| Conservation | Enhancing or protecting natural enemies |
| | Ecological |
| | Long-term |
| Broad-spectrum | Polyphagous natural enemies used specifically |
| | Small numbers of grazing herbivores confined in limited spaces |
| | Technological |
| | Short-term |

conservation biological control; and in broad-spectrum biological control, grazing herbivores are confined in some way, making it safe to use these polyphagous agents.

Classical biological control is essentially ecological, relying on the inherent characteristics of natural enemies for establishment and spread after release (Table 1). Similarly, conservation biological control relies on ecological manipulation of the environment to enhance or protect natural enemies. Augmentative and broad-spectrum biological control are essentially technological and generally short-term, whereby large numbers of biological control agents are released at a time when they are predicted to have the greatest impact on the target pest. Establishment and self-perpetuation is not a requirement for an augmentative biological control agent, and can be detrimental if the agent is to be used as a commercial product, such as a mycoherbicide (a plant pathogen used augmentatively).

Classical, augmentative, and conservation biological control use host-specific natural enemies, either monophagous (restricted to one host), or acceptably oligophagous (attacking only a small number of closely related hosts), and therefore there is no requirement (and usually no possibility) to confine their movement to protect potential nontarget hosts (see "The Ecology of Biological Control," this volume). However, broad-spectrum biological control deliberately uses polyphagous (many hosts) natural enemies such as goats, sheep, or grass carp, and uses technology (such as fences to confine grazing livestock in a pasture, barriers to keep grass carp localized in aquatic systems, etc.), so they can be employed safely. Grazing livestock have the added advantage of being able to control specific vegetation classes (e.g., shrubs or weedy forbs) to maintain desired plants (e.g., grasses for cattle).

Several common themes occur in all types of biological control. The four developed below are safety, risk, regulation, and measures of success.

## Safety of Biological Control

Biological control of weeds has had an outstanding record of safely managing invasive weeds for more than 100 years. In this time, 133 weed species have been targeted, and more than 350 biological control agents have been introduced into 70 countries (Julien and Griffiths 1999). Waterhouse (1999) noted that only eight agent species worldwide have been recorded as damaging nontarget species; none of the economic damage was significant or long-term and none of the ecological damage was at population level. These figures are sometimes challenged (e.g., Follett and Duan 2000, Howarth 1991, Louda et al. 1997, Louda et al. 2003a, b, Simberloff and Stiling 1996a, b, Strong and Pemberton 2000), and it is certain that additional nontarget species have been attacked by some agent species since the data in Julien and Griffiths (1999) were compiled. However:

1) It is important to distinguish incidental, exploratory nibbling on a few leaves of a nontarget species, or development of a few pathogen spores causing very low infection, from long-term, population-level impacts, which have *not* been documented from biological control agents (although research is in process to document potential long-term impacts of *Rhinocyllus conicus*—a weevil introduced to control *Carduus nutans*—on native *Cirsium* spp.; Louda et al. 1997). If a plant on which minor attack from a biological control agent was recorded during physiological (laboratory) host specificity testing is present in the field in the area where the agent is released, it should be expected that some minor events may occur on individual plants, such as nibbling, oviposition, some larval development, spore germination, small numbers of lesions, etc., depending on the type of agent. However, it is important to note that temporary and minor damage *to a plant* is far different from long-term and extensive damage *to a plant population.*

2) Transitory damage in the field on plants attacked in physiological host specificity testing has occurred in rare instances. This can happen when there are a very high number of individuals of an agent in the field, but poor quality or quantity of the targeted weed due to significant attack by the agent, and sympatric availability of a secondary host. One of the best examples of this is *Cactoblastis cactorum* in Australia, which attacked secondary hosts in the field for a short period of time when the targeted weeds, *Opuntia* spp., were heavily damaged. This damage was not significant in terms of long-term damage to the nontarget species, and was predicted by the physiological host specificity testing (see "Host Specificity Testing of Biological Control Agents of Weeds," this volume).

3) One of the key roles of government regulators is to represent society in matters where significant individual citizen participation would be impractical (see "Technical Advisory Group for Biological Control Agents of Weeds," this volume). It follows that issuing a release permit for an agent represents a societal decision to accept potential short-term risk (including the "acceptable risk" of some damage to a nontarget species; see below), weighed against potential long-term gains. Thus, an indication of potential minor damage to nontarget species in physiological host specificity testing is often appropriately rated as an acceptable risk through the regulatory process. However, it should be accepted that a release anywhere in the United States represents a release for North America, so the plants selected for use in host specificity testing must reflect the potential ecological (field) host range of the agent. One of the hypotheses that should be tested once an agent is released in the field is that the physiological host specificity testing accurately predicts the ecological host range of an agent. If damage occurs as

3

predicted, the hypothesis is supported. The need for long-term post-release monitoring should be obvious in this context (see "Monitoring in Weed Biological Control Programs," this volume). Field records of the ecological host range can be used to improve selection of plants to be used for host specificity testing for future agents. It is also important to note that a biological control agent can be an outstanding success in one geographic region of the world, but its use is unacceptable in another part of the world, usually due to too many plants that are closely related to the target weed.

## Biological Control and Risk

Risk is a measure of uncertainty, the product of hazard x exposure (R=H x E, and see "Plant Pathology and Biological Control of Invasive Weeds," this volume) (Delfosse 2003). The concept of "the risk from biological control" has been discussed extensively, particularly over the last decade (Delfosse 2001a, b, 2004, Goeden and Andres 1999, Louda et al. 2003a, b, Simberloff and Stiling 1996a, b, Strong and Pemberton 2000, Wajnberg et al. 2001). However, this is a misleading phrase unless qualified. There is not "a [single] risk" from biological control. If there were, then the risk (to nontarget species) from a polyphagous parasitoid, predator, or herbivore would be the same as risk from a monophagous arthropod or from a one-host pathogen. Clearly, release of a monophagous natural enemy in an area where there are no closely related congeners carries much less risk than the release of an oligophagous natural enemy in a region with many closely related species, some of which were attacked during physiological host specificity testing, which in turn carries much less risk than use of a polyphagous natural enemy. The difficulty for biological control practitioners is that there are relatively few monophagous natural enemies in nature, but many polyphagous species (Fig. 1). Assessment of risk must take these levels of risk into account; it is extremely difficult and misleading to compare across different types of risk.

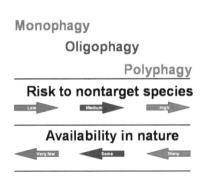

Figure 1.

Risk to nontarget species from biological control can be estimated through an ecological risk analysis process (Delfosse 2001a, b, 2003). Five highly related components of risk should be considered when predicting risk from a biological control agent: awareness, perception, assessment, management, and communication.

### • *Risk awareness and perception*

Risk awareness leads to risk perception. However, because information is always incomplete, the perception of risk from a biological control agent tends toward "the worst case scenario," and the actual risk is often overestimated.

Risk is a relative concept commonly used when uncertainty can be quantified. Uncertainty is measured by the deviation from "expected values" (i.e., risk perception) which may also be difficult to quantify. It is difficult to isolate the exclusive impact of agents on potential nontarget species when dealing with physiological host specificity

4

results. Other environmental factors influence risk and determine the ecological host range, such as number and distribution of species in the same genus as the target weed, closely related species in genera in the same family as the target weed, phenology of the target weed and potential nontarget hosts, bridging species, weather, ecology of agent species, etc. When these environmental factors are omitted from consideration, the risk attributed to biological control agents may be overestimated. Thus, when probabilities of different outcomes are unknown, uncertainty is transformed into risk, where probabilities of outcomes are subjectively weighted according to their likelihood of occurrence.

Incomplete information complicates objective estimates of risk because the nature and extent of actual risk is not understood, so the subjective valuation, or perception, of risk is biased and usually overstated. "Acceptable risk" is used when uncertainty is quantified to the subjective satisfaction of a viewer—but varies widely depending upon the perspective of the viewer!

Probabilities of possible outcomes can be estimated (e.g., risk of impact on nontarget species from a biological control agent). The crucial variable of risk (and most fears) from biological control is measured by *assumption of potential damage to nontarget species*.

### • *Risk assessment*

The risk assessment component for biological control is measured by host specificity testing, during which the ecological (field) host range is predicted from the results of physiological (laboratory) host testing. Post-release monitoring should focus on determining the ecological host range and on gathering the data used to improve testing and prediction procedures. This does not always happen. Risk assessments for biological control are normally qualitative—a "low," "medium," or "high" risk to nontarget species.

### • *Risk management*

Risk management is the component of risk analysis that is usually not well developed for biological control. Risk management can take many forms, but for biological control, it is sometimes limited to testing pesticides to find what will kill the agent "if something goes wrong," so is often a response to a requirement of a gatekeeping regulatory system (see below). The reality is that once released, it is nearly impossible to recall a biological control agent. Clearly, better risk management procedures should be developed for biological control.

### • *Risk communication*

Risk communication should take place at every step of the biological control process, and should involve customers and stakeholders as well as regulators and scientists. However, risk communication often occurs only *after* problems, such as the involvement of a threatened or endangered species, have been discovered. A good example of the need for early communication in biological control is illustrated by the saltcedar biological control program (DeLoach et al. 2004).

The first natural enemy released against saltcedar was the leaf-feeding beetle, *Diorhabda elongata*. Before releases were initiated, an endangered bird, the Southwestern willow flycatcher (*Empidonax traillii extimus*) was found to use saltcedar as a nesting substrate in parts of Arizona, New Mexico, and Nevada. The flycatcher had adopted saltcedar because this invasive tree displaced the native trees and shrubs that are its preferred nesting habitat. To ensure minimal impact on the flycatcher, the United States Department of the Interior Fish and Wildlife Service was involved in selecting release sites, and is participating in post-release monitoring of *D. elongata*. Unfortunately, the situation with the flycatcher did not come to light until late in the program planning, and a significant amount of time was lost resolving this situation. Early and effective risk communication could have saved years and considerable costs in implementing this program.

## Regulation of Biological Control

The greatest impediment to research and implementation of biological control is a regulatory system that does not facilitate a decision to import or release a natural enemy for a major pest species. Most countries regulate biological control by a combination of old quarantine laws and newer environmental laws, often using a pesticide testing paradigm.

Globally, there are two diametrically opposed philosophies of regulation of natural enemies for biological control: gatekeeping and facilitating. No country's regulations are completely one way or the other, and the following is presented to show how these different regulatory attitudes can theoretically impede or accelerate biological control.

The gatekeeper model is characterized by a "command and control" attitude, and there is often a lack of transparent standards upon which regulations are based. Regulations are not based on science and are not proportional to the potential risk to nontarget species. There is little meaningful peer review, turnover times for review of petitions for the introduction of weed biological control agents are long, and the Precautionary Principle ("guilty until proven innocent") is used uncritically. Precautionary Principle procedures ask scientists to prove negatives (i.e., "prove that the natural enemy will not attack other plants"), and regulatory decisions are delayed until "all information is available." The proper question to ask petitioners is posed in a risk analysis context (see below). The gatekeeper model results in poor service to petitioners in the short term, and to citizens who ask for biological control in the longer term.

The facilitator model is based on entrepreneurial brokering, whereby regulators facilitate rapid decisions on petitions for importation and release. Standards are transparent and science-based and regulation is proportional to risk of potential impact to *populations of nontarget species*. The facilitator model is based on peer review by qualified partners and stakeholders, resulting in shorter turnover times for review compared to the gatekeeper model. Potential agents are considered neither "guilty until proven innocent" or "innocent until proven guilty." Rather, the primary regulatory agency ensures that the petition for release is seen by all other regulatory agencies with responsibility in the area, and ensures an objective, science-based review by sending the application to external reviewers with expertise in the taxa. Instead of being asked to prove a negative, the regulatory agency asks a petitioner to pose the question in a risk analysis context; i.e., "estimate the risk to other plants if the agent was released." A high level of customer service can be delivered from a facilitator system.

## Success in Biological Control

Most success in biological control has been judged based on a subjective assessment of "control" of the target weed (Julien and Griffiths 1999). Factors such as "agent establishment," "some percent reduction [in a phenological stage]," "considerable reduction," "good control," "substantial control," "no significant control," "complete control," etc., are used. These are important factors to measure, but are too fuzzy and do not fully measure success. Thus, a degree of success is often assumed in biological control of weeds programs, but rarely defined, and even more rarely quantified.

I have proposed seven measures of success that biological control practitioners should consider: biological, ecological, scientific, economic, social, legal, and political (Delfosse 2004). Perception of success varies depending upon the point of view (for example, a rancher vs. a scientist), the scale (control in a pasture vs. in several states), and the time ("How long will other strategies have to be used until biological control is successful?"). All seven components of success are important, but they must feed into political success or the long-term benefits of the other components will be lost. A brief discussion of the seven components follows; see Anderson et al. (2003) for how this model was used for leafy spurge (*Euphorbia esula*).

### • *Biological success*

Biological success is a measure of management of the target weed. This is the component that is estimated for the examples in Julien and Griffiths (1999). It is often the only component of success that is measured or reported and is certainly very important, but more quantification is needed. Statements such as "negligible success" or "partial success," while better than nothing, do not convey the detail needed. "Complete" biological success has been achieved when the invasive weed is reduced to the point that no other inputs are needed for its management. Establishment of agents is a very important feature of success, but by itself does not measure the success of a program (McFadyen 1998). The other components of success discussed below can be as important as biological success.

### • *Ecological success*

A long-term, biologically based, integrated weed management program that is sustainable, low-input, and energy-conserving would be rated an ecological success. The goal is to maintain or increase natural biological diversity. A decrease in risk from pesticide use is also a possible outcome. Has recovery of native, threatened, or endangered plant species been documented? Similarly, if nontarget damage occurs, is it documented, analyzed, and published so that the results can be discussed by peers? (See "The Ecology of Biological Control," this volume).

### • *Economic success*

To measure economic success of biological control programs, benefit-cost ratios and similar factors should be calculated. It is imperative to determine whether less cash was expended for the same or better weed control compared to other strategies, how much money is returned to the producer, etc. Ideally, this requires economists to be involved *early* in program development, but this is not often the case. Careful documentation is

7

important to be able to determine whether the program is an economic success (see "Economic Benefits of Biological Control," this volume).

### • *Social success*

Social success is measured by a large set of complex factors involving individuals across society. Scientific programs must identify and address social needs. Is there an increase in the awareness of the problem weed? Do average citizens accept individual responsibility for weed management? Do they understand biological control and other integrated weed management strategies? Does the degree of control that is achieved meet their perceived needs? Does it increase their enjoyment of open spaces? Broad agreement can be reached on goals of a biological control program when it is discussed with customers and partners before a program is initiated. Appropriate societal groups should be involved in program development (scientists, environmentalists, federal agencies, states, tribal governments, the private sector, and other special interests). Extension personnel, sociologists, and the media should be involved early in program development. Technology transfer leading to increased implementation of biological control should be one of the key goals. It is essential to identify the customers, stakeholders, and beneficiaries of a proposed biological control program. Action committees, coalitions, etc., composed of customer groups generally concerned with a single target weed or resource can help focus the need and obtain funding, but can also leverage core resources away from ongoing projects.

### • *Legal success*

Legal success includes enactment of laws and regulations that facilitate biological control research and implementation. Knowledgeable lawyers and regulators should be involved early in the development of laws and regulations. It is critical that laws, regulations, and implementing procedures are transparent and science-based (see "Regulation of Biological Control," above). Enabling legislation for biological control should be considered to empower regulators to become entrepreneurial facilitators, to include customers in the peer review process, and to establish a single primary point of contact for regulating biological control in the United States.

### • *Scientific success*

Knowledge that increases our understanding of ecological systems is among the most important and enduring long-term gains from biological control programs. For example, the biological control process should be viewed as a series of ecosystem-level experiments in applied ecology based on testing refutable hypotheses; however, this often is not the case. Results of these ecological experiments should be used to improve predictability of the host specificity testing process, and particularly in understanding the relationship between the physiological and ecological host range of biological control agents. Biological control experiments can contribute to the development of ecological theory.

Biological control scientists conform to the highest ethical standards when determining the safety of natural enemies. Adoption of the International Code of Best Practices for Classical Biological Control of Weeds (see "International Code of Best Practices for Classical Biological Control of Weeds," this volume) by all practitioners will help demonstrate and communicate the ethical nature of the process. For example, it is unethical to release a classical biological control agent without post-release monitoring, and most biological control of weeds programs now emphasize this factor. In fact, monitoring and evaluation should always be the key post-release activities of biological control programs, to determine the efficacy and impact of past and continuing projects.

Biological control scientists should take more advantage of the techniques of metapopulation biology (Hanski and Gilpin 1997), including landscape ecology (patch quality, boundary effects, patch context, and connectivity) and metapopulation dynamics (interpatch movement, extinction of local populations, and recolonization of the patches), to increase understanding of the impacts of releasing herbivores and plant pathogens into the environment. Increased support for systematics is essential, and the role of systematics should be elucidated and emphasized whenever appropriate.

### • *Political success*

Long-term and increasing support (philosophical and monetary) for biological control as the base strategy of integrated weed management is the key measure of political success. This can best be done by ensuring that results from the other six types of "success" are funneled into political success. How many new programs are initiated? An oddity of biological control of weeds is that successful programs reduce targeted weeds to nonweedy levels, thus making the problem "invisible" and resulting in the "what have you done for me lately" syndrome. Are the positive results from the programs captured and transferred to customers and stakeholders? Are economic analyses available that show success or otherwise? Has the post-release monitoring shown not only a decrease in the target weed, but an increase in native plant species, possibly threatened or endangered species, that were out-competed by the invasive weed? Is nontarget damage discussed openly and objectively in nonscientific arenas? Buy-in from traditional and nontraditional customers and stakeholders, following on from social success, will help achieve the goal of increasing support for biological control of weeds.

## Future Challenges and Opportunities

Biological control faces many challenges and opportunities.

The primary challenge is to develop and implement regulatory systems that are transparent and based on science, set and use achievable standards, use risk analysis, regulate in proportion to risk, and pose answerable questions about the risk to nontarget species from release of biological control agents. The horrific events of September 11, 2001, inevitably led to a tightening of biological control regulation (Knott and Flanders 2003), but biological control practitioners should welcome some of these restrictions. For example, a better tracking process and increased personal responsibility by scientists who import biological control agents from overseas will increase the safety of the process and demonstrate the high ethical standards that have characterized the biological control community for over 100 years.

Biological control practitioners must not act defensively when biological control safety and efficacy are challenged. Nontarget damage has occurred and will continue to occur. Prerelease procedures have evolved from a scientist making a decision based primarily on ecological observations in the native range of a weed, to the crop-testing method to release an agent, or the centrifugal phylogenetic and relatedness testing procedures. At each step, the prerelease procedures have become increasingly science-based and sophisticated. Testing protocols will continue to evolve as new challenges emerge. This evolution of testing in response to societal needs and concerns is one of the strongest elements of biological control.

One of the key jobs of the biological control community is to design experiments in the post-release monitoring phase that will produce data that quantify and explain nontarget damage and allow a more objective assessment of "success." It is important that biological control research is hypothesis-driven. We should publish the "zeros" (no attack on plants tested in physiological host specificity testing), as well as data that indicate nontarget impacts.

Expectations should not be raised beyond what biological control can deliver; there are few silver bullets and many blanks. It must also be realized that there may not be effective biological control agents for some weeds. Education of citizens, politicians, resource managers, regulators, and sponsors about realistic expectations—capabilities and pitfalls—of biological control is a continuing challenge.

In summary, biological control is not a panacea and is not risk-free. It is a key strategy of integrated weed management, but it should not be implemented without a consideration of all the pest management options available for a given weed; often a biologically based weed management program is the appropriate response. The safety record of biological control is clear, and far better than any other strategy used to manage weeds. The new political awareness of invasive species has already significantly increased the support for biological control of weeds, and with no slowdown in the rate of pest invasion in sight, this support is likely to continue for many years.

The future of biological control of weeds is very bright if we act responsibly, continue to establish and maintain effective partnerships, and if important regulatory and environmental safety issues can be resolved.

## *References*

Anderson, G. L., E. S. Delfosse, N. R. Spencer, C. W. Prosser, and R. D. Richard. 2003. Lessons in developing successful invasive weed control programs. J. Range Manage. 56: 2-12.

Barbosa, P. 1998. Conservation Biological Control. Academic Press, San Diego, CA.

Bellows, T. S., and T. W. Fisher. 1999. Handbook of Biological Control. Principles and Applications of Biological Control. Academic Press, NY.

DeBach, P. 1964. Biological Control of Insect Pests and Weeds. Chapman and Hall, London.

Delfosse, E. S. 2001a. History of the risk assessment process: A North American perspective on risk assessment for biological control introductions. XXI International Congress of Entomology, August 20-26 2000, Iguassu Falls, Brazil (abstract).

Delfosse, E. S. 2001b. Management of invasive species. Proc. 10th Pacific Science Intercongress, 1-6 June 2001, Guam (abstract).

Delfosse, E. S. 2003. Overview of the ecological risk analysis process. Proc. of the 14th USDA Interagency Research Forum on Gypsy Moth and Other Invasive Species. Symposium entitled Risk Analysis, It's Not Just Risk Assessment. Organizer, E.S. Delfosse. 16 January 2003, Annapolis, MD. (*in press*).

Delfosse, E. S. 2004. What is "success" in weed biological control?–Thinking outside the plant. Page 605 *in* J. M. Cullen, D. T. Briese, W. M. Lonsdale, L. Morin, and J. K. Scott, eds. Proc. XI Symp. Biol. Contr. Weeds, April 27 - May 2, 2003, Australian National University, ACT, Canberra, Australia.

DeLoach, C. J., R. I. Carruthers, T. Dudley, D. Eberts, D. Kazmer, A. Knutson, D. Bean, J. Tracy, J. Knight, J. Herr, G. Abbott, G. Adams, I. Mityaev, R. Jashenko, B. Li, R. Sobhian, A. Kirk, and E. S. Delfosse. 2004. First results for control of saltcedar (*Tamarix* sp.) in the open field in the Western United States. Pages 505-13 *in* J. M. Cullen, D. T. Briese, W. M. Lonsdale, L. Morin, and J. K. Scott, eds. Proc. XI Symp. Biol. Contr. Weeds, April 27 - May 2, 2003, Australian National University, ACT, Canberra, Australia.

Follett, P. A., and J. J. Duan. 2000. Nontarget Effects of Biological Control. Kluwer Academic Publishers, Boston, MA.

Goeden, R. D., and L. A. Andres. 1999. Chapter 34: Biological control of weeds in terrestrial and aquatic environments. Pages 871-90 *in* T. S. Bellows and T. W. Fisher, eds. Handbook of Biological Control. Principles and Applications of Biological Control. Academic Press, NY.

Hanski, I. A., and M. E. Gilpin. 1997. Metapopulation Biology. Ecology, Genetics, and Evolution. Academic Press, NY.

Howarth, F. G. 1991. Environmental impacts of classical biological control. Annu. Rev. Entomol. 36: 485-509.

Hoy, M. A., and D. C. Herzog. 1984. Biological Control in Agricultural IPM Systems. Academic Press, NY.

Huffaker, C. B., and P. S. Messenger. 1976. Theory and Practice of Biological Control. Academic Press, NY.

Julien, M. H., and M. W. Griffiths. 1999. Biological Control of Weeds. A World Catalogue of Agents and their Target Weeds. Fourth Edition. CABI Publishing, Wallingford, Oxon, UK.

Knott, D., and R. Flanders. 2003. Letter to permit holders, 10 October 2003. USDA-APHIS-Plant Protection and Quarantine (available at www.aphis.usda.gov).

Louda, S. M., A. E. Arnett, T. A. Rand, and F. L. Russell. 2003a. Invasiveness of some biological control insects and adequacy of their ecological risk assessment and regulation. Conserv. Biol. 17(1): 73-82.

Louda, S. M., D. Kendall, J. Connor, and D. Simberloff. 1997. Ecological effects of an insect introduced for the biological control of weeds. Science 277: 1088-90.

Louda, S. M., R. W. Pemberton, M. T. Johnson, and P. A. Follett. 2003b. Nontarget effects - The Achilles' Heel of biological control? Retrospective analyses to reduce risk associated with biocontrol introductions. Annu. Rev. Entomol. 48: 365-96.

McFadyen, R. E. C. 1998. Biological control of weeds. Annu. Rev. Entomol. 43: 369-93.

Simberloff, D., and P. Stiling. 1996a. How risky is biological control? Ecology 77: 1965-74.

Simberloff, D., and P. Stiling. 1996b. Risks of species introduced for biological control. Biol. Conserv. 78: 185-92.

Strong, D. R., and R. W. Pemberton. 2000. Biological control of invading species-risk and reform. Science 288: 1969-70.

Wapshere, A. J., E. S. Delfosse, and J. M. Cullen 1989. Recent developments in biological control of weeds. Crop Prot. 8: 227-50.

Waterhouse, D. F. 1999. Foreword *in*: M. H. Julien and M. W. Griffiths, eds. Biological Control of Weeds. A World Catalogue of Agents and their Target Weeds. Fourth Edition. CABI Publishing, Wallingford, Oxon, UK.

Wajnberg, E.W., J.K. Scott, and P.C. Quimby. 2001. Evaluating Indirect Ecological Effects of Biological Control. CABI Publishing, Wallingham, UK.

*Section I*

# The Theory and Practice of Biological Control

# The Ecology of Biological Control

Shon S. Schooler, Peter B. McEvoy, and Eric M. Coombs

## Introduction: Ecology and Biological Control

Ecology is the scientific study of the interactions between organisms and their environment. Ecologists seek to identify, understand, and then predict the outcomes of interactions between the individuals within a species (populations), between organisms of different species (communities), and between organisms and the surrounding environment (ecosystems). Our level of ecological understanding is measured by our ability to predict the future of organism populations based on current environmental conditions.

Biological control is the attempt to control the population of one organism (target species) with that of another (biological control agent). A successful biological control agent will reduce the density of a target species to a desired level and maintain it there (effectiveness) while minimizing the probability of negative impacts to nontarget species (risk). Since ecology is the study of such interactions, biological control practitioners who employ scientific methodology are applied ecologists.

Ecological theory should aid us in predicting the risk and effectiveness of biological control agents. Risk is assessed primarily through studies of host specificity that aim to predict the impact of the agent on the communities and ecosystems of its new range. Effectiveness is often defined as the decrease in the abundance of the target organism; however, usually the true goal is a change toward a desired plant community composition and function.

Studies of biological control programs can also help to advance the science of ecology. In most cases it is unethical to increase or redistribute populations of harmful invading organisms in order to scientifically examine their impacts. However, biological control programs allow ecologists to manipulate and study the progress of controlled introductions, which in turn increases our ability to understand, predict, and manage invasive species.

## Ecology in the Service of Biological Control

### Can we predict which introduced species will become invasive?

Moreover, can we do so while prevention is still feasible and before they become a problem? Many plants are purposefully or accidentally introduced into the United States each year. Few of these become established and fewer yet become a problem. Should we allow a new exotic nursery species to be propagated, sold, and dispersed within the United States? What will be the impact of an aquatic plant if it is released from an aquarium into a nearby lake or stream? If we can predict potential invasiveness and subsequent impact, we have the option to prevent the introduction or eradicate small populations before they become entrenched.

Currently our best predictor of the potential invasiveness of a particular plant species is whether it is already recognized as causing harm somewhere else (Reichard and Hamilton 1997). This is important information and is used to judge the invasive risk of many nursery plants. However, in itself this knowledge does little to predict the invasiveness of most newly introduced plant species. A more inclusive predictive strategy is to identify plant traits that are associated with invasiveness. In a study of 24 pine species, several researchers found that three of the 10 life history traits they measured were associated with invasiveness (Rejmanek and Richardson 1996). Pine species that had a short juvenile period, a short interval between seed crops, and a small seed mass were more invasive. This evidence supports prior hypotheses that plants with high reproductive rates have a greater potential to become invasive (Elton 1958).

Nevertheless, our ability to predict plant invasions is poor (Enserink 1999). This is because invasions depend not simply on attributes of the invader, but also on attributes of the recipient environment. Kudzu ("the plant that ate the South" or *Pueraria lobata*) planted into arboreal regions of Alaska is unlikely to become invasive, whereas kudzu planted in the Willamette Valley of Oregon is cause for great concern. Thus, to truly predict whether a plant will become invasive, we also must understand where it might be introduced, either purposefully or accidentally. Ecology may help to predict purposeful introductions, but is little use in predicting accidents. Ecology can help determine what sorts of environments are likely to be invaded by particular types of plants, but the social and economic aspects of plant spread also need to be considered.

Intact communities of native plants may be better able to repel invasive plants. Fewer resources (water, sunlight, or nutrients) may be available in undisturbed communities so that some type of disturbance event is necessary to make some of the limited resources available and allow an invasive plant to become established. A study of invasive plants in Australia provided some evidence supporting this claim (Hobbs 1989, 1991). A similar line of research asks whether more diverse plant communities are less likely to become invaded. Theoretically, this should be the case; however, several studies provide evidence that question this hypothesis (Levine 2000).

## Can we predict the impact of an invasive species?

Clearly our control efforts should target species with greater potential impact over those with minimal impact. Again, ecology can help us with a framework of how to understand negative effects and prioritize control programs. Negative impacts can occur at the population, community, and ecosystem levels of ecological organization (Parker et al. 1999). If the invasive species only impacts the population of a single other species, it is generally a low priority for control efforts unless the impacted species is considered to be of ecological (e.g., threatened or endangered species) or economic (e.g., crop or horticultural plants) value. For example, the justification for biological control of tansy ragwort (*Senecio jacobaea*) was based primarily on its impact on livestock (Coombs et al. 1996). A species that negatively affects many other species is considered a greater threat. Purple loosestrife (*Lythrum salicaria*) is considered a noxious weed because it displaces a range of native species from sedges to waterfowl (Blossey et al. 2001). Finally, those species that alter ecosystem processes—such as saltcedars ( *Tamarix* spp.), which modify river morphology and water levels in the desert southwest (Loope et al. 1988); fayatree (*Myrica faya*), which alters nitrogen concentrations on volcanic Hawaiian soils (Vitousek and Walker 1989); or cheatgrass (*Bromus tectorum*), which alters fire regimes throughout the Great Plains (D'Antonio and Vitousek 1992)—have the potential to cause the most harm and should be priorities for control efforts.

Ecologists can devise and refine models that predict impact by understanding the biology and ecology of the invasive organism in its native range and by monitoring its impacts in the new location. For example, if the invasive species is a nitrogen-fixing legume, we know from the biology of the plant and research on prior invasions that it may increase soil nitrogen concentrations. We can then measure soil nutrient levels and the associated impact on local organisms. Once we understand the nature and magnitude of the impact, we can create estimates of environmental harm and economic loss to guide our prioritization and management (Parker et al. 1999). Not all impacts will necessarily be negative; in many instances, the plant may have benefits, such as herbal remedies (St. Johnswort, *Hypericum perforatum*), nectar for honey production (purple loosestrife), or forage for domesticated livestock (reed canarygrass, *Phalaris arundinacea*). We can use ecology and economics to create benefit-cost analyses that will provide the scientific basis for informed management decisions (Radtke and Davis 2000).

## Is biological control the most appropriate method?

Unlike many other control strategies, biological control of weeds is initially expensive (Andres 1977, Harris 1979). A complete biological control program is likely to cost $1.2 to $1.5 million (it will, on average, require 18.8 to 23.7 scientist-years) and thus is only economical for use against major weeds (Harris 1979). In addition to economic cost, risk to nontarget organisms continues from immediate new associations into the indefinite ecological and evolutionary future (McEvoy 1996, McEvoy and Coombs 1999).

Biological control does not aim to eradicate the invader from the entire invaded area, but to regulate it at lower densities. Clearly other strategies, including eradication through physical and chemical removal and altering cultural management practices, should be weighed before introducing expensive and risky organisms into new habitats (McEvoy and Coombs 1999, Ruesink et al. 1995).

Ecology can help estimate the benefits of diverse and integrated control strategies and these can be weighed against the costs of management and risk. Early identification of harmful invasive plants may make total eradication possible in some cases. Cultural controls, such as increasing the competitive ability of desirable vegetation by altering management practices, could reduce the impact of the invasive species without the cost and risk of biological control. Regardless of their impact, plants closely related to native and horticultural species may not make good targets for biological control because of the higher probability of risk to nontargets.

## What is the risk of the control organisms to nontarget species?

Biological control agents can impact nontarget organisms directly or indirectly (Lonsdale et al. 2001). Direct damage through herbivory is generally the greatest concern. Indirect damage may occur when agents modify the local environment or enhance the populations of insectivores, thus changing the structure of food webs. Ecologists studying and managing weed biological control programs have, since 1902, been successful in predicting the risk of direct impacts of herbivorous control agents on nontarget plants (Pemberton 2000). By using our understanding of ecological and evolutionary relationships, biological control practitioners can select and test plants that are most likely at risk of direct effects before candidate agents are released. However, indirect impacts are much more difficult to predict and the best way to reduce this risk may be to limit the number of introductions to those clearly shown to be necessary, effective, and safe (McEvoy and Coombs 2000).

## Can we predict the effectiveness of biological control agents?

Along with predicting risk, this has been a primary focus of biological control practitioners. If we can predict the effectiveness of a control agent, we can reduce unnecessary introductions and spend fewer resources on a given problem, freeing more for the control of other species. An effective biological program is linked to traits of the target plant, the biological control agent, and the local environment; interactions among these three factors; and strategies for the release of the biological control agent.

### • Traits of the target plant

Some plants may be more easily controlled than others. In a 1981 study, researchers analyzed the case histories of 45 plants and found that biological control was more successful against asexually reproducing plants (Burdon and Marshall 1981). They hypothesized that this was due to a lack of genetic variation that reduces the plants' ability to develop defenses against herbivores. Subsequent analyses confirmed this conclusion (Lawton 1990). However, 14 years later a different research group examined an updated set of data and found that four new case histories had shifted the result and plant reproductive mode was no longer significant in predicting success (Chaboudez and Sheppard 1995). They concluded that the current database was "too small and unreliable to be of real value in making predictions for selecting weeds that would be easier to control." This is a good example of the merits of the scientific approach. Nothing is ever considered "proven" in science and, therefore, there is always room for improvement in light of new data. New data may reinforce or question widely held beliefs, which can then be modified to reflect our current state of knowledge.

### • Traits of the biological control agent

Ecologists have focused primarily on the traits of the control agents when studying effectiveness. A weighted scoring system to predict effective agents has been proposed (Goeden 1983, Harris 1973). The revised system scores each potential agent using 16 criteria ranging from the type of damage inflicted to its prior history of effectiveness. According to this system, the most effective agents: 1) destroy the vascular tissues of the plant; 2) vector a virulent pathogen (assuming the pathogen has been approved by USDA-APHIS Plant Protection and Quarantine); 3) attack the plant throughout the growing season; 4) have multiple generations per year; 5) have high reproductive rates; 6) are immune from generalist predators; 7) feed gregariously; and 8) cover the full range of the target plant. Critics of this approach comment that the scoring system is difficult to apply because necessary information is often lacking (Lawton 1990), may ignore important plant traits, and may unnecessarily reject effective agents (Crawley 1989a, Cullen 1995, Lawton 1990).

Taxonomic relationships have also been used to predict effectiveness. In analyses of prior biological control programs (Julien et al. 1984, Lawton 1990), releases of bugs (Hemiptera) and beetles (Coleoptera) tend to be more effective than flies (Diptera) and moths (Lepidoptera). However, these correlations do not tell us what makes bugs and beetles more effective than flies and moths. Perhaps bugs and beetles tend to do greater damage to individual plants because they aggregate in higher densities or maybe they tend to preferentially feed upon plant parts that are more important for survival and reproduction. They also may be less susceptible to acquired natural enemies such as predators and pathogens. Selecting bugs and beetles over flies and moths without understanding the underlying mechanisms of success will likely lead to missed opportunities. For example, prickly pear cactus (*Opuntia* spp.) might never have been controlled across vast stretches of Australia and other arid environments if it were not for the pyralid moth *Cactoblastis cactorum*.

Not surprisingly, traits linked with successful biological control are similar to those conferring invasiveness as observed with some of our worst pests. Again, the most predictive trait is whether the agent has had an impact elsewhere (Crawley 1989b). Other similarities are traits that confer quick population growth through high reproduction rates (Crawley 1986, Lawton 1990).

Despite all the research devoted to identifying factors that are correlated with success, there is no panacea. No single trait, when held up to scientific scrutiny, has been found to accurately predict general effectiveness. Perhaps the most notable rule is that there are always exceptions. These correlations provide general guidelines for selection of effective agents, but a potential agent should not be dismissed simply because it is a moth or has slow population growth (Cullen 1995). Multivariate analyses combining traits of control organisms, target organisms, and recipient environments may lead to a general set of factors that can be used to screen potential agents. However, more experimental research and environmental monitoring need to be done before this is feasible.

### • Traits of the local environment

It is obvious that the surrounding environment—including both abiotic and biotic components—will impact the effectiveness of biological control organisms. Abiotic components include topographic variation, elevation, climate (precipitation and temperature), and disturbances such as floods and fires. Biotic components include vegetation structure, host-plant phenology, microclimate (moisture, shade, etc., as influenced by vegetation), and natural enemies (predators, pathogens, and parasitoids).

Perhaps the most obvious factor that may influence establishment and effectiveness of the biological control agents is the local climate, though some reviewers find little evidence to support this hypothesis. It has been suggested that the lack of evidence is due to biological control practitioners' sensitivity to climate matching, so that few failures can be directly attributed to this factor (Lawton 1990).

Rather than generating *post hoc* correlations and explanations through database analyses, we might consider a question-based approach (Cullen 1995). It has been argued that the scoring system/checklist approach previously described may include factors that are not relevant to a particular weed-agent interaction and might miss factors that are particularly important. Researchers have suggested an approach based on a set of questions organized similar to the dichotomous keys used in taxonomy. This would allow consideration of specific traits of agents in response to specific plant and environment traits.

Others advocate more holistic approaches to agent selection including consideration of targeted disruption of pest life cycles and combinatorial ecology (McEvoy and Coombs 1999). Targeted disruption identifies the most sensitive transitions (seed to juvenile, juvenile to adult, etc.) of the plant's life cycle and finds agents that will specifically reduce those transitions. In this way, even small amounts of damage will presumably have large impacts. Combinatorial ecology considers coordinated manipulation of disturbance, plant competition, and natural enemies in biological control programs. Rather than requiring the agent to do all the work, this approach emphasizes that the three aspects that influence the populations of the invasive plant can be used together. Both of these methods demand less of the biological control agents and may require fewer (and safer) agents to be introduced for the same or better control. However, either method requires more prerelease study to model the invasive plant's life cycle and more collaboration with land managers to manipulate factors that affect disturbance and plant competition such as grazing pressure, urban growth, fertilizer application, and timber harvesting.

## • *Release strategies of the biological control practitioners*

In addition to predicting the performance of biological control agents, we can also improve existing programs using scientific methodology. Passive adaptive management uses post-release monitoring to guide future management decisions and active adaptive management manipulates agent releases as scientific experiments to answer predetermined questions (Shea et al. 2002). Results are translated into recommendations that are incorporated into subsequent release strategies. For example, practitioners might initially release a range of population sizes and then use the relationship between establishment and release size to guide subsequent release and redistribution efforts.

A good example of the active adaptive management approach is a study involving purple loosestrife (Grevstad 1999a, b). The study examined the probability of establishment of the black-margined and golden loosestrife beetles (*Galerucella calmariensis* and *G. pusilla*) when purposefully released in differing numbers (20, 60, 180, and 540 individuals) across 36 release sites. Using mathematical models coupled with field data it was found that: a) releases of 180 individuals exhibited only 30% success in establishment, whereas releasing 540 beetles increased the rate to 90%; and b) in more variable or disturbed environments it is better to make many small releases, whereas in environments with less environmental variability it is better to make fewer but larger releases. System-specific studies like this help generate data that can be used to hypothesize and test generalizations that might benefit biological control as a whole.

Ecological theory can also be used to determine which of multiple introduced agents are most effective at controlling the target. An Oregon study coupled a field experiment with population matrix models to investigate whether the cinnabar moth (*Tyria jacobaeae*) or the ragwort flea beetle (*Longitarsus jacobaeae*) was more effective against tansy ragwort, after controlling for other environmental factors such as disturbance and competition (McEvoy et al. 1993). Researchers found that the flea beetle was the more effective agent, while the moth provided minimal additional control. This result is particularly important because the cinnabar moth is known to attack nontarget native species. This study provides support for ending redistribution efforts of the cinnabar moth in favor of the flea beetle.

## Is the biological control program successful?

Once agents are released, we would like to know how effective they are at reducing the populations of the target plant and whether the reduction results in a more desirable plant community. Photographs of the site before and after control are commonly used to demonstrate success. They are easy to obtain and are often visually dramatic. However, they may be misleading. A site that was covered by a sea of flowering purple loosestrife one year is shown to be devoid of the invasive wetland plant four years after the introduction of the biological control agents. Was the reduction caused by the control organisms, herbicide drift from a nearby cornfield, physical removal by concerned citizens, competition from another invasive plant species (e.g., reed canarygrass), or foraging by deer, geese, or muskrats? The answer is that we don't know unless we monitor the site to establish the relationship between causes and effects.

Ecology provides a framework that helps us understand and monitor the progress of a biological control program. We can examine a program as a series of discrete steps: 1) the establishment of the agent; 2) the increase of the agent population; 3) the spread of the agent; 4) the damage to individual target plants; 5) the response of the target plant population; and 6) the change in the local plant community or ecosystem. If the release is not successful, this approach allows us to detect at which stage the problems occurred and to respond with a revised strategy (adaptive management). If one agent establishes but does not reach high enough densities to damage the target plant, another agent may be chosen for release at that site. Alternatively, if an agent significantly reduces the population of the weed but the desired plant community is not achieved, perhaps other restoration efforts are necessary. Observed success and then subsequent failure of a biological control agent may be caused by interference from a predator (Pratt et al. 2003). By using standardized measures we can also compare performance across release sites and between biological control systems. Note that both the adaptive management approach and cross-study comparisons depend upon post-release monitoring.

Information gained through monitoring can identify nontarget impacts, generate hypotheses both of what makes a good control agent and of why a program is successful, and lead to more effective management strategies (McEvoy and Coombs 1999). Knowledge of agent population growth and spread can aid in decisions of when and where to redistribute the agent and how many to release at a given site. The intensity, timing, and type of damage may elucidate why some control agents are more effective than others and help determine which agents to prioritize in redistribution efforts. Increasing damage to the target weed resulting in declines in abundance strongly suggests that the biological control agent is effective. Measuring the decline in biomass of the target species at different times and places provides a reliable first measure of effectiveness. The final evidence for success lies in the change in composition and function of the plant community. Identifying and sampling the abundance of plant species can be a daunting task and many land managers may not have the resources for this final piece

21

of the puzzle, but the prize is worth it. Those who have sampled have gained greater understanding of the true effectiveness of the control (McEvoy et al. 1991), including the return of threatened native species (Gruber and Whytemare 1997) and desired forage species (Huffaker and Kennett 1959).

## Biological Control in the Service of Ecology

We have seen how ecology can help us understand and predict the invasiveness and impact of exotic plants and the risk and effectiveness of biological control programs. These systems also make ideal testing grounds for ecological theory. Biological control systems offer ecologists opportunities to test theories of invasion, community assembly, and population dynamics as the populations of the agents go through phases of colonization, establishment, growth, spread, impact, genetic changes, and new associations.

### • *Populations*

Because biological control systems introduce a known number of organisms to a known location, we can examine the growth and spread of the agent populations across landscapes and generate generalized models of dispersion. These can then be used to build and test general theories of organism spread for biological control agents and invasive species (Andow et al. 1990, Shigesada and Kawasaki 1997).

Using biological control, we can also examine how populations are controlled. There is an ongoing debate as to whether populations are controlled by a limiting resource (bottom-up) or by predators and pathogens (top-down). Some general ecological theory suggests that plant populations are usually regulated by resources (nutrients, sunlight, water) that are limited through competition with other plants, and that animal populations are often regulated by predators. However, the nature of classical biological control presumes that exotic weeds often become invasive because they are freed from the negative impact of their natural herbivores. Therefore, general ecological and biological control theories of population regulation are at odds. Most likely both types of regulation interact through space and time to generate the diversity of plants and animals we see in nature. Successful biological control programs are suggestive of strong top-down control, although plant competition may play an important but less visible role. Biological control failures may be due to reduced plant competition caused by tilling, grazing, fire, or other disturbances that serve to release limiting resources.

A closely related debate reframes the issue in terms of the degree of symmetry in the interactions between plants and herbivorous insects (McEvoy 2002). The symmetrical view holds that insects and plants strongly impact each other's ecology and evolution, whereas the asymmetrical view contends that plants impact the evolution and ecology of insects but not vice versa. Again, biological control may help resolve the debate by providing simplified systems that are readily manipulated and within which herbivores are expected to impact plant populations. If the asymmetrical view is correct, we should expect to see low success rates in classical weed biological control programs, especially where local plant competition is reduced.

## • *Communities*

Often the goal of weed biological control is the return of a more diverse plant community. Research in this field is therefore a fertile ground for investigating theories of species coexistence and hypotheses of how diversity is maintained. The theory of competitive exclusion states that two species cannot coexist when feeding on the same limiting resource because one will eventually gain an advantage and eliminate the other through competition. In a study of two leaf-skeletonizing beetles in the same genus (*Galerucella*) introduced for the control of purple loosestrife, biological control researchers found no indication that the beetles were subdividing the resource (Blossey 1995). Further study will reveal whether these species can coexist in North America when purple loosestrife is controlled and their food supply is limited. This case study may challenge the theory of competitive exclusion and force us to reexamine a long-standing paradigm in theoretical ecology.

Biological control programs provide simplified food webs for examining how energy flows through biotic systems. Weed biological control agents are host-specific, which helps us to accurately identify and measure their populations and resource base. We can then examine how populations of these herbivores affect the populations of generalist predators and detritivores and compare this with a system in which the herbivore is absent. These programs also give us discrete systems to study how stability is maintained in an environment of constantly fluctuating resources and predators. What happens if one link in the web becomes locally extinct? Will the entire web collapse or will the population of another organism compensate for the loss?

## • *Ecosystems*

Studies of invasive plants and biological control programs help us understand the connection between organisms and the local environment. Changing environmental conditions impact species composition, but individual species can also impact the environment. The introduction of fayatree increases soil nitrogen levels and therefore can alter the local soil conditions for others (Vitousek and Walker 1989). Plants that rely on their ability to efficiently gather and use nitrogen to outcompete other plants may be displaced, which may cause a cascade of community and ecosystem changes. Successful biological control programs allow us to naturally manipulate densities of invasive species and observe the consequences. If nitrogen-enriching plants are reduced to lower densities, do soil nitrogen concentrations return to previous levels? Will a reversion to prior soil conditions be followed by the return of the prior plant and animal communities? Does water flow increase with the removal of saltcedar trees? How do the aquatic plants and animals survive water-level fluctuations?

## Conclusion

The relationship between ecology and biological control is mutually beneficial. Ecological theory can help us organize and measure the specifics of biological control programs to better predict and monitor the impact of the invasive species and the risk and effectiveness of the agents. We can do this by separating seemingly complex processes into a series of biologically discrete steps. We can measure performance at each step in standard ecological terms and then contrast it within and between biological control systems. We can identify specific stages where failure has occurred and try to remedy problems. In addition, we can learn more about systems by setting up biological control releases with active adaptive management in mind. This method emphasizes viewing releases as an experiment in which the agents' species, number, and release location are manipulated, replicated, and monitored. These studies can then be used to improve effectiveness of the specific program and may lead to generalizations that will increase the effectiveness and safety of future programs.

Ecology gains from this relationship as well. Using the data generated from biological control programs, ecologists can create new hypotheses of how the natural world is structured and test theories with real data. General theories can then be built that, in turn, can be used to increase our understanding of invasive species and biological control.

## *References*

Andow, A. A., P. M. Kareiva, S. A. Levin, and A. Okubo. 1990. Spread of invading organisms. Landscape Ecol. 4: 177-88.

Andres, L. A. 1977. The economics of biological control of weeds. Aquat. Bot. 3: 111-23.

Blossey, B. 1995. Coexistence of two leaf-beetles in the same fundamental niche. Distribution, adult phenology, and oviposition. Oikos 74: 225-34.

Blossey, B., L. C. Skinner, and J. Taylor. 2001. Impact and management of purple loosestrife (*Lythrum salicaria*) in North America. Biodiversity Conserv. 10: 1787-1807.

Burdon, J. J., and D. R. Marshall. 1981. Biological control and the reproductive mode of weeds. J. Appl. Ecol. 18: 649-58.

Chaboudez, P., and A. W. Sheppard. 1995. Are particular weeds more amenable to biological control?—A reanalysis of mode of reproduction and life history. Pages 95-102 *in* E. S. Delfosse and R. R. Scott, eds. Proc. VIII Int. Symp. Biol. Contr. Weeds, 2-7 February 1992, Lincoln Univ., Canterbury, New Zealand. DSIR/CSIRO, Melbourne, Australia.

Coombs, E. M., H. Radtke, D. L. Isaacson, and S. P. Snyder. 1996. Economic and regional benefits from the biological control of tansy ragwort, *Senecio jacobaea*, in Oregon. Pages 489-94 *in* V. C. Moran and J. H. Hoffman, eds. Proc. IX Int. Symp. Biol. Contr. Weeds, 19-26 January 1996, Stellenbosch, South Africa. Univ. of Cape Town.

Crawley, M. J. 1986. The population biology of invaders. Phil. Trans. Royal Soc. London, Series B. 314: 711-31.

Crawley, M. J. 1989a. Insect herbivores and plant population dynamics. Annu. Rev. Entomol. 34: 531-64.

Crawley, M. J. 1989b. The successes and failures of weed biocontrol using insects. Biocontrol News Info. 10: 213-23.

Cullen, J. M. 1995. Predicting effectiveness: fact and fantasy. Pages 103-9 *in* E. S. Delfosse and R. R. Scott, eds. Proc. VIII Int. Symp. Biol. Contr. Weeds, 2-7 February 1992, Lincoln Univ., Canterbury, New Zealand. DSIR/CSIRO, Melbourne, Australia.

D'Antonio, C. M., and P. M. Vitousek. 1992. Biological invasion by exotic grasses, the grass/fire cycles and global change. Annu. Rev. Ecol. System. 23: 63-87.

Elton, C. S. 1958. The Ecology of Invasions by Plants and Animals. Chapman and Hall, London.

Enserink, M. 1999. Biological invaders sweep in. Science 285: 1834-36.

Goeden, R. D. 1983. Critique and revision of Harris' scoring system for selection of insect agents in biological control of weeds. Prot. Ecol. 5: 287-301.

Grevstad, F. S. 1999a. Experimental invasions using biological control introductions: The influence of release size on the chance of population establishment. Biol. Invasions 1: 1313-23.

Grevstad, F. S. 1999b. Factors influencing the chance of population establishment: implications for release strategies in biocontrol. Ecol. Appl. 9: 1439-47.

Gruber, E., and A. Whytemare. 1997. The return of the native? *Sidalcea hirtipes* in coastal Oregon. Pages 121-24 *in* T. N. Kaye, A. Liston, R. M. Love, D. L. Luoma, R. J. Meinke, and M. V. Wilson, eds. Conservation and Management of Native Plants and Fungi. Native Plant Soc., La Grande, OR.

Harris, P. 1973. The selection of effective agents for the biological control of weeds. Can. Entomol. 105: 1495-1503.

Harris, P. 1979. Cost of biological control of weeds by insects in Canada. Weed Sci. 27: 242-50.

Hobbs, R. J. 1989. The nature and effects of disturbance relative to invasions. Pages 389-405 *in* J. A. Drake, H. A. Mooney, F. di Castri, R. H. Groves, F. J. Kruger, M. Rejmanek, and M. Williamson, eds. Biological Invasions: A Global Perspective. John Wiley, NY.

Hobbs, R. J. 1991. Disturbance a precursor to weed invasion in native vegetation. Plant Prot. Q. 6: 99-104.

Huffaker, C. B., and C. E. Kennett. 1959. A ten-year study of vegetational changes associated with biological control of Klamath weed. J. Range Manage. 12: 69-82.

Julien, M. H., J. D. Kerr, and R. R. Chan. 1984. Biological control of weeds: an evaluation. Prot. Ecol. 7: 3-25.

Lawton, J. H. 1990. Biological control of plants: a review of generalisations, rules, and principles using insects as agents. Pages 3-17 *in* C. Bassett, L. J. Whitehouse, and J. A. Zabkiewcz, eds. Alternatives to the Chemical Control of Weeds. Proc. Int. Conf. Ministry of Forestry, FRI Bull.

Levine, J. M. 2000. Species diversity and biological invasions: relating local process to community pattern. Science 288: 851-54.

Lonsdale, W. M., D. T. Briese, and J. M. Cullen. 2001. Risk analysis and weed biological control. Pages 185-210 *in* E. Wajnberg, J. K. Scott, and P. C. Quimby, eds. Evaluating Indirect Effects of Biological Control, CABI Publishing, NY.

Loope, L. L., P. G. Sanchez, P. W. Tarr, W. L. Loope, and R. L. Anderson. 1988. Biological invasions of arid land nature reserves. Biol. Conserv. 44: 95-118.

McEvoy, P. B. 1996. Host specificity and biological pest control. BioScience 46: 401-5.

McEvoy, P. B. 2002. Insect-plant interactions on a planet of weeds. Entomol. Exp. Appl. 104: 165-79.

McEvoy, P. B., and E. M. Coombs. 1999. Biological control of plant invaders: regional patterns, field experiments, and structured population models. Ecol. Appl. 9: 387-401.

McEvoy, P. B., and E. M. Coombs. 2000. Why things bite back: unintended consequences of biological weed control. Pages 167-94 *in* P. A. Follett and J. J. Duan, eds. Nontarget Effects of Biological Control. Kluwer Academic Publishers, Boston, MA.

McEvoy, P. B., C. Cox, and E. Coombs. 1991. Successful biological control of ragwort, *Senecio jacobaea*, by introduced insects in Oregon. Ecol. Appl. 1: 430-42.

McEvoy, P. B., N. T. Rudd, C. S. Cox, and M. Huso. 1993. Disturbance, competition, and herbivory effects on ragwort *Senecio jacobaea* populations. Ecol. Monogr. 63: 55-75.

Parker, I. M., D. Simberloff, W. M. Lonsdale, K. Goodell, M. Wonham, P. M. Karieva, M. H. Williamson, B. Von Holle, P. B. Moyle, J. E. Byers, and L. Goldwasser. 1999. Impact: toward a framework for understanding the ecological effects of invaders. Biol. Invasions 1: 13-19.

Pemberton, R. W. 2000. Predictable risk to native plants in weed biological control. Oecologia 125: 489-94.

Pratt, P. D., E. M. Coombs, and B. A. Croft. 2003. Predation by phytoseiid mites on *Tetranychus lintearius* (Acari: Tetranychidae), an established weed biological control agent of gorse (*Ulex europaeus*). Biol. Control 26: 40-47.

Radtke, H., and S. Davis. 2000. Economic analysis of containment programs, damages, and production losses from noxious weeds in Oregon. Oregon Dept. Agric., Salem, OR.

Reichard, S. H., and C. W. Hamilton. 1997. Predicting invasions of woody plants introduced into North America. Conserv. Biol. 11: 193-203.

**25**

Rejmanek, M., and D. M. Richardson. 1996. What attributes make some plant species more invasive? Ecology 77: 1655-61.

Ruesink, J., I. M. Parker, M. J. Groom, and P. Kareiva. 1995. Guilty until proven innocent: reducing the risk of non-indigenous species introductions. BioScience 45: 465-77.

Shea, K., H. P. Possingham, W. W. Murdoch, and R. Roush. 2002. Active adaptive management in insect pest and weed control: Intervention with a plan for learning. Ecol. Appl. 12: 927-36.

Shigesada, N., and K. Kawasaki. 1997. Biological Invasions: Theory and Practice. Oxford University Press, NY.

Vitousek, P. M., and L. R. Walker. 1989. Biological invasion by *Myrica faya* in Hawai'i: plant demography, nitrogen fixation, ecosystem effects. Ecol. Monogr. 59: 247-65.

# Foreign Exploration

## Matthew F. Purcell, John A. Goolsby, and Wendy Forno

Classical biological control of introduced weeds is based on the use of herbivores and pathogens from the weed's native range, where it is typical to find a large suite of herbivores and pathogens, called "natural enemies." In this setting, natural enemy species have had thousands or even millions of years to adapt to the plant. Exploration for natural enemies to control an introduced target weed is the first active step in a biological control program. The job of the foreign explorer is to sift through the many natural enemies and find species with narrow host ranges and strong potentials to control the weed.

Before the foreign explorer starts work, conflicts of interest regarding the commercial or environmental value of the target plant have been resolved. The plant has been declared a noxious weed through the state and/or federal process. A clear consensus among stakeholders, regulatory officials, and weed biological control researchers has emerged that the weed should be targeted for biological control. This consensus ensures that the target weed's potential impact justifies the release of nonendemic agents and that the program has multi-agency approval. This process is consistent with the International Code of Best Practices for Classical Biological Control of Weeds (see "International Code of Best Practices for Biological Control of Weeds," this volume).

Exploration programs typically involve scientists and land managers in the target plant's invasive range as well as scientists and cooperators in its native range. Often a team of scientists is assembled to conduct the exploration. The skills required are mainly those of an entomologist, botanist, and mycologist. Foreign explorers should also be self-reliant, resourceful, capable of networking, good field biologists, and have a knowledge of the local language and culture where the plant is native.

It is common for foreign exploration to continue several years or even longer. During this time the goals are to survey the entire native range of the target plant and to collect, identify, and determine the host range of associated herbivores and pathogens. Frequently, the first biological control agent candidate is identified within the first year of exploration. Each candidate species undergoes quarantine testing (conducted overseas or at a U.S. quarantine laboratory) to assess its environmental safety. Agents can be approved for release within 18 months of discovery, although the research and approval process often takes longer depending on the complexity of the biology of the candidate species, the availability of nontarget plant species to be tested, and the permitting process.

Funding for foreign exploration generally comes from a coalition of stakeholders, as well as state and federal agencies. Several years of funding is required to complete exploration and preliminary host range studies. The most effective exploratory programs have research stations in or near the native range of the target weed where year-round exploration and testing of agents can be conducted without quarantine restrictions. A budget of $250,000 per year per target weed is optimum for foreign exploration.

The U.S. Department of Agriculture's Agricultural Research Service maintains biological control research laboratories in Beijing, China; Buenos Aires, Argentina; Brisbane, Australia; and Montpellier, France. U.S. scientists based at these laboratories are able to field-test potential biological control agents in their native habitats. In addition, research is conducted for U.S. organizations by many institutions including Commonwealth Agricultural Bureau International (CABI) Bioscience (Switzerland Centre, Delémont) and Commonwealth Scientific and Industrial Research Organization (CSIRO) Entomology (Canberra, Australia).

## Process of Exploration

Before exploration begins, a thorough survey of all available literature is conducted. Valuable information on the plant species and its associated herbivores and pathogens may already be published and may influence survey plans. Comparisons are made of the climate where the plant is a problem and its native range. Climatically similar areas are prioritized for exploration because the natural enemies from these areas are usually better adapted to the climate in the introduced range. The climate-matching program CLIMEX is often used to make these comparisons. CLIMEX calculates an *ecoclimatic index* based on rainfall, degree-days, growth, temperature, moisture, cold stress, heat stress, drought stress, and water stress indices for each location. CLIMEX has a worldwide climate database, so comparisons can be easily made for many locations.

Exploration must be carried out over the whole region where the target plant is believed to be native; explorers proceed along transects traversing altitude, temperature, rainfall, and other ecological gradients. It is hypothesized that the greatest number of natural enemies occurs near the center of diversification or evolution of the weed genus or subgenus. Some of the natural enemies are likely to have evolved with the weed and have a host range specific to the weed, or to the weed and a few species within the same genus. When the center of diversification is known, intensive surveys are made in that region. Molecular methods are also used to match the target population with a corresponding genotype in the native range. Searches in the center of diversification are particularly important when dealing with highly specialized herbivores such as eriophyid mites.

Surveys are designed to take account of season, habitat, and soil or water type, and follow a predetermined plan. Many natural enemies are seasonal in abundance, even in the tropics. It is desirable to conduct preliminary surveys at a time when the weed and its natural enemies are abundant. Night searches are often incorporated into the surveys. A different suite of nocturnal herbivores may be revealed and the activity of some species may be stimulated.

Surveying methods are determined by the type and habitat of the target weed. At each locality, care is taken to collect natural enemies from different phenotypes of the target weed, all parts of the plant, and from closely related plant species growing in the same habitat. Plants are searched directly for natural enemies at field sites, or indirectly from plant samples collected for examination at the field station. Field searches are beneficial for obtaining data on the distribution, seasonal abundance, behavior, and habitat of herbivores, and for collecting large numbers of agents for laboratory studies, colonization, or exportation. Any evidence of damage on the target plant, including defoliation, gall formation, flower and fruit damage, or dead tissue, is carefully investigated for the causative agents. Potential biological control candidates are collected directly from the plants by hand or by using an aspirator. The niche and feeding habits of all herbivores are recorded. Specimens not needed for rearing or laboratory trials are placed into killing jars, then preserved.

Preserved specimens are forwarded to taxonomists for identification and to molecular geneticists for characterization. Additional specimens lodged in museum collections are called vouchers and are representative of the population under evaluation. If the organisms are described as new species, an individual specimen is designated as the holotype. Arthropod species that show potential as biological control agents are often characterized using molecular methods, i.e., gene sequencing, using multiple specimens to insure consistency in the results. All life stages of herbivores can be used for gene sequencing. Molecular characterization is an interim method of assessing species diversity, identifying cryptic species, and matching immature stages with adults. It is also used to assess the genetic diversity of the target weed species and its close relatives. Sequence data are posted on GenBank (http://www.ncbi.nlm.nih.gov/) and serve as vouchers for molecular diagnostics.

Field samples of terrestrial weeds are made using hand-pruners or loppers attached to extension poles for taller trees. Aquatic weeds are collected from shore by hand, while wading, or by using a rake or grappling hook attached to a rope. In deeper water, a boat or equivalent vessel is used. The plant material is placed into labeled collection bags for transportation back to the laboratory. Both plant material and live insect specimens are protected from temperature extremes during transit.

Field data sheets are completed for each collection and record site descriptions (e.g., name, locality, latitude/longitude, topography, vegetation), collection date, time, physical parameters (e.g., temperature, humidity, weather), weed parameters (height, reproductive state, sample weight), herbivores/pathogens observed and collected, as well as general notes. Each collection is assigned a unique number for specimen records and entry into a database.

Collections processed in the laboratory are searched for natural enemies by hand or by using insect extraction devices (e.g., Berlese funnels). Extraction devices are especially useful for collecting internal feeders such as stem-borers and leaf-miners, and for processing large amounts of plant material. Specimens of the host plant, including reproductive parts, collected from each region are labeled and stored or sent to specialist botanists for determination.

Immature and adult stages of each herbivore species are collected. Ideally, at least 10 specimens representing all regions surveyed are preserved; if there are different forms (polymorphic species), then specimens of each form are collected. Immature phytophagous insects are reared and the hosts of predators and parasitoids noted. Records of the feeding habits and life histories of each insect are maintained. Digital images of specimens are taken, which can be forwarded to taxonomists. Usually, adult specimens are sent to specialist taxonomists because immature stages are generally difficult or impossible to determine, although they are sometimes required. Collection records of the same species held in national collections are obtained. These records may contain valuable information on hosts, niche, feeding habits, seasonality, and distribution. Pathogenic fungi are collected, dried, and sent to specialists for determination.

Laboratory colonies of promising biological control agents are established. These colonies are used for detailed studies of biology and host range, impact assessments, and exportation to U.S. quarantine facilities. Detailed studies determine which agents have the greatest potential as candidates for introduction against the weed in its introduced range. Selection is based largely on host specificity, but there is increasing emphasis on predicting effectiveness. Impact is assessed more accurately when specialist parasitoids and predators, which will not be present in the introduced range of the weed, are excluded. Most agents are selected because they have a narrow host range and are very damaging to the target plant in its native range.

Studies document a natural enemy's seasonal abundance, distribution, and biology and provide an indication of host range. The latter may be achieved through field observations and by screening a few plants closely related to the weedy species. When sufficient resources exist, candidate agents are host-tested in the field in the native range of the weed. Studies are conducted throughout the year and involve intensive searching for candidate agents on both the target weed species and all other potential hosts along transects or in a defined area where potential agents occur. The abundance and seasonality of these agents, and the damage they cause, is then determined for each plant species. Trap gardens are sometimes planted to draw in potential agents for collection and study. Plants that are not present but are required for host testing are placed or planted in the survey areas whenever possible. Critical related nontarget plant species can be planted near the trap plants for field host range studies.

Seasonality surveys involve regular monitoring of field populations of potential agents, ideally over a two- to three-year period. These surveys not only account for climatic variation, but also the various growth phases of the target weed. This information is invaluable in determining timing of field releases of the biological control agent and also improving its chances of successful establishment in the introduced range.

Studies may include pesticidal exclusion experiments in which individual plants of the target weed are divided into two groups under natural conditions. One group is sprayed with a pesticide that excludes herbivore attack but does not damage the plants, while the other group is untreated and therefore exposed to attack. The growth rates or biomass of each group are then compared to gauge herbivore impact. This method is often useful when the damage of the candidate agent is subtle or the effect is cumulative over a long period of time.

In summary, the process for a candidate agent proceeds as follows: 1) field host range observations; 2) establishment of laboratory rearing colonies; 3) exclusion of parasitoids and pathogens from the laboratory colony; 4) preliminary laboratory host range evaluation; 5) biological studies; 6) target weed impact studies; and 7) taxonomy of the organism. Once these steps are completed and USDA Animal and Plant Health Inspection Service Plant Protection and Quarantine (USDA-APHIS-PPQ) has approved the importation, preparations are made to ship the agent to U.S. quarantine facilities for final host range testing. It should be noted that permits are also needed from the country of export to comply with international law under the Convention on International Trade in Endangered Wildlife and Flora (CITES). The U.S. Department of Interior Fish and Wildlife Service checks importations for CITES compliance.

To ensure that biological control agents arrive in the United States in the best possible condition, the fastest and most direct route (usually by air) is determined. Survival is directly related to the time the agents spend in transit. Shipments must be sent directly to the containment facility from the country of origin.

Correct packaging is crucial for the survival of insects and ensures that strict quarantine safeguards are maintained. Packaging techniques vary widely according to the type and habit of the insects. Insects are usually held on host plant material within a gauze or cloth bag tied off with string. The bags are then placed into sealed plastic containers. Excess moisture within the containers is controlled by lining the interior with absorbent paper and/or by ventilation (e.g., gauze-covered windows or small ventilation holes). The containers are put into an appropriately sized box (preferably insulated), sealed, and then placed into a larger carton surrounded by sufficient packaging to insulate them from external temperature and physical damage. Designated labels provided by USDA-APHIS (PPQ Form 599) must be affixed to all packages shipped under permit.

## References

Balciunas, J. K. 1999. Code of best practices for classical biological control of weeds. Page 435 *in* N. R. Spencer, ed. Proc. X Int. Symp. Biol. Contr. Weeds, 4-14 July 1999, Montana State Univ., Bozeman, MT.

Forno, I. W., and M. F. Purcell. 1997. Exploration for agents. Pages 51-55 *in* M. Julien and G. White, eds. Biological Control of Weeds: Theory and Practical Application. ACIAR Monogr. No. 49.

Goolsby, J. A., A. D. Wright, and R. W. Pemberton. 2003. Exploratory surveys in Australia and Asia for natural enemies of Old World climbing fern, *Lygodium microphyllum*: Lygodiaceae. Biol. Control 28: 33-46.

Goolsby, J. A., C. J. Burwell, J. Makinson, and F. Driver. 2001. Investigation of the biology of Hymenoptera associated with *Fergusonina* sp. (Diptera: Fergusoninidae), a gall fly of *Melaleuca quinquenervia*, integrating molecular techniques. J. Hymenoptera Res. 2: 172-200.

Goolsby, J. A., J. R. Makinson, D. M. Hartley, R. Zonneveld, and A. D. Wright. 2004. Pre-release evaluation and host range testing of *Floracarus perrepae* (Eriophyidae) genotypes for biological control of Old World climbing fern. Pages 113-16 *in* J. M. Cullen, D. T. Briese, W. M. Lonsdale, L. Morin, and J. K. Scott, eds. Proc. XI Int. Symp. Biol. Contr. Weeds, April 27-May 2, 2003, Australian Nat. Univ., Canberra, ACT, Australia.

Sutherst, R. W., G. F. Maywald, T. Yonow, and P. M. Stevens. 1999. CLIMEX. Predicting the Effects of Climate on Plants and Animals. CD-ROM and User Guide. DSIR/CSIRO, Melbourne, Australia.

# Host Specificity Testing of Biological Control Agents of Weeds

Jeffrey L. Littlefield and Gary R. Buckingham

An important selection criterion for potential biological control agents for weeds is that the agents will be reasonably host-specific and not become pests. The determination or estimation of an agent's host range and host specificity is used primarily to identify potential risks and prevent unacceptable impacts on nontarget plants. In addition, excluding nonspecific agents early in the screening process reduces time, costs, and effort. A potential biological control agent's host range must be determined prior to the release of the organism in a new environment. In this chapter we present a general overview, based upon our experiences and those of other researchers, of various testing strategies that are utilized for this determination.

Many factors play a role in the selection and utilization of a plant by its natural enemy. Phylogenetic, genetic, physiological, behavioral, and ecological constraints determine the breadth of the organism's host range. Host selection and utilization is a set of complex behaviors. Not only does the organism need to locate the host plant and often specific host tissues, in many cases it also has to locate the habitat of the host. The host, once found, must be suitable for feeding, reproduction, and progeny development. Because of the potential complexity of the selection process, the researcher's task of determining agent specificity has to be somewhat simplified. Typically, we consider only a limited portion of the organism's selection process, i.e., the selection of the plant for oviposition and feeding, and its suitability for the development of reproductive offspring. To determine the potential host range of an herbivore, we often ask a series of questions.

## How specific?

How specific an agent must be is largely dependent upon the target host and its phylogenetic relationship with plants of economic or environmental importance. A genus-specific organism may be suitable if the genus containing the target weed is only represented by exotic species of no economic or environmental importance. However, if the target weed has closely related native species or exotic species of environmental or economic concern, a more species-specific biological control organism is desirable. Also the potential effectiveness of the agent must be weighed against its potential host range. In certain cases, a highly effective but less specific agent may be more desirable than a less effective but more host-specific agent.

## Can the organism be rejected quickly?

It is advantageous to quickly eliminate unwanted agents from the screening process. A quick method is simply to review the literature for references regarding possible host plants of the agent. Although this may quickly identify nonspecific organisms, especially those that are of a pestiferous nature, the literature may lack references to less common insect species or host plants. Insect collections—generally those associated with universities and government agencies—are often inspected to determine host records, although in many cases a specimen's host information may not exist. While these sources of information are valuable, care must be exercised to prevent prematurely eliminating a suitable candidate because of incorrect data or the misidentification of either the herbivore or its host plant. Although the literature may provide information about the host range of the organism, this information is of limited use in predicting the risk to plant species that have never come into contact with the proposed agent. Thus, experimental testing of potential nontarget hosts is required to identify potential risks.

## How do we test for specificity?

Much has been written regarding host specificity testing (see References). While different approaches have been used or suggested, we will present a simplification or generalization of the testing procedures that typically have been employed. Often there is no "correct" testing method because the nature of the herbivore-plant interaction differs among species; therefore testing procedures are tailored to the specific species or to similar groups of herbivores, for example, defoliating moths, root-boring beetles, seed head gall flies, etc.

### • Test plant lists

The number of plant species that may be exposed to a new biological control agent could be immense. It is impractical, both financially and procedurally, to test every plant species. Therefore, biological control workers have sought methods of selecting test plants that are practicable and will provide an adequate level of safety and numerous researchers have addressed how to make the selection of test plants more efficient (see "Technical Advisory Group for Biological Control Agents of Weeds," this volume). Recommendations for choosing a test plant list can be found in USDA-APHIS (1998). These recommendations are based partially on the phylogenetic approach by which closely related species are theorized to be at greater risk of attack than are distantly related species. Seven general categories of plants are considered for testing. These include: 1) genetic types of the target weed, 2) species within the same genus, 3) species of other genera—same plant family, 4) threatened and endangered plants, 5) species found in other families—same plant order, 6) species found in other plant orders, and 7) plants on which the agent or its close relatives feed or reproduce.

### • Selecting representative populations

Herbivores and test plants are carefully selected so that they are representative of field populations and do not bias the results of the tests. The test population of the candidate agent should be genetically diverse to provide for a "typical" expression of host acceptance and utilization; but care should be taken in selecting these populations because potential biotypes or races of the organism may exhibit differences in host ranges. Representative individuals are often selected from a single or several specific locations for testing and possible future field release. In choosing individual organisms for testing,

33

it is recognized that results may be influenced by inherited differences in the herbivore's response due to age, life stage, or sex, as well as differences due to previous feeding by the herbivore or the presence of disease.

When plants are selected for testing, similar concerns must be considered. Differences in plant age, tissue type, overall plant quality, biotypes, the variability or presence of plant defense compounds caused by disease or damage, or differences in the use of intact versus excised tissues may alter the herbivore's acceptance of the plant for oviposition or feeding. Environmental conditions, especially light, temperature, and humidity, also affect herbivore-plant interactions during testing.

### • *Types of tests*

Several tests are usually performed to assess the potential host range of a herbivore: 1) oviposition tests to determine the acceptability of plants for egg laying; and 2) feeding tests to determine the suitability of test plants for successful development from egg to adult and for the emergence or development of viable individuals. Oviposition and feeding tests may be divided into two general categories: 1) no-choice tests in which the individual is forced to feed on the test plant or starve, or to either lay eggs or not; and 2) multiple-choice tests in which the organism is exposed to several potential hosts. No-choice tests provide a simple way to ascertain which plant species are definitely not suitable hosts and to quickly eliminate them from further consideration. These tests are highly artificial because they do not consider the mechanisms of host selection by the insect and, if used alone, may lead to the rejection of possible host-specific agents. Multiple-choice tests are often better predictors of the potential host range of the organism because they introduce the element of choice, which is more typical of natural conditions. These tests may be more complex to set up compared with no-choice tests, and therefore fewer plants are usually tested. Multiple-choice tests are often used to further delineate the host range after no-choice tests have been conducted, usually when positive feeding responses on nontarget plants have been observed. These tests may be conducted either in the laboratory or in outdoor gardens at overseas sites.

### • *Testing location*

Testing may be conducted either in the laboratory or under field conditions. Both have advantages and limitations. Sometimes the complete screening process is conducted at a U.S. quarantine facility, especially if the country of origin lacks adequate facilities or trained personnel to conduct field or laboratory tests. Equipment, personnel, or test plants may be more readily available in the United States, or economic or political concerns may not favor performing the work overseas. Laboratory or greenhouse tests may allow a large number of test plants to be evaluated, although in a somewhat artificial situation. In some cases this screening may be adequate, but with other herbivores, field tests may better differentiate their potential host range.

Testing in the native range of the biological control agent has several advantages: 1) the agent does not have to be reared under laboratory conditions (which can sometimes prove difficult); 2) fresh agents can be collected from the field should problems arise in maintaining a culture; and 3) more realistic tests can be performed under field conditions. However, field tests in an herbivore's native range may be more difficult to set up, especially with large numbers of test plants or when countries where the field tests are occurring restrict the importation of field test plants from the United States. Under field conditions, agents may be exposed to natural parasitoids, predators, or pathogens that may decimate test populations of both insects and plants. In addition, tests that require the confinement of insects to the test plants are faced with the same problems in the interpretation of results as are laboratory tests. Despite these concerns, field tests are normally used for safety reasons (e.g., preliminary tests of key plant species with little-known herbivores) or to supplement laboratory tests in which feeding on nontarget plants is observed. In some cases, observations of the biological control agents are also made in its native range to determine feeding and ovipositional behaviors that may restrict its host range, or to observe the utilization of other plant species under "natural" conditions.

### • *Testing aquatic biological control agents*

Host range testing of aquatic weed agents is similar to that of terrestrial weed agents, but with some differences—mostly in techniques. Choice of test plants varies somewhat from many terrestrial programs. Many submersed plant families are relatively small and the genera often include only a few species. Also, host plant records for aquatic herbivores are poorly documented. Thus, testing protocols for aquatic biological control agents often include a larger number of plant families and genera than do many terrestrial programs. Because some aquatic insects hibernate on shore and might feed in spring before flying back to the waterways, cultivated terrestrial crops and ornamentals that might be exposed to them are also considered for testing.

It is obvious that even the best-planned host range tests will fail without proper techniques (see "Post-Release Procedures for Biological Control Agents of Aquatic and Wetland Weeds," this volume). In testing herbivores on aquatic plants, sufficient oxygenation of water must be provided under laboratory conditions. Cages must not be overfilled with plants, especially when working with agents that obtain air from the water with plastrons, such as many aquatic caterpillars and weevils. Excess submersed plant biomass can become deadly at night when the plants respire and use up oxygen. Even leaf-mining fly larvae, which can live in low-oxygenated conditions, are affected. When testing internal feeders on plants in water, stem length is also important. Short stems often become waterlogged killing the immature insects, especially the immobile pupae. Some larvae that feed in the stems do not pupate until the plant is exposed by receding water. This is suspected when mature larvae do not pupate in submersed stems.

Techniques for testing and rearing insects on floating and immersed plant species are more similar to the terrestrial techniques. In some instances, testing may be conducted without water, e.g., most beetle larvae that feed internally and pupate either internally or in a cocoon. These insects are always surrounded by a plastron of air, not water, and therefore can be reared and tested on plants, including submersed species, that are held out of water but in a moist environment. This solves the problem of stem waterlogging and air deficit, although it might produce a broader laboratory host range. Selected plants may be retested later in water.

Sterilization of cages is extremely important when working with aquatics. Plants often quickly decompose without a bleach treatment of the cage prior to each change of plant material, especially with submersed plants. Because aquatic insects require high humidity, insect pathogens, especially fungi, are more bothersome in these programs. Plants and insects sometimes are dipped in a fungicide harmless to insects. Insects are also examined repeatedly for protozoans that can be introduced on field-collected plants.

• *Interpretation—assessing host range versus specificity*

The host range of a herbivore may vary depending upon conditions. The fundamental or physiological host range, which is the potential number of hosts that could be utilized based upon laboratory tests, may be restricted under field conditions. Ideally, biological control practitioners would like the insect to feed and develop only upon the target species. Nevertheless, many of our biological control agents are not strictly monophagous in their native range and may utilize closely related species, either in the same genus or in related genera. When other plant species are used by the herbivore during testing, researchers must decide whether feeding is an artifact of the testing procedure or truly indicates the potential host range of the organism—and if so, is the damage significant? To further delimit the potential host range of a herbivore, the ecological context in which the organism will interact with its potential hosts must also be considered. Such ecological criteria or questions may be: Does the phenology of the nontarget species overlap with the activity of the insect? Is the herbivore constrained by specific ecological or physiological factors, such as habitat, elevation, moisture, nutritional requirements, etc.? Are potential nontarget species geographically or ecologically isolated from target species? Can the organism maintain itself on nontarget plants? Does feeding significantly damage these plant species or their populations? These are difficult questions to answer or to test, and often we can only estimate what the limitations of the agent will be prior to its introduction. However, careful evaluation of agents already released—their ecological limitations, spread, interactions with other plant species, etc.—may allow us to better address these questions for future agents, or perhaps lead to additional questions to be answered.

## How can we improve our predictions?

Defining an organism's host specificity is often a complex task, but one that is essential prior to the agent's release into a new environment. Predictions of the potential host range of an organism should be based upon biological, behavioral, ecological, and taxonomic information or considerations, as well as laboratory and field experimentation. Researchers should be aware of the limitations and shortcomings of traditional testing procedures so they do not reject host-specific organisms, yet still maintain a high level of safety. In the future, a greater emphasis will be placed on demonstrating the safety of biological control organisms prior to their introduction. To make host specificity testing more reliable, we must still rely on traditional testing techniques, but we must also have a greater understanding of the key elements of the host selection and utilization process, of the taxonomic and phylogenetic relationships of the organisms involved, of evolutionary processes, and of the ecological context in which the organisms will be placed.

## References

Cate, J. R., and J. V. Maddox. 1994. Host Specificity in Biological Control Agents: Report of a Workshop Sponsored by the National Audubon Society. National Audubon Society, Washington, DC.

Harley, K. L. S., and I. W. Forno. 1992. Biological Control of Weeds, a Handbook for Practitioners and Students. Inkata Press, Melbourne, Australia.

Jacob, H. S., and D. T. Briese. 2003. Improving the selection, testing and evaluation of weed biological control agents. Proc. CRC for Australian Weed Management Biol. Contr. Weeds Symp. and Workshop, 13 Sept. 2002, Perth, West Australia. CRC for Australian Weed Manage. Tech. Series No. 7.

McEvoy, P. B. 1996. Host specificity and biological pest control. Bioscience 46: 401-5.

Paynter, Q., and J. L. Littlefield. 1998. Safety in weed biological control: host specificity screening of insect herbivores. Pages 12-14 *in* D. Isaacson and M. H. Brookes, eds. Weed Biocontrol: Extended Abstracts from the 1997 Interagency Noxious-Weed Symposium. USDA Forest Service Forest Health Technology Enterprise Team, Publ. FHTET-98-12. Morgantown, WV.

Schaffner, U. 2001. Host range testing of insects for biological weed control: how can it be better interpreted. Bioscience 51: 951-59.

USDA-APHIS. 1998. Reviewer's Manual for the Technical Advisory Group for Biological Control Agents of Weeds: Guidelines for Evaluating the Safety of Candidate Biological Control Agents. USDA-APHIS-PPQ, Marketing and Regulatory Programs. 03/98-01.

Van Driesche, R., T. Heard, A. McClay, and R. Reardon. 2000. Host Specificity Testing of Exotic Arthropod Biological Control Agents – The Biological Basis for Improvement in Safety, 8 July 1999, Bozeman, MT. U.S. Forest Service. Forest Health Technology Enterprise Team FHTET-99-1. Morgantown, WV.

Wapshere, A. J. 1974. A strategy for evaluating the safety of organisms for biological weed control. Ann. Appl. Biol. 77: 201-11.

Withers, T. M., L. B. Browne, and L. Stanley. 1999. Host Specificity Testing in Australasia: Towards Improved Assays for Biological Control. Scientific Publishing, Queensland Dept. Nat. Res., Brisbane, Australia.

# Technical Advisory Group for Biological Control Agents of Weeds

Alfred F. Cofrancesco, Jr. and Judy F. Shearer

Agencies and organizations represented on TAG are:

USDA Cooperative State Research, Education and Extension Service

USDA Forest Service

USDA Agricultural Research Service (ARS)

USDA APHIS

USDA ARS Biological Control Documentation Center

USDI Geological Survey

USDI National Park Service

USDI US Fish and Wildlife Service (USFWS)

USDI Bureau of Land Management

USDI Bureau of Indian Affairs

USDI Bureau of Reclamation

Department of Defense U.S. Army Corps of Engineers

US Environmental Protection Agency

Weed Science Society of America

National Plant Board

Canadian representative

Mexican representative

The Technical Advisory Group (TAG) for Biological Control Agents of Weeds was established in 1987. Its inception dates back to 1957 when the Subcommittee on Biological Control of Weeds was established by the U.S. Departments of Agriculture (USDA) and Interior (USDI) as an unbiased advisory group to communicate with researchers and regulatory agencies on the safety of organisms as weed biological control agents (Klingman and Coulson 1983). In 1971 the group was changed to the Working Group on Biological Control of Weeds; however, the focus of both of these organizations revolved around the unbiased assessment of organism safety. The composition and specific roles of the organization have been changed and modified over time, but the ultimate goal has always been to ensure that only safe biological control organisms are introduced (Drea 1991, USDA-APHIS 1998).

During the existence of the first two organizations, methods were developed for information exchange between researchers and regulatory agencies on the safety of potential agents. It was quickly realized that the release of an organism in the United States could have direct or indirect impacts on all of North America. Reciprocal reviews were established with Canada and Mexico for collaboration and comments on petitions submitted by biological control researchers (Coulson 1992). The composition of the organizations was also expanded to include specialists in fields such as plant taxonomy and plant quarantine research. The first 30 years provided a strong foundation for the development of the current organization (Coulson 1992).

In 1987 a formal charter was developed to establish the TAG. This organization provides recommendations to the USDA Animal and Plant Health Inspection Service, Plant Protection and Quarantine (APHIS-PPQ) and researchers on the safety of organisms as potential biological control agents (USDA-APHIS 1998). Currently, TAG is composed of 15 governmental agencies representing the United States, Canada, and Mexico. Membership on the working group continues until individuals are replaced by their agency. The TAG is administered by an executive secretary from USDA-APHIS-PPQ who is a nonvoting member. The chair is elected for a three-year renewable term by the membership.

**Petitioner**
- Consults with U.S. Fish and Wildlife Service
- Prepares petition for release or test plant list
- Sends to APHIS-PPQ
↓

**TAG Executive Secretary**
- Establishes timelines
- Sends to TAG members
↓

**TAG members**
- Review and evaluate (Evaluation Guidelines)
- Synthesize comments from subject-matter specialists
- Submit comments and recommendations
↓

**Subject-matter specialists** evaluate

**TAG Executive Secretary**
- Logs and files comments and recommendations
- Sends to Chair
↓

**TAG Chair**
- Consolidates recommendations
- Submits TAG recommendations to APHIS-PPQ, Petitioner, TAG members, and other interested parties
- Files petition and recommendation with ARS, BCDC
↓

**Does TAG recommend release?**  →  No  →

**Petitioner**
- Conducts more research and
- Resubmits petition or test plant list, or
- Discontinues effort, or
- Elects to submit permit application to APHIS anyway

↓
Yes
↓

**Petitioner** submits a permit application to APHIS-PPQ
↓

**APHIS-PPQ consults with the State Plant Regulatory Official (SPRO)**
- APHIS-PPQ sends permit application to the SPRO
- SPRO returns it with comments
↓

**APHIS-PPQ**
- Prepares an EA
- Notifies TAG of results
↓

**APHIS-PPQ** advises Petitioner that an environmental impact statement (EIS) is needed
↓

**Does APHIS-PPQ reach a finding of no significant impact (FONSI)?**  →  No  →

**EIS is prepared.** Based on the EIS, the Petitioner may
- Receive a permit, or
- Discontinue effort

↓
Yes

**APHIS-PPQ issues a permit**

Figure 1. Review process for TAG petitions.

The TAG reviews two types of documents: Test Plant List Petitions and Petitions to Release an Agent. Each petition undergoes a number of simple and direct steps in the review process (Fig. 1). The researcher submits the petition to the TAG executive secretary. The secretary establishes timelines for review (usually 60 to 90 days) and distributes the petition to the members. The members review the petitions and can also solicit subject-matter experts to review them. Once the members have finished their reviews and consolidated the responses provided by the subject-matter experts, they make a recommendation to the executive secretary. The secretary logs and files the comments and recommendations and provides a total packet to the chair. The chair reviews the recommendations submitted by the members and consolidates the information into a single recommendation that is sent to the USDA-APHIS-PPQ, all TAG members, and to the researcher that submitted the petition.

In reviewing the Test Plant List Petitions TAG members examine the proposed list to ensure that the full range of critical plants will be included in the tests. Reviewers can accept the list or suggest modifications to the researcher. This list will then become part of the Petition to Release an Agent. When examining the release petition the TAG usually makes one of three recommendations to the USDA-APHIS-PPQ: 1) the agent should be released; 2) more information is needed; or 3) the agent should not be released at this time. The USDA-APHIS-PPQ then decides on the next step. If the TAG has recommended that an agent be released, the researcher must follow established procedures in submitting a request for a permit (PPQ Form 526) to the USDA-APHIS-PPQ for the release of the agent. It is then the responsibility of the USDA-APHIS-PPQ to determine whether they will issue a release permit. If the TAG has recommended that more information is needed or that the agent should not be released, the reason for the recommendation and concerns of the TAG members are provided to the researcher. The researcher can then address the concerns of the reviewers and resubmit the petition with additional information.

When submitting a petition, the researcher is encouraged to provide an address and phone number to allow direct contact between the reviewer and the researcher. Direct contact during the review process can often eliminate potential concerns. It is also recommended that the researcher involve local or regional USFWS representatives early in the investigation of potential biological control agents to ensure that threatened and endangered species are considered throughout the research efforts (see "Permitting," this volume).

TAG reviewers concentrate on four main areas when examining a petition.

### • Taxonomy

Reviewers examine the taxonomies of both the target plant and the agent. Information on the taxonomy of the target plant is presented so that reviewers can be certain which plant is being targeted for biological control and which closely related species occur in North America. The region of the world from which the potential agent originates is indicated so there is no confusion over which geographical population is being examined.

### • Test plant list

Test plant lists are critically examined to determine whether the researcher has included a full range of potential alternate hosts that exist within the proposed release range. Care should be taken to ensure that any concerns about the agent previously identified by the TAG are addressed. Information should be presented if the approved TAG list has been modified; this may occur because the researcher encounters difficulties in obtaining test plant species that are rare, threatened, or endangered.

### • *Host range tests*

Host range tests are conducted using choice or no-choice tests (see "Host Specificity Testing of Biological Control Agents of Weeds," this volume). All life stages of the agent that impact the target plant are examined. The researcher presents a wide range of information on the ability of arthropods to oviposit, feed, and develop, while pathogen information documents an agent's epidemiology and disease symptoms on all test plants.

### • *Impact to nontarget plants*

Any impact to nontarget plants must be fully discussed because it raises serious concerns about the safety of the agent. The researcher must explain why the impact that occurred will not cause problems if the agent is approved for release. On rare occasions, agents that impact nontarget plants have been recommended for release when researchers have demonstrated that the agent died or could not develop or that seedlings outgrew an infection of a pathogen.

The Reviewer's Manual for the Technical Advisory Group for Biological Control Agents of Weeds (USDA-APHIS 1998) was developed specifically to assist reviewers (it can be found online at http://www.aphis.usda.gov/ppq/manuals/). It is also an extremely useful document for scientists conducting biological control research for weeds. The manual lays out the format that a petition should follow and gives detailed descriptions of all information that is required in a petition.

## References

Coulson, J. R. 1992. The TAG: development, functions, procedures, and problems. Pages 53-60 *in* R. Charudattan and H. W. Browning, eds. Regulations and Guidelines: Critical Issues in Biological Control, Proceedings of a USDA/CSRES National Workshop. Inst. Food Agric. Sci., Univ. Florida, Gainesville.

Drea, J. J., Jr. 1991. The philosophy, procedures, and cost of developing a classical biological control of weeds project. Natural Areas J. 11: 143-47.

Klingman, D. L., and J. R. Coulson. 1983. Guidelines for introducing foreign organisms into the United States for the biological control of weeds. Bull. Entomol. Soc. Am. 29: 55-61.

USDA-APHIS 1998. Reviewer's Manual for the Technical Advisory Group for Biological Control Agents of Weeds: Guidelines for Evaluating the Safety of Candidate Biological Control Agents. USDA-APHIS-PPQ, Marketing and Regulatory Programs. 03/98-01.

# Permitting

Tracy Horner

The Plant Protection Act (7 United States Code (U.S.C.) 7701 et seq.) provides the Secretary of Agriculture with the authority to regulate any enemy, antagonist, or competitor used to control a plant pest or noxious weed. However, other legislation such as the National Environmental Policy Act (NEPA) 1969, as amended (42 U.S.C. 4321 et seq.), the Endangered Species Act (ESA) 1973, as amended (16 U.S.C. 1531 et seq.), the Coastal Zone Management Act (16 U.S.C. 1451 et seq.), and Executive Order 13112 (64 Federal Register 6183) for Invasive Species may affect the decision to release a nonindigenous weed biological control agent into the environment. The approval process can be very complicated and difficult to navigate without guidance.

## Early Input

Whether or not a candidate biological control agent has yet been identified, researchers should submit a proposed test plant list to the Technical Advisory Group for Biological Control Agents of Weeds (TAG) (see "Technical Advisory Group for Biological Control Agents of Weeds," this volume). At this early stage of the approval process, the TAG comments on the target weed choice and makes suggestions for the proposed test plant list that the researcher plans to use for host specificity testing. The test plant list petition is not required. Two TAG petitions are submitted, but the other one is submitted later in the process after host specificity testing is complete. In addition to submitting a test plant list to the TAG, researchers should be sure that threatened and endangered species are considered in the test plant list. Proposed and candidate species for listing should also be considered because they may be listed at any time. The appropriate agency to contact for information regarding listed species is usually the U.S. Department of the Interior U.S. Fish and Wildlife Service (USFWS), but sometimes the U.S. Department of Commerce National Marine Fisheries Service (NMFS), depending on the nature of the proposed action. Both these agencies have the responsibility of enforcing the Endangered Species Act (ESA). Although a USFWS representative participates on the TAG, this does not substitute for the ESA consultation process. Biological control researchers' separate and direct contact with these agencies facilitates the consultation process. Although not required at this stage, receiving early input on a weed biological control project from the TAG, USFWS, and NMFS reveals problems or concerns that can be addressed, potentially saving years of delays in the biological control approval process.

## Permits for United States Importation

Once potential biological control organisms have been discovered in a foreign country, the researcher must apply for a U.S. Department of Agriculture, Animal and Plant Health

Inspection Service, Plant Protection and Quarantine (USDA-APHIS-PPQ) permit (PPQ Form 526) to import them into the United States for further host specificity testing in a containment facility. USDA-APHIS-PPQ issues plant pest and noxious weed permits under the authority of the Plant Protection Act, the federal plant pest regulations (7 Code of Federal Regulations (CFR) § 330.200–212), and the noxious weed regulations (7 CFR § 360.300). Importation and/or interstate movement of plant pests (including pest insects and mites, butterflies, earthworms, and plant pathogens), noxious weeds, and biological control organisms must occur only in accordance with permit conditions.

Shipments of weed biological control organisms may contain harmful plant pests that might be included accidentally or as a result of improper species identification. Field-collected, imported biological control agents may also be contaminated with exotic parasitoids, or pathogens. In addition, biological control agents may not be acceptably host specific, resulting in harmful impacts to native plants or wildlife. Therefore, the unrestricted importation and distribution of biological control organisms may be detrimental to the environment, agriculture, and commerce of the United States. USDA-APHIS-PPQ oversight of importation, containment, inspection, release, and distribution of weed biological control organisms plays an important role in protecting the natural resources of the United States.

It usually takes six to eight weeks from submission of the application to receive a permit for importation of biological control agents into a containment facility. Once the USDA-APHIS-PPQ permit evaluation unit has reviewed the application and prepared proposed conditions, these are sent to the appropriate state plant regulatory official for review. For applications that are likely to be approved, USDA-APHIS-PPQ will send proposed conditions to the applicant. Permit applicants are required to place their initials next to each condition of the permit and sign a verification letter indicating that they agree to the conditions under which the permit is issued. Once the applicant returns both the initialed conditions and signed verification letter, USDA-APHIS-PPQ will issue the permit.

Permitted agents can be imported only into an adequate containment facility in the United States. Permittees may not hand-carry permitted organisms from the port of entry in the United States to the containment facility. Designated labels provided by USDA-APHIS (PPQ Form 599) must be affixed to all packages entering the United States under permit. The label identifies the package as entering the United States under permit, facilitates its movement through the port of entry, and directs it to USDA-APHIS-PPQ officers for inspection, permit verification, and random sampling of contents. USDA-APHIS-PPQ assigns a unique number to each label and requires that the permittees keep account of each label issued to them.

## Maintaining the permit for importation

Permittees are responsible for safeguarding an authorized organism throughout the duration of the permit, as specified by the permit condition requirements. The permit must be kept valid as long as the organism is in his or her possession, whether or not more movement of the organism takes place. If the permittee plans to leave the institution where the organism is maintained, he or she must either: 1) designate a qualified individual to assume responsibility for the continued maintenance of the organism, and the designee must then obtain a new permit prior to the permittee's departure; 2)

43

apply for a new permit to move the organism to a new facility; or 3) destroy the organism. In any case, the permittee must notify USDA-APHIS-PPQ and the original permit will be revoked.

## Permit for Environmental Release

### Environmental documentation in support

Issuance of permits by USDA-APHIS-PPQ for the environmental release of nonindigenous weed biological control organisms is considered a federal action and triggers compliance with NEPA and the ESA. Although USDA-APHIS-PPQ is ultimately responsible for compliance with those environmental statutes, the applicant can reduce the turnaround time by preparing draft documents for USDA-APHIS-PPQ to finalize. Documentation may be initiated during the host specificity testing phase of the project.

The document required for NEPA compliance is the Environmental Assessment (EA), a concise public document that provides sufficient evidence and analysis to determine whether a Finding of No Significant Impact (FONSI) can be reached or whether a more in-depth Environmental Impact Statement (EIS) must be prepared. The EA provides the public with the potential positive and negative  (direct and indirect) environmental impacts that may occur. Applicants from other federal agencies must also consider their own NEPA implementing procedures specific to any proposed actions.

The document required for compliance with the ESA is the Biological Evaluation (BE). This document is usually submitted to the USFWS. The BE should include several elements: 1) a description of the action to be considered; 2) a description of the specific area that may be affected by the action; 3) a description of any listed species or critical habitat that may be affected by the action; 4) a description of the manner in which the action may affect any listed species or critical habitat and an analysis of any cumulative effects; 5) relevant reports, including any EIS or EA; and 6) other relevant information on the action, affected listed species, or critical habitat.

### TAG recommendation

After host specificity testing for a weed biological control agent has been completed, the researcher must submit a petition to the TAG requesting recommendation for release of the agent into the environment (see "Technical Advisory Group for Biological Control Agents of Weeds," this volume). Unlike the test plant list petition, which is not required, it is USDA-APHIS-PPQ policy that all proposed first-time releases of nonindigenous weed biological control agents be reviewed and recommended by the TAG before the approval process for the agent will proceed.

### Application

When a recommendation from the TAG for the release of a weed biological control organism has been received, the researcher must submit an application (PPQ Form 526) to USDA-APHIS-PPQ requesting a permit to release the biological control agent into the environment. The draft EA and BE prepared by the researcher should also be submitted to USDA-APHIS-PPQ at this time. To speed the review process, it is important to submit documents that are as close to complete as possible.

## Section 7 consultation

According to the ESA, any action that is authorized, funded, or carried out by a federal agency must comply with the consultation requirements of Section 7 of the ESA. This compliance may be achieved through formal or informal consultation. Although the researcher should have been in contact with USFWS and/or NMFS from the beginning, USDA-APHIS-PPQ determines whether formal consultation with those agencies must be conducted at this point in the process. Formal consultation involves the submission of the BE to USFWS and/or NMFS and is required when there are concerns that the proposed release may adversely affect endangered, threatened, or candidate species or designated critical habitat. For weed biological control releases, both formal and informal consultations are conducted between USFWS and/or NMFS and USDA-APHIS-PPQ. However, applicants from any federal agency are strongly encouraged to conduct and complete the consultation prior to applying to USDA-APHIS-PPQ for an environmental release permit. Nonfederal applicants may conduct informal consultations, but first must be designated as a nonfederal representative by USDA-APHIS-PPQ. In any case, early, open communication between the applicant, USDA-APHIS-PPQ, USFWS and/or NMFS is essential to ensure efficient movement through this portion of the approval process.

## Public comment

Once the Section 7 consultation is complete, USDA-APHIS-PPQ incorporates the response from USFWS and/or NMFS (either a Letter of Concurrence or Biological Opinion) into the EA and makes any final changes necessary. The USDA Office of General Counsel reviews the EA to ensure that it meets all legal standards. Once the EA has been approved by the USDA Office of General Counsel, USDA-APHIS-PPQ publishes a 30-day (or longer) notice of availability of the EA in the Federal Register to allow the public to comment on the proposed action. After considering the comments, USDA-APHIS-PPQ either reaches a FONSI and issues the release permit, advises the applicant that an EIS must be prepared, or advises the applicant to discontinue the project. If USDA-APHIS-PPQ issues a permit for environmental release, specific conditions are placed on the permit, including limitations on the release area, notification requirements, submission of monitoring reports, etc. States may impose additional conditions on the release.

## Environmental Protection Agency

Researchers working with pathogens for weed biological control work through the U.S. Environmental Protection Agency (US-EPA), Office of Pesticide Programs, Biopesticides and Pollution Prevention Division. All microbial pathogens for weed biological control proceed through the approval process described previously and USDA-APHIS- PPQ maintains regulatory authority over all interstate movement and release of these organisms. However, the US-EPA regulates microbial pathogens as biological pesticides under the Federal Insecticide, Fungicide, and Rodenticide Act of 1972 (7 U.S.C. § 136 et seq.). The US-EPA approval process for pathogens for biological control of weeds is in addition to and not a substitute for the USDA-APHIS-PPQ approval process. The US-EPA has authority to regulate releases of these organisms in areas greater than 4 ha (10 acres) and less than 20 ha (50 acres) (cumulative in the United States). A release area greater than 20 ha requires an experimental use permit.

The US-EPA registers biological pesticide products intended for commercial use. USDA-APHIS-PPQ regulations for weed biological control agents that are commercialized under US-EPA registration are reexamined on a case-by-case basis.

## Interstate Movement of Approved Weed Biological Control Agents

Once a weed biological control organism has been approved for environmental release, land managers are often interested in distributing the agent into new states where the target weed occurs. Currently, interstate movements of all arthropods and noncommercial pathogens for weed biological control must be authorized by a USDA-APHIS-PPQ permit. However, permits for environmental release are only approved for states that have been covered under an EA and consultation with USFWS and/or NMFS. A supplemental EA and another Section 7 consultation must be conducted before releases into additional states can be approved. Although generally not as time consuming as the original approval process, supplemental approvals can create a considerable delay for land managers anxious to implement a release program. Therefore, it is wise for weed biological control researchers to consider broad areas for release when preparing the TAG release petition and environmental documentation for initial approval. The permitting policy changes frequently, so please check with USDA-APHIS-PPQ for current requirements.

*Acknowledgments*

The author thanks Charles Divan, Erich Rudyj, Walter Gould, and Robert Flanders, USDA-APHIS-PPQ; and Michael Oraze, US Department of Homeland Security, for their helpful editorial comments during the preparation of this manuscript.

*References*

TAG Website <http://www.aphis.usda.gov/ppq/permits/tag>

USDA-APHIS-PPQ permit Websites <http://www.aphis.usda.gov/ppq/permits/> and <http://www.aphis.usda.gov/ppq/permits/biological/weedbio.html>

# Documentation

Jack R. Coulson, Eric M. Coombs, and Baldo Villegas

In the field of biological control of weeds, we need to know what agents are placed where, when, what for, and by whom. The answers to these questions help us to document success, learn from failures, improve techniques, reduce redundancy and costs, and improve monitoring, implementation, and safety.

Generally, releases of biological control agents are documented according to political units (country, state, county, local, etc.) because funding, regulation, and jurisdiction of the biological control process are coordinated by agencies associated with those entities. The need for accurate data on biological control that can be shared across multiple jurisdictions has been acknowledged by a number of agencies. Guidelines to determine the core elements of biological control release and monitoring records began to be developed in 2001.

Shipments of biological control agents from foreign countries into the United States and across state lines require proper documentation through a permitting process (see "Permitting," this volume). Nonquarantine shipments of biological control agents within a state generally do not require federal permits, but it is very helpful to provide documentation to a center for biological control information, usually the state Department of Agriculture.

## National-level Documentation

Documentation of a classical biological control program—and an important step in monitoring—begins at the initial stage of the program. Information concerning the origin of a biological control agent should be provided, in addition to documentation of its release, dispersal, and effects (Coulson 1992). Geographical, ecological, taxonomic, and biological data recorded during foreign collection provide important information affecting future release sites, target hosts, and effectiveness of the introduced agent.

The next stage in a classical biological control program is quarantine clearance: the shipment of foreign material through quarantine facilities (Fisher and Andres 1999). Here any contaminant organisms are destroyed, the identity of the introduced beneficial organism is confirmed, the organism may be cultured, and it is finally consigned or shipped for field release or further culture at nonquarantine facilities. The third stage includes the actual release of the introduced organism into the field. A collateral stage is the recolonization of the organism, i.e., collection from a previously established population for release elsewhere. A final stage, after the organism becomes established, is the monitoring of its effects on target and nontarget organisms.

Each of these three stages is documented in a variety of ways. Prior to 1980, each U.S. agency and quarantine involved in classical biological control programs had its own method of documentation. As the number of quarantine facilities in the United States began to grow, the possibility of a standardized documentation system was explored.

The USDA Agricultural Research Service Biological Control Documentation Center (USDA-ARS-BCDC), established in 1982, developed paper forms for recording each of the three stages. Based on these, a computerized program titled "Releases of Beneficial Organisms in the United States and Territories"—dubbed "ROBO"—was developed. Three annual reports of releases were published—for 1981, 1982, and 1983.

In the 1990s, ROBO documentation data and procedures were posted on the Internet (http://www.ars-grin.gov/nigrp/robo.html). By the end of 2002, importation data for the years 1981-1997 and release data for 1981-1985 were available from ROBO. The online form "Invertebrate Shipment Record-Foreign/Overseas Source" for recording importation data was finalized in 2002 for use by the USDA cooperating overseas and domestic quarantine facilities. Online procedures for direct recording of quarantine consignments and field releases will be developed in the next few years, as will forms for recording importation and release of pathogens for classical biological control.

Information on the release, establishment, and U.S. distribution of many introduced biological control agents is also included in the USDA-APHIS National Agricultural Pest Information System (NAPIS) database (http://ceris.purdue.edu/napis/index.html). This system is used to track information on biological control agents because it is already functioning and uses many of the same data that are required in pest surveys. Data entry forms are available.

> The address for obtaining USDA-APHIS-PPQ Form 526 is:
>
> USDA-APHIS-PPQ, Biological Assessment and Taxonomic Support, 4700 River Road, Unit 113, Riverdale, MD 20737.
>
> The address for obtaining USDA-ARS-BCDC Forms 941, 942, and 943 is:
>
> Biological Control Documentation Center, USDA-ARS National Agricultural Library, 14th Floor, Beltsville, MD 20708.

Everyone involved in releasing biological control agents is solicited to help document the release, establishment, and dispersal of introduced agents by submitting the USDA-ARS Biological Shipment Record—Non-Quarantine (Form AD943) to the ROBO database and/or using the NAPIS data entry forms. It is particularly useful to record otherwise undocumented releases of commercially supplied biological control agents.

Any time biological control agents are shipped across state lines, the receiving state Department of Agriculture should request from the originator a year-end report of shipments made into the state as a part of the conditions set in Section 2 of USDA-APHIS PPQ Form 526—Application and Permit to Move Live Plant Pests or Noxious Weeds. The PPQ Form 526 can be downloaded from the Internet at http://www.aphis.usda.gov/ppq/permits. Upon approval of PPQ Form 526 by USDA-APHIS, an Interstate Shipment Authorization Label (PPQ Form 549) is issued to the applicant and affixed to the outside of shipping containers prior to shipment. Copies of forms for recording nonquarantine releases or recolonizations (and other ROBO-related forms) can be obtained from USDA-ARS-BCDC.

Copies of paper forms for nonquarantine releases or recolonization (and of other forms) can be obtained from (and data returned to) USDA-ARS-BCDC or local, federal, or state offices involved in biological control programs.

## State-level Documentation

In addition to national documentation programs, a number of state and local programs document field releases and establishments of classical biological control organisms. Two of the most developed state programs are in Oregon and California. Oregon has more than 12,000 biological control release records that span more than 56 years and document an estimated 80 to 90% of the releases made in the state. These records are very useful in documenting the history of biological control, creating distribution maps, evaluating successes and failures, and monitoring the impacts of biological control projects. These records form the foundation of information for the important but often neglected aspect of monitoring releases over time.

Anyone releasing biological control agents should record a minimum amount of data: date, target weed, biological control agent species, number of agents released, state, county, land owner/manager, location description (latitude and longitude, Township, Range & Section, Meets and Bounds, UTM, etc.), source of agents, and the name of the person making the release. Proper site information will allow accurate portrayal and analysis using Geographic Information Systems (GIS). Collection and redistribution of biological control agents can be hastened by using GIS analysis to prioritize release locations and reduce redundancy (Isaacson et al. 1995). GIS and historical records of biological control releases on the target weed can be used to identify areas where nontarget impacts are likely to occur and to identify native species at risk.

Releases of biological control agents within a state can be documented with a generic release form, by copying USDA-ARS-BCDC Form 943, or with forms provided by a state Department of Agriculture, university, or other responsible agency.

*References*

Coulson, J. R. 1992. Documentation of classical biological control introductions. Crop Prot. 11: 195-205.

Fisher, T. W., and L. A. Andres. 1999. Chapter 6: Quarantine; Concepts, Facilities, and Procedures. Pages 103-24 *in* T. S. Bellows and T. W. Fisher, eds. Handbook of Biological Control. Academic Press, San Diego and London.

Isaacson, D. L., G. A. Miller, and E. M. Coombs. 1995. Use of geographic systems (GIS) distance measures in managed dispersal of *Apion fuscirostre* for control of Scotch broom (*Cytisus scoparius*). Pages 695-99 *in* E. S. Delfosse and R. R. Scott, eds. Proc. VIII Int. Symp. Biol. Contr. Weeds, 2-7 February 1992, Lincoln Univ., Canterbury, New Zealand. DSIR/CSIRO, Melbourne, Australia.

# Plant Pathology and Biological Control of Invasive Weeds

William L. Bruckart, III, Dana K. Berner, Anthony J. Caesar, and Timothy L. Widmer

The potential for plant diseases to control weeds has been recognized for more than a century; observations about damage to Canada thistle by a rust fungus were described in 1893. The active, deliberate development of pathogens for weed control, however, has been pursued only since the 1960s; there have been several successful cases of weed control since then.

Disease-causing organisms have unique characteristics that make them appealing and useful for weed control. By definition from a general plant pathology text, disease is a "series of invisible and visible responses of plant cells and tissues … that results in adverse changes in the form, function, or integrity of the plant and [that] may lead to partial impairment or death of the plant or its parts." Disease is caused either by living organisms (i.e., biotic or infectious) or nonliving factors (i.e., abiotic stresses) like nutrient deficiency. Biotic diseases, which are caused by living organisms and are the focus of this chapter, result from the combination of three factors: 1) a pathogenic organism; 2) a susceptible host; and 3) a favorable environment. If any of these factors is lacking, then there will be no disease.

Disease is different from injury. Injury is damage from a temporary (noncontinuous) or physical event that also may change the form, function, or integrity of a plant and lead to partial impairment or death. Damage from most insect feeding is considered injury rather than disease. Certain arthropods, such as eriophyid mites, may induce gall formation that mimics disease. For the sake of this discussion on the biological control of weeds, only disease caused by microorganisms will be considered.

## Plant Pathogens

What organisms cause disease in plants? Organisms in every category have been identified as causing disease, including bacteria, fungi, higher plants, mollicutes, nematodes, protozoa, and viruses (Agrios 1997). Not every plant is susceptible to all of these organisms. Organisms that cause diseases in plants are known as "plant pathogens." Some of these organisms require a living host in which to grow; these are "obligate" pathogens. Others can grow without the plant, with other sources of nutrients; these are "facultative saprophytes."

Generally, fungi, bacteria, and viruses are the most commonly studied plant pathogens and, therefore, they are best understood for their role as disease incitants. Pathogens in these three groups of organisms also are the most commonly considered for biological weed control. Some individual organisms are host-specific, while others are capable of infecting several species. Each group of pathogens has characteristics that may be useful or problematic in its application in biological weed control.

## • *Bacteria*

These organisms require a wound or other opening (e.g., hydathode or stomate) to get into the plant and they are spread passively, i.e., they move in rain, wind, or running water, or they are moved by vectors such as humans or insects. Most bacteria can be easily cultured, although some require a living host for survival.

## • *Fungi*

Most of these organisms are capable of making their own way into susceptible plants, but some require insect vectors or wounds. Many fungi produce large numbers of spores that can be blown long distances or splashed in water from plant to plant. Many can be cultured, but many others grow only on living plants. Some cause only local infections, while others can move through the plant systemically.

## • *Viruses*

These are pathogens that need a living host and require either insects, nematodes, or wounding for transfer.

## Infection and Disease

Infection depends upon the three factors necessary for disease: the pathogen, a susceptible host, and a favorable environment. Under optimal conditions, disease can be very severe, leading to the death of the host plant. The amount of damage lessens as each of these factors declines from the optimum condition. In some cases, a pathogen can be applied or spread artificially at a time when the plant is susceptible and the environment favors disease. This is an important component of biological control.

All plant parts can become infected and diseased. Some infections occur throughout the plant (i.e., are systemic) while others are local, occurring only on one or a few plant parts or occurring as spots or cankers on leaves or stems (Agrios 1997).

Many plant pathogens produce plant toxins or enzymes that cause cells to leak nutrients that can then be used by the invading organism. Others, particularly viruses, use the plant's DNA replication system to make more of the pathogen, while many of the obligate pathogens form an intimate relationship with the host that allows the pathogen to use the host nutrients without killing the cells. In some cases, a plant can be infected without showing symptoms or damage. But generally, the amount of plant damage is proportional to the amount of disease, unless a vital part of the plant such as the root system is damaged or destroyed. In that case, the plant may be stunted or killed with limited amounts of infection.

There are a number of cases in which plant pathogens interact either with other plant pathogens or other organisms. Commonly, insects are important components of these interactions. Insects are essential as vectors of some pathogens, usually viruses, moving them from diseased plants to healthy plants. In other cases, insects wound plants while feeding, and microorganisms can invade these sites (Agrios 1997). This combination often results in disease and can lead to increased plant damage and death. Plant pathogens and other microbes also can interact in ways that increase the amount of disease. It has been shown that *Colletotrichum* diseases of several plants are enhanced by nonpathogenic bacteria and yeasts that grow on the leaf surface (Fernando et al.

1994, Shisler et al. 1991). Another study indicated that two pathogens, *Puccinia xanthii* and *Colletotrichum orbiculare*, together cause more disease on Noogoora burr (*Xanthium occidentale*) than either one alone (Morin et al. 1993).

## Biological Control of Weeds with Plant Pathogens

Two considerations determine the potential application of a pathogen for biological control of weeds in the United States. The first is whether the pathogen is present in the United States. Pathogens from foreign sources must be evaluated carefully and approved before they can be used in the field. This is an involved process conducted in special containment facilities. The second consideration is whether the pathogen is obligate or whether it can be cultured. Pathogens that can be cultured have the potential to be increased artificially and used in large quantities like herbicides. The process of growing, packaging, and applying a pathogen is relatively expensive. Obligate pathogens are not suitable for large-scale application, but they can increase in the right environment and spread great distances on their own. A small amount of inoculum does not cost much to produce and release, and the rest of the process is free.

There are two general options for using plant pathogens in biological weed control (TeBeest 1991).

### • *Classical approach*

The inoculative or classical approach involves release of a pathogen into an area where it has not been before. Many invasive plant species are of foreign origin and, as such, do not have the full complement of natural enemies that affect their survival where they are native. Pathogens are collected in the country of origin, evaluated, and released into populations of the target plant species, thus reuniting the weed with natural enemies that may facilitate control. The impact of a release depends upon the agent becoming established and increasing to the point that it damages the target plant. The agent is expected to spread to new areas on its own, creating further damage to the target weed and impact on target weed densities. For this reason, host specificity of the biological control agent is of paramount importance, since other plants will be exposed to the pathogen as it spreads. The classical approach has been used against most invasive species, particularly in rangelands and pastures. It is very cost effective in terms of labor, implementation, and permanence, but it is expensive in time—progress or impact is often measured in years or even decades. In some cases, however, decline in pest populations has been noted within two to three years of release.

### • *Bioherbicidal approach*

An option that has not been pursued actively for control of invasive rangeland or pasture weeds is the inundative or bioherbicidal approach (Boyetchko 1997). It is similar to the use of chemical herbicides except the "active ingredient" is a living organism. This approach involves growing a large quantity of the pathogen, then adding materials to the pathogen preparation to improve storage (up to a year) and enable application with conventional agricultural equipment. The inundative approach is more expensive to develop and use because the pathogen must be grown, formulated, and applied in order to be effective. These characteristics are generally not suitable to weed control over vast acreages of low-income rangeland, or natural or recreational areas. However, this approach should not be overlooked as an option for weed control.

## Testing, Quarantine, and Safety

Plant pathogens have caused significant damage in the United States and have impacted U.S. history (Agrios 1997). Many Irish died or immigrated to the United States as a result of the Irish Potato Famine (1845-1846) caused by the fungus *Phytophthora infestans*, which killed potatoes that were food for the working class. The chestnut blight, caused by the fungus *Cryphonectria parasitica*, decimated stands of the most important lumber tree in the eastern United States and reduced chestnut to an understory plant. Dutch elm disease, caused by *Ophiostoma ulmi*, changed the landscape of America by killing a large number of American elms, a popular shade tree in towns and cities. Each of these pathogens was accidentally introduced into the United States with devastating results (Agrios 1997).

Clearly, it is important that pathogens selected for introduction are safe and will not damage important and valuable plant species. All pathogens evaluated thus far for biological control of invasive weeds have been of foreign origin. Because they come from plants that are not of commercial or ornamental value in their native ranges, very little is known about the pathogens or the diseases they may cause. For this reason, foreign organisms are studied either in a containment facility or at locations overseas until they are known to be damaging to the target plant and not to plants of commercial, ornamental, or ecological importance in the United States. As of 2003, evaluations of foreign plant pathogens are conducted at the United States Department of Agriculture Agricultural Research Service (USDA-ARS) laboratory and quarantine in Frederick, MD, at Montana State University in Bozeman, MT, at the USDA-ARS facility in Montpellier, France, or through cooperators in countries where the target plants are native.

### • $R = H \times E$

There are two steps to the evaluation of any candidate pathogen. First, it is necessary to show it damages the target plant. That is, it must be both pathogenic (infects the target) and virulent (damages the target). It is then necessary to learn whether the candidate pathogen can infect and damage other plants that are of value in the United States.

Risk (i.e., the chance of an undesired event) is the product of hazard (what can go wrong) and exposure (the meeting of pathogen and host where the environment is favorable), or $R = H \times E$ (Alexander 1988). If there is no hazard ($H = 0$ because the pathogen does not infect nontarget plants), then there is no risk. Likewise, if there is no exposure (i.e., $E = 0$ because the pathogen and the nontarget plant never meet), then there is no risk. Because a foreign biological control agent cannot be controlled once it is released, exposure is assumed, it is tested for the one thing that can be controlled, which is hazard. If the candidate pathogen does not infect and damage nontarget species under optimal greenhouse conditions, then it will not infect them in the field.

Tests to determine potential hazard involve inoculation of plants that are related either to the target host or are of commercial or ecological importance in the United States (Bruckart and Dowler 1986, Bruckart and Shishkoff 1993). Environmental conditions used in these tests are optimal for infection. Rarely, important plants in the United States have been infected but not damaged in these tests. In such cases, comparative studies that include indigenous U.S. pathogens of the nontarget species have been tested under equivalent conditions in the greenhouse (Bruckart et al. 1996).

## Evaluations of Pathogens for Biological Control of Weeds in the United States

Several recent examples follow of pathogens used for biological control:

### • *Puccinia carduorum* on musk thistle (*Carduus nutans*)

*Puccinia carduorum* is a rust fungus from Europe, Asia, and North Africa. Significant infection of and damage to musk thistle occurred in the greenhouse, and isolates of the pathogen did not infect other species of *Carduus* in host tests. During the risk assessment, however, some minor infections occurred on a few species of *Cirsium* (mostly native North American thistles) and artichoke (*Cynara scolymus*). These were regarded as potential hazards and more tests were run to learn more about the possibility of risk to nontarget species. In every case, infection of nontargets was weak and *P. carduorum* could not be kept alive on these plants, even under ideal greenhouse conditions (Bruckart and Peterson 1991, Bruckart et al. 1996). One critical test in containment involved slenderflower thistle (*Carduus tenuiflorus*), a relative of musk thistle that is naturally infected with a strain of *P. carduorum* in California (Watson and Brunetti 1984). This plant also grows next to artichoke in California, but the artichokes do not get the rust disease; however, under containment greenhouse conditions, artichokes from California could be infected. This supported conclusions that the foreign strain of musk thistle rust, which infected artichokes in containment, would not likely infect or damage them in the field. Field tests in Virginia, approved by the USDA Animal and Plant Health Inspection Service Plant Protection and Quarantine (APHIS-PPQ), showed that nontarget species were not infected, despite the fact that *P. carduorum* readily infected experimental and volunteer musk thistle plants in and around the study site (Baudoin and Bruckart 1996, Baudoin et al. 1993, Bruckart et al. 1996). Following establishment of *P. carduorum* in Virginia, the fungus spread and now occurs on musk thistle as far west as California (Baudoin et al. 1993, Littlefield et al. 1998, Woods et al. 2002).

### • *Rhinocyllus conicus* plus *Puccinia carduorum* on musk thistle

During a field evaluation of *P. carduorum* in Virginia, it was clear that there was a positive indirect interaction between *Rhinocyllus conicus* (a thistle seed head weevil) and damage by the rust fungus. The weevil caused an average reduction in seed production of 69.3%, compared to controls, in a three-year study. Combinations of the weevil and the rust disease reduced seed production from the controls by 83.9%. In one wet year, reductions in seed production were 77.2% for the insect alone and 97.0% for the combination of the two agents (Baudoin et al. 1993). In another USDA study, there was no evidence that the rust infection reduced impact by any of the insect biological control agents attacking musk thistle (Kok et al. 1996).

### • *Puccinia jaceae* on yellow starthistle (*Centaurea solstitialis*)

Isolates of this rust fungus (primarily from Greece and Turkey) damage yellow starthistle in the greenhouse. Infection of safflower (*Carthamus tinctorius*) by *P. jaceae* was identified as a potential hazard (Bruckart 1989). In this case, greenhouse comparisons were made between *P. jaceae* and *P. carthami* (the cause of safflower rust) from California and Mexico. The results indicated that the yellow starthistle rust, although it infected safflower, did not damage it, and that resistance in safflower to *P. carthami* seems to convey resistance to *P. jaceae*. In contrast, all cultivars of safflower were easily infected by *P. carthami*. Additional tests looked into the potential for *P. jaceae* teliospores to infect safflower seedlings in a manner similar to that caused by *P. carthami* (Bruckart 1989, Shishkoff and Bruckart 1993). No infections have been observed from *P. jaceae* teliospore inoculations, even though teliospores are viable and applied at rates similar to *P. carthami* teliospores that cause severe damage on inoculated plants (Bruckart and Eskandari 2002). The permit application to release *P. jaceae* in California was approved by USDA-APHIS-PPQ and release was made in early July 2003. Infection on yellow starthistle was verified three weeks after inoculation.

### • *Aphthona* spp. beetles plus soil-borne fungi on leafy spurge (*Euphorbia esula*)

Reductions in leafy spurge stands have been associated with releases of flea beetles in the genus *Aphthona*. There is a direct interaction between beetle larvae, which feed on roots of spurge, and soil-borne fungi in the United States, particularly species of *Rhizoctonia* and *Fusarium*. This interaction between the fungi and insect larvae causes significantly greater damage than that caused by the larvae alone, leading more quickly to reductions in stand densities (Caesar 1994, Caesar et al. 1993, 1998, 1999).

### • *Mixtures of pathogens*

A limiting factor in the development of biological control agents is the desire for host specificity. Although host specificity is one way to insure the candidate agent is safe to use around valuable nontarget plants, host specificity may severely limit the practical application of the agent. However, it may be possible to combine host-specific pathogens into a product that has a broader range of targets. For example, three fungi—*Drechslera gigantea*, *Exserhilum longirostratum*, and *E. rostratum*—can be combined to control seven weedy grasses in Florida citrus (Chandramohan et al. 2002). Similarly, two isolates of *Colletotrichum gloeosporioides* were combined for effective control of northern jointvetch (*Aeschynomene virginica*) and winged waterprimrose (*Ludwigia decurrens*) (Boyette et al. 1979).

In some cases, damage to an individual target weed can be increased greatly by the use of additional organisms. These usually are other pathogens, as in the case of synergism between the rust fungus *P. xanthii* and *C. orbiculare* that infects leaves of cocklebur (*Xanthium*) species. Much higher levels of damage occurred to leaves when these two fungi were used together (Morin et al. 1993). However, enhancement of infection of hemp sesbania (*Sesbania exaltata*) occurred from *Colletotrichum truncatum* that was combined with nonpathogenic leaf-surface bacteria (Schisler et al. 1991).

## Future Considerations

### • *Foreign candidates*

Foreign candidates remain the focus of evaluations for biological control of invasive weeds. New discoveries of pathogens and progress with nonobligate fungi that grow on media, in particular, may provide options for the future of weed biological control. Several new obligate pathogens, which only grow in the living host, have been collected in Europe, Asia, and Africa. These cause significant damage to target species in greenhouse tests and are likely to be host specific. Among these candidates are smut fungi, obligate pathogens that attack the seeds. The classical approach remains the primary option for developing agents for biological control of invasive weeds.

In bioherbicidal developments, several leaf-spotting fungi have been collected overseas on target species, which they significantly damage in greenhouse tests. One fungus, *Ramularia crupinae*, was damaging to common crupina (*Crupina vulgaris*) in field tests and caused a 30% reduction in root biomass after a single greenhouse inoculation (Eskandari and Bruckart 2002, Hasan et al. 1999). The use of leaf-spotting fungi has not been proposed in the past because application as a bioherbicide is expected to be relatively costly compared with the strictly classical approach. Some of these pathogens may function in a classical sense without the manipulations normally associated with facultative pathogens. In any event, to be viable candidates these organisms must be very damaging to target weeds and useful at costs competitive with other approaches.

### • *Indigenous candidates*

Indigenous species may also be candidates for biological control of invasive weeds. Development and application of these organisms will require inclusion of known pathogens into practical control approaches, and others only await their discovery and the imagination of researchers to bring them into play.

Several species of *Uromyces* cause rust on alfalfa (*Medicago sativa*), sweet clovers (*Melilotus* spp.), and clovers (*Trifolium* sp.) in the United States. These fungi have alternate hosts in the Euphorbiaceae, possibly including leafy spurge. Planting these legumes into stands of leafy spurge may facilitate completion of the life cycle of the pathogen and cause additional damage or death of the target weed (Stack and Statler 1989). Also, strains of *Agrobacterium* spp. have been tested and proposed as candidates for biological control of leafy spurge as well as Russian, spotted, and diffuse knapweed (*Acroptilon repens, Centaurea stoebe* ssp. *micranthos*, and *C. diffusa*) (Caesar 1994b).

A number of bacteria that grow on roots have been shown to reduce growth of plants, including leafy spurge and other invasive weeds (Kremer and Kennedy 1996). Limitations of efficacy and delivery may be overcome with additional research that includes the use of other agents (e.g., insects that feed on plant crowns and roots) to move the bacteria to desirable sites.

Host specificity and limited efficacy are two frequently cited shortcomings of the biological control approach. Evidence presented earlier suggests that both combinations of pathogens and interactions with other biological control agents (e.g., insects or nematodes) may enhance the usefulness of this strategy, either in terms of the number of weeds controlled or from improved efficacy against a single target (Boyette et al. 1979, Chandramohan et al. 2002, Morin et al. 1993, Schisler et al. 1991).

Pathogens have great potential for biological control of invasive weeds. Several foreign and endemic candidates have properties useful for adaptation to invasive weed control programs. Obligate fungi are the candidates of choice to date, but the flexibility of culturing and combining facultative organisms also may be useful and practical for weed control with biological agents.

## References

Agrios, G. N. 1997. Plant Pathology. Academic Press, San Diego, CA.

Boyetchko, S. M. 1997. Principles of biological weed control with microorganisms. HortScience 32: 201-05.

Boyette, C. D., G. E. Templeton, and R. J. Smith. 1979. Control of winged water primrose (*Jussiaea decurrens*) and northern jointvetch (*Aeschynomene virginica*) with fungal pathogens. Weed Sci. 27: 49-51.

Bruckart, W. L. 1989. Host range determination of *Puccinia jaceae* from yellow starthistle. Plant Dis. 73: 155-60.

Bruckart, W. L. 1999. A simple quantitative procedure for inoculation of safflower with teliospores of the rust fungus *Puccinia carthami*. Plant Dis. 83: 181-85.

Bruckart, W. L., and F. Eskandari. 2002. Factors affecting germination of *Puccinia jaceae* var. *solstitialis* teliospores from yellow starthistle. Phytopathology 92: 355-60.

Bruckart, W. L., and N. Shishkoff. 1993. Foreign plant pathogens for environmentally safe biological control of weeds. Pages 224-30 *in* R. D. Lumsden and J. L. Vaughn, eds. Pest Management: Biologically Based Technologies, Proc. Beltsville Symp. XVIII, ARS, USDA, Beltsville, MD. May 2-6, 1993. Am. Chem. Soc., Washington, DC.

Bruzzese, E. 1992. Present status of biological control of European blackberry (*Rubus fruticosus* Aggregate) in Australia. Pages 297-99 *in* E. S. Delfosse and R. R. Scott, eds. Proc. VIII Int. Symp. Biol. Contr. Weeds, 2-7 February 1992, Lincoln Univ., Canterbury, New Zealand. DSIR/CSIRO, Melbourne, Australia.

Caesar, A. J. 1996. Identity, pathogenicity, and comparative virulence of *Fusarium* spp. related to stand declines of leafy spurge (*Euphorbia esula*) in the Northern Plains. Plant Dis. 80: 1395-98.

Caesar, A. J., G. Campobasso, and G. Terraglitti. 1998. Identification, pathogenicity and comparative virulence of *Fusarium* spp. associated with diseased *Euphorbia* spp. in Europe. Biocontrol Sci. Technol. 8: 313-19.

Chandramohan, S., R. Charudattan, R. M. Sonoda, and M. Singh. 2002. Field evaluation of a fungal pathogen mixture for the control of seven weedy grasses. Weed Sci. 50: 204-13.

Cullen, J. M. 1985. Bringing the cost benefit analysis of biological control of *Chondrilla juncea* up to date. Pages 145-52 *in* E. S. Delfosse, ed. Proc. VI Int. Symp. Biol. Contr. Weeds, 19-25 August 1984, Vancouver, Canada. Agric. Canada, Ottawa.

Fernando, W. G. D., A. K. Watson, and T. C. Paulitz. 1994. Phylloplane *Pseudomonas* spp. enhance disease caused by *Colletotrichum coccodes* on velvetleaf. Biol. Control 4: 125-31.

Frohlich, J., S. V. Fowler, A. Gianotti, R. L. Hill, E. Killgore, L. Morin, L. Sugiyama, and C. Winks. 1999. Biological control of mist flower (*Ageratina riparia*, Asteraceae): Transferring a successful program from Hawai'i to New Zealand. Pages 51-57 *in* N.R. Spencer, ed. Proc. X Int. Symp. Biol. Contr. Weeds, 4-14 July 1999, Montana State Univ., Bozeman, MT.

Hasan, S., R. Sobhian, and L. Knutson. 1999. Preliminary studies on *Ramularia crupinae* sp. nov. as a potential biological control agent for common crupina (*Crupina vulgaris*) in the USA. Ann. Appl. Biol. 135: 489-94.

Kremer, R. J., and A. C. Kennedy. 1996. Rhizobacteria as biocontrol agents of weeds. Weed Technol. 10: 601-09.

Morin, L., B. A. Auld, and J. F. Brown. 1993. Synergy between *Puccinia xanthii* and *Colletotrichum orbiculare* on *Xanthium occidentale*. Biol. Control 3: 296-310.

Morris, M. J. 1999. The contribution of the gall-forming rust fungus *Uromycladium tepperianum* (Sacc.) McAlp. to the biological control of *Acacia saligna* (Labill.) Wendl. (Fabaceae) in South Africa. Afr. Entomol. Mem. 1: 125-28.

Schisler, D. A., K. M. Howard, and R. J. Bothast. 1991. Enhancement of disease caused by *Colletotrichum truncatum* in *Sesbania exaltata* by coinoculating with epiphytic bacteria. Biol. Control 1: 261-68.

Shishkoff, N., and W. Bruckart. 1993. Evaluation of infection of target and nontarget hosts by isolates of the potential biocontrol agent *Puccinia jaceae* that infect *Centaurea* spp. Phytopathology 83: 894-98.

Stack, R. W., and G. D. Statler. 1989. Unexplained noncurrent distribution of leafy spurge and alfalfa in noncropped areas of eastern North Dakota. Proc. N.D. Acad. Sci. 43: 86.

Supkoff, D. M., D. B. Joley, and J. J. Marois. 1988. Effect of introduced biological control organisms on the density of *Chondrilla juncea* in California. J. Appl. Ecol. 25: 1089-95.

TeBeest, D. O. 1991. Microbial Control of Weeds. Chapman and Hall, NY.

Trujillo, E. E. 1985. Biological control of Hamakua Pa-Makani with *Cercosporella* spp. in Hawaii. Pages 661-71 *in* E. S. Delfosse, ed. Proc. VI Int. Symp. Biol. Contr. Weeds, 19-25 August 1984, Vancouver, Canada. Agric. Canada, Ottawa.

Wilson, C. L. 1969. Use of pathogens in weed control. Annu. Rev. Phytopath. 7: 411-34.

# Handling Insects for Use as Terrestrial Biological Control Agents

Richard W. Hansen

After promising biological control agents have been approved for introduction into the United States, the process of introducing these agents into the country begins. This process starts with the collection of the agent within its native range, its shipment to the United States, and agent release at field insectary sites. The process is completed when field insectary populations are established and they become large enough to permit collections for redistribution. These domestically collected agents are then used to initiate general releases throughout the range of the target weed. A general scenario for introduction, establishment, and redistribution of a weed biological control agent is illustrated in Fig. 1.

## Initial Release and Recovery

### Importing agents

New biological control agents must first be collected from one or more sites within their native range. Generally, collection locations are selected based on the relative ease of collection and agent population size. However, other factors should be considered. Matching host plant species and/or plant biotypes at collection sites to those likely to be encountered in the area of introduction and, generally, some degree of climate and site matching between collection and release locations may be attempted, based on available knowledge. Additionally, potential collection sites where parasitoids or pathogens are known to be relatively abundant should be avoided, if possible. Finally, agents must be collected from the same general geographic range as agents used in the host specificity experiments and other investigations on which a United States Department of Agriculture Animal and Plant Health Inspection Service Plant Protection and Quarantine (USDA-APHIS-PPQ) release permit is based, thus ensuring compliance with the terms of the permit.

Collections should target the agent's life stage or stages that are most appropriate for long-distance shipment. Of course, this decision depends on the insect species involved. In general, nonfeeding or inactive stages (pupae, eggs, or adults) are better suited to transport than are active, feeding stages (larvae or nymphs). Insect adults may be easy to collect in fairly large numbers, but shipment of this life stage should be attempted only among those groups with comparatively robust, long-lived adults (for example, beetles) rather than insects with more delicate, short-lived adults (such as flies and moths). Eggs may require fairly strict temperature or humidity regimes to survive during shipment.

Shipping boxes that cushion and protect their contents while providing temperature and humidity control are essential. In general, polystyrene foam containers within

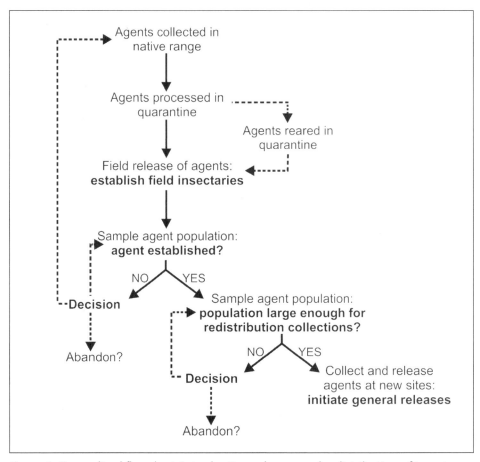

Figure 1. Generalized flowchart: introduction, releases, and redistribution of insects used as weed biocontrol agents.

cardboard boxes are useful for shipping live insects; these containers are commonly used to ship fresh or frozen food, biomedical specimens, and other temperature-sensitive materials. Within shipping containers, insects should be confined in unbreakable cartons that allow some movement of air and water vapor; unlined paper cartons and vented plastic containers may be employed. If host plant material must be provided as a food source during transit, this should be as free as possible of contaminants (soil, other insects, pathogens, etc.) and weed propagules (seeds, rhizomes, etc.). Insect containers may also be provided with a substrate for the insects to rest or hide during shipment; inert, nonliving materials such as shredded paper, wood shavings, or vermiculite are better suited than host plant tissues for this purpose. Excessive moisture is usually more of a problem than desiccation during long-distance shipments, especially if plant material is included. In general, excessive heat is also more likely to affect insect health or survival than cold temperatures during transport, so synthetic ice packs should be included, though they should not come into direct contact with insect containers.

Export permits from the country of origin as well as United States and state import permits (if required) need to be acquired well before collection and shipment of agents (see "Permitting," this volume). Copies of these permits must be provided with the shipping container but must be accessible without opening the portion of the container holding the insects. Appropriate labels, typically as directed by import permits, are affixed prominently on the outside of the shipping container.

Transport of agents should be accomplished by the quickest means possible. Ideally, shipment from the collector to the receiver in quarantine should be completed within 48 hours; even the hardiest of insects is unlikely to survive five or more days in transit. All permitted organisms must be shipped from the country of origin directly to the containment facility. Designated labels provided by USDA-APHIS (PPQ Form 599) must be affixed to all packages shipped under permit. Unless the arrival of biological control agents is a routine event, it is usually a good idea to contact U.S. Department of Homeland Security's Division of Border and Transportation Security personnel beforehand to explain what is arriving and when.

A quarantine facility is a specially designed building in which exotic and/or potentially dangerous or undesirable organisms are held with minimal risk of escape into the environment. Imported insects intended as weed biological control agents must be considered potentially undesirable (e.g., possible crop pests) until their beneficial identity is confirmed. Thus, they must be held in quarantine immediately after their arrival from a foreign location. Shipping containers must be opened only in an approved quarantine facility upon receipt. Any parasitoids, predators, or additional insects or other animal hitchhikers found in shipping boxes or insect containers should be immediately killed and held for later identification. Actively feeding insect agents should immediately be provided with fresh plant material, while inactive stages should be placed in growth chambers under conditions that maintain or break dormancy, as desired. After agents are removed, any plant material and soil must be sterilized and disposed of in an appropriate manner.

Some of the insects received in quarantine should be sent to an appropriate taxonomic specialist so that the identity of the species in question may be confirmed. Arthropod taxonomists may be located at federal or state laboratories or at universities, and many are willing to identify biological control samples. Arrangements should be made with the appropriate specialist well before the agents arrive in quarantine so that identifications may be made as quickly as possible, especially if agents are to be released in the field. If at least some of the sampled organisms are identified as something other than the permitted agent, it may be possible to separate the approved agents from the contaminants. Otherwise, the organisms must not leave quarantine and the colony should be destroyed.

A sample of the agents received in quarantine should also be examined for the presence of internal pathogens and parasitoids. This process typically involves specialized microscopic examinations that may be performed by personnel affiliated with some quarantine institutions; otherwise, the assistance of specialists at various federal, state, or university laboratories may be required. Additionally, several private entities provide these diagnostic services. Again, arrangements should be made before the agents arrive so that samples may be processed as quickly as possible.

Detection of pathogens or parasitoids may require some difficult decisions. By themselves, parasitoids and pathogens do not appear to preclude establishment among field-released agent populations, although the effectiveness of these populations may be reduced. However, it is also possible that an insect parasitoid is an exotic organism

61

not yet established in the United States. For this reason, agent colonies that support exotic parasitoids should not be released into the field unless the identity of the parasitoid can be confirmed and the species is known to occur in the country. This determination may require rearing of parasitoid-infested insects in quarantine until adult parasitoids can be collected and identified by appropriate specialists. Pathogens may be selectively eliminated after selective matings over several generations in quarantine; however, this process requires considerable time, especially for univoltine agents. Thus, a prudent response may be to destroy pathogen- or parasitoid-infected colonies and attempt new collections of uninfected agents.

## Releasing agents in the field

A typical strategy for releasing new insects for use as biological control agents in the United States involves the use of "field insectary" or "field nursery" sites. These sites are weed-infested locations that possess those biological and abiotic characteristics that are believed, given available knowledge, to optimize agent survival, reproduction, and population growth. New agents are released at a relatively few insectary sites and left relatively undisturbed. Once populations have increased to high enough levels (generally, three to five years), surplus agents are harvested for redistribution throughout weed-infested regions.

Initial agent releases may also be made from laboratory or greenhouse colonies established from foreign collections and maintained on the target weed in quarantine; additional USDA-APHIS-PPQ and state permits may be required before agents from quarantine colonies can be released into the field. These colonies provide increased numbers of agents for eventual field release and also permit more control over the timing of field release. Quarantine colonies may be practical for foliage-feeding arthropods, especially those that are multivoltine (e.g., mites, some moths). However, quarantine colonies are much more difficult and time-consuming to establish among univoltine agents and agents that feed in plant stems, on or in flowers, fruits, and roots. With these agents, field releases are usually made directly with individuals collected in and shipped from foreign sources.

Many factors may ultimately determine whether or not agents released at field insectaries survive and become successfully established. Most biotic and abiotic environmental factors that influence establishment cannot be controlled by biological control practitioners, but one important controlling factor is under their direct control—how many agents are released. Established agent populations have been initiated with anywhere from a single gravid female to several hundred or several thousand individuals. Ecological theory, retrospective analyses, and limited experimental evidence suggest that the likelihood of successful establishment increases with larger numbers of insects released and greater numbers of releases. Unfortunately, only a limited number of agents are typically available for initial field releases, so both factors are rarely optimized. Thus, initial release strategies must involve some compromise between the number of releases and the size of each release.

Larger releases increase the probability of successful agent establishment, primarily by reducing the impacts of intrinsic genetic effects and population changes that may impede survival among small populations. Increasing the number of releases spreads the risk by reducing the chance of local extinctions in a highly variable environment.

This is important since we often have limited knowledge of what makes a given site favorable or unfavorable for agent establishment at the start of a release program. For example, most initial introductions involve several hundred to, at most, several thousand individuals. Assuming an initial population of 500 insects, an initial field insectary strategy could involve the release of all 500 at one location, the release of 250 at two locations, or the release of 100 at five locations. Releasing large numbers at one or two sites maximizes the intrinsic likelihood of population establishment but increases the risks of extinction due to catastrophic environmental events (e.g., fire, flood, unacceptable host plants, mowing, insecticide application) at one or both sites. Releasing relatively few agents at many locations reduces the intrinsic probability that any one release will become established, but increases the likelihood that at least one or a few releases will survive in the face of environmental variability; catastrophic events are more likely to affect a few sites than many.

Additional factors must also be considered in determining how many field insectary releases to make. Chief among these are the costs (in time and money) inherent in initiating and subsequently monitoring release locations, including travel, equipment, and sampling and processing time. These costs will be based on the distance between personnel and release locations, relative accessibility of sites, and the comparative complexity of release and monitoring protocols. Obviously, such costs are minimized with one or a few local releases and maximized with many scattered releases.

Thus, determining how many field insectary releases to make and how many agents to assign to each release is an exercise in risk and cost management. Making an intermediate number of intermediate-sized initial releases is an intuitive, reasonably conservative approach that is often pursued by biological control practitioners. In the preceding example, such an approach might involve the release of 100 agents at each of five sites. However, there is no reason why all field insectary releases have to be the same size. Some researchers suggest that a mixed release strategy, employing some large, some intermediate, and some small releases, may be a sound tactic, spreading risk while providing experimental data that may then be used to determine an optimal release size.

Once the number and size of field insectary releases has been at least tentatively settled, the next step is to determine where to make the releases. The distribution and relative abundance of the target weed will obviously control this decision. Field insectary sites should be located where abiotic and biotic factors appear well suited for agent survival and reproduction. Analysis of the impacts of site factors on agent abundance within its native range may offer guidance. However, such guidelines may ultimately have limited value since classical biological control introductions involve completely novel interactions among agent biology and environmental factors that may not at all resemble interactions within the native range. In general, it appears that large, contiguous weed infestations may be more favorable for agent establishment than patchy, discontinuous infestations. Sites that have not been subjected to other weed management tactics (e.g., herbicides, mowing, or grazing) in the recent past and that are unlikely to experience such treatments in the future are probably best suited for the establishment of agent insectaries. Similarly, sites where other disturbances (e.g., seasonal flooding, fires, or road construction) are likely should be avoided. Field insectaries have been successfully established on public and privately owned land; in either case, arrangements ensuring site access, avoidance of site disturbance, and management consistency for at least several years after release should be formalized.

Agent releases at field insectaries may be open releases or may be enclosed within cages of various sizes. Cages confine agents, reducing dispersal and making them easier to find, and may offer protection from natural enemies. On the other hand, cages may affect agent biology if dispersal is a normal behavioral component during the life cycle. They also alter the enclosed microclimate, which can have impacts on agent and host plant biology, limit resources for the agents (especially root-feeders), and make them easy prey for predators trapped inside the cage. Field cages represent an added cost for biological control projects and may be subject to weather damage or vandalism. Ultimately, the decision to cage field insectary releases should be made after careful consideration of agent biology, conditions at the field insectary site, and available project resources. In general, cages may be appropriate for agents with actively flying adults that readily disperse on their own or are easily moved by wind (for example, flies and moths), while agents with comparatively sedentary adults (such as beetles) are better suited for open releases.

Cages are best constructed of ultraviolet-stabilized nylon or other synthetic materials and should consist of a mesh size fine enough to confine the agent in question while large enough to allow movement of air and precipitation and limit shading. Light-colored cages are preferred to reduce solar heating within the cage. Cages may be supported by internal or external frames constructed of steel, aluminum, plastic, or other materials; internal frames are generally preferred for larger cages and for cages left in place year-round where snow commonly occurs. Cage sizes employed in weed biological control projects are highly variable. Small cages are easier to set up, considerably less expensive, and usually less vulnerable to weather damage; larger cages provide easier access and can support larger agent populations, but are more expensive, require stronger frames, and offer greater wind resistance. Cages measuring 3 by 3 m and 1.8 m high, supported by an internal framework of galvanized steel pipes, are used in many USDA-APHIS-PPQ projects throughout the northern United States.

In the first few years after initial release, many insect agents disperse only short distances and their populations increase quite slowly. Thus, the exact release point must often be found to assess the status of populations over time. Open field insectary releases should be marked with a permanent stake, pole, or other vertical object constructed of any of a variety of weather-resistant materials (not wood). Ideally, the marker should be conspicuously colored and tall enough to be visible above the plant canopy from some distance away; however, smaller, more discrete markers should be used if the release site is close to human traffic where vandalism may become an issue. Similarly, shorter markers may be better suited in areas where livestock occur so they will not rub against nor be injured by larger stakes or poles. Caged releases obviously provide their own markers, but it is probably a good idea to add a small stake or other marker near a cage corner in case the cage is lost to weather events or vandalism.

Precise documentation of agent releases is a crucial component of all classical biological control programs. Maps showing the location of the open or caged insectary releases should be drawn at the time of release. These should show routes from major roads, appropriate vehicle parking spots, access trails or roads, relevant landmarks (e.g., fences, trees, bodies of water, etc.), and distances from landmarks to the point of agent release. These maps should permit anyone to locate the exact release point or very close to it, even if markers are removed. Latitude and longitude coordinates for the release, derived from global positioning system (GPS) units, can assist in locating the release site but they should be used with, rather than instead of, a drawn map.

The collection of accurate information at the time of release and the subsequent analysis of this information serve many purposes in a weed biological control program. Agents to be released in the field should be accompanied by USDA-ARS Form AD-942 (for agents collected from foreign populations and processed through a quarantine facility) or USDA-ARS Form AD-943 (for agents collected from domestic or, possibly, Canadian populations). These forms are self-explanatory and record information about the source population, the release location, numbers of agents involved, collection and release dates, and individuals participating in the entire process. A copy of the completed form is sent to a central USDA-ARS Biological Control Documentation Center so that the information can be incorporated into a database of biological control agent introductions in the United States. Additional but similar forms may be supplied by various states or agencies for interstate or local redistribution.

Many weed biological control practitioners collect additional data at the time of field release. Typically, these include more detailed site information, including characteristics of the weed population (infestation size, density, plant height, etc.), characteristics of the overall plant community, soil conditions, slope and aspect, current land uses and management practices, and details about land ownership and access. Site information sheets may also record weather conditions at the time of agent release and may provide space to draw release site maps.

The types of qualitative and quantitative site data gathered, and the ways these data are collected and recorded, are as varied as the biological control agents utilized and the personnel and agencies involved. Though different site data collection protocols may be developed for different purposes, it is probably a good idea to attempt some standardization for classical biological control programs targeting a particular weed, in general, and release programs for a specific agent, in particular. This allows data collected by different agencies and/or individuals to be pooled for retrospective analyses (see "Monitoring in Weed Biological Control Programs," this volume).

## Monitoring established populations

Population establishment may be defined as the persistence of a biological control agent population in a given area without additional human intervention. From a practical standpoint, establishment has a number of definitions. Some practitioners believe that an agent is established when it successfully completes at least one generation in the field and is collected in the year following initial release (i.e., the agent successfully overwinters). Others feel that establishment is not confirmed unless the agent is collected for at least two or three years following initial release. One possible scheme for classifying establishment is as follows: 1) *established population*—agent has survived for at least two years after initial release, and the population is increasing and at least beginning to spread beyond the initial release area; 2) *surviving population, but establishment uncertain*— insect has successfully survived for at least one year after initial release (i.e., at least one generation successfully completed), but there is no spread or only very limited spread beyond the initial release point; and 3) *failure to establish (tentative)*—no agents

65

recovered after initial release (see "Factors That Affect Successful Establishment of Biological Control Agents," this volume).

Most unsuccessful biological control efforts appear to be the result of failure to establish after initial release rather than lack of efficacy, failure to spread, or poor long-term survival. Many factors can influence whether or not an agent becomes established after initial release. These include biotic factors (condition of agents released, genetic characteristics of agent populations, actions of natural enemies, quality of host plants, agent migration, etc.), abiotic factors (weather and climate conditions, soil characteristics, slope and aspect, fires and other disturbances, etc.), "procedural" factors (improper site selection, poor handling of agents during shipment, lack of food while in quarantine, release at wrong time of year, etc.); and combinations of all three.

Monitoring is the process by which the biological control agent population, and its impact on the target weed, is measured and evaluated over time. For field insectary sites, the primary information desired from a monitoring program concerns the status of the biological control agent population—is it established, how large is it, is it spreading, and will it support redistribution collections? Agent populations are best monitored using a predetermined sampling program developed for the agent in question. Sampling is the collection and analysis of a small portion of the population in a systematic way so that reliable estimates about the population as a whole may be made.

Sampling protocols are highly variable because they must be based on the life cycle and biology of the particular agent; thus, generalizations are difficult. Generally, an active aboveground life stage, often the adult, is targeted for sampling, particularly among insects that spend most of their life cycle in the soil or within host plant tissues. Agents may be sampled using a variety of techniques appropriate to the sampled life stage; these may involve visual counts, sweep netting, various active or passive traps, or suction devices. Occasionally, the products of the insect's attack on the target weed (galls, feeding damage, or frass) rather than the insect itself may be sampled. The timing of sampling visits, both seasonally and diurnally, is based on a thorough knowledge of agent biology. Seasonal scheduling is determined by the agent's life history—when during the year the sampled life stage occurs. Daily scheduling is a function of the agent's behavior—is it active during the day, evening, or night, and is it affected by sunny or cloudy weather or precipitation?

Initially, sampling protocols developed for field insectary sites will be designed primarily to confirm establishment and will usually be fairly simple for the first few years after agent release, determining presence or absence of the agent. With time, it may be desirable to collect quantitative data about the agent population (i.e., how many are there?), so more elaborate sampling procedures may need to be devised.

A standardized form may be prepared on which results of post-release visits to field insectary sites can be recorded. Information to be recorded on such a form may include date and time of visit, weather conditions, name and affiliation of person(s) visiting the site, number of agents collected or observed, and notes about any target weed impacts observed. This form should also provide a brief summary of sampling procedures for the agent(s) to be monitored.

Since agent establishment and population growth are the goals of a field insectary site, agents collected by various means should be immediately returned to the field after examination. However, a small sample should be collected to confirm identification of the agent by taxonomic specialists. Typically, the assistance of specialists employed for identification of quarantined, foreign-collected agents prior to release should also be sought at this time. Collected agents may be preserved in alcohol (typically, 70% ethanol in distilled water) or frozen after collection until they can be processed for identification.

## Collection and Redistribution

Once establishment of the agent is confirmed at one or more field insectaries, the next step in agent management is deciding whether or not each established population is large enough to permit redistribution collections. Redistribution is the movement of agents from a relatively small number of field insectaries to a relatively large number of general releases throughout weed-infested areas. It is these general releases that will effect large-scale weed control over time.

### Collectable sites

A collectable agent population is an established population that is large enough to permit collection of agents without jeopardizing the population. Agents may be harvested once, at intervals of several years, annually, or several times during a given year depending on the biology of the agent and the size of the insectary population.

As mentioned previously, agent sampling protocols are designed to acquire reliable estimates about the entire population. To determine whether an agent population is collectable, sampling is employed to estimate the size of the population. Sampling procedures used to determine establishment can be quite simple because they are usually designed to detect only the presence or absence of an agent. Sampling protocols used to decide whether or not the population is collectable, however, are usually more complex because they must provide absolute estimates of population size. Absolute population estimates describe the numbers of agents per unit area, per plant, or per plant part (e.g., stems or seed heads). When the area occupied by the agent or the number of plants or plant parts in a given area can be quantified, absolute population estimates can be used to estimate how many agents are present at a site.

Agent-specific sampling protocols developed to generate absolute estimates of agent population size may already be described in the scientific literature, may be adapted from sampling protocols developed for closely related species, or may need to be developed anew through experimentation. Sampling protocols developed to assess whether an agent population is collectable may have little or considerable overlap with those used to determine whether the same population is established. Again, sampling protocols are based on the life cycle and biology of each biological control agent, making generalizations problematic. However, basic concepts previously described, such as the selection of an appropriate life stage to be sampled, use of a technique for agent collection, and the seasonal and diurnal timing of sampling visits, also apply here.

To determine whether an insect population is collectable, population estimates based on sampling data must be compared to a decision threshold. This threshold represents the population size at or above which collections are feasible. Thresholds may be derived from experimental data, but more often they are estimates designed to allow conservative collections of agents while maintaining the field insectary population.

*Collecting agents*

The agent life stage that is most readily collected in large numbers is the appropriate target for redistribution collections. The adult stage is often the life stage harvested among beetles and other insects with complete metamorphosis. However, insect larvae or pupae may be collected if adults are fragile or difficult to collect. Once the appropriate life stage is targeted, collections should be scheduled seasonally and diurnally to coincide with maximum abundance of that life stage, which is determined by agent phenology and behavior. Hopefully, information collected during post-release monitoring efforts will have provided quantitative information on agent seasonality.

Obviously, collection efforts at field insectary sites should be concentrated in those areas where agents congregate or are otherwise most abundant. If weed mortality has occurred at the site, agents may be most numerous on the margins of the impacted area. Feeding damage may also be a good indicator of agent abundance. Agents may be collected from field insectary sites by hand or with sweep nets, beating trays, and vacuum devices, depending on the behavior and fragility of the collected life stage. Many of these collection methods are not selective and may also harvest nontarget arthropods and plant material. Contaminants may not be an issue if collected agents are to be redistributed locally (same or adjacent counties). However, agents must be separated from other organisms and materials if they are to be distributed to many environmentally sensitive areas (e.g., parks or preserves) or moved within or among states. Collected insects may be sorted from contaminants by hand or with an aspirator; chilling collected material before processing slows down arthropod activity and may assist the sorting effort. Collected agents may also be sorted using a variety of passive or active sorting devices. Processing and sorting of agents may be done at the field site, or collected material may be processed at a laboratory or other facility, particularly if specialized equipment and electricity are required.

Collected and sorted agents should then be packaged for transportation to general release locations. The sophistication of agent containers largely depends on the biology of the agent and the distance that they must be shipped before release. Containers need to allow movement of air and moisture, particularly if plant materials are provided as food for the biological control agents. Plastic or coated-paper containers should be avoided unless ventilation holes can be provided. In USDA-APHIS-PPQ redistribution programs, it has been found that unlined cylindrical cardboard containers of various sizes are a useful (though somewhat expensive) container for agent distribution. These are rigid enough to protect the agents and are air- and moisture-permeable.

Once packaged, some agents may be stored at cool but nonfreezing temperatures (about 4 to 10° C) for various lengths of time prior to shipment, though other agents may not tolerate cold storage and must be shipped immediately. The shipping method used will, of course, depend on the distance to the general release sites. Transport to local release sites can usually be accomplished by car; packaged agents should be kept cool and away from exposure to direct sunlight during transit. Shipment to more distant sites usually requires postal or commercial courier services; senders should check with those receiving the agents to determine what next-day shipment options are available for each location. Agents should be shipped to these locations in insulated foam containers, as described previously. Shipping containers should once again be provided with synthetic ice to keep contents cool during transit. Again, to avoid freezing, agent containers should not directly contact synthetic ice.

A USDA-APHIS permit (PPQ Form 526) is required for interstate shipment and release of weed biological control agents; this must be acquired before agents are transported. Shipments of collected biological control agents should be accompanied by USDA Form AD-943, which is completed at the time of field release. Other forms for describing the release process and/or biotic and abiotic conditions at the release site may also be provided.

## Distributing and releasing agents

The location of release sites is generally determined by the distribution of the target weed in a state or region. Within these broad areas, specific sites are chosen based on factors such as the nature of weed infestations, abiotic and biotic site factors, likelihood of other weed management treatments, and the relative proximity of cooperators who will make and monitor releases. Data collected at field insectary sites may identify those site conditions favoring agent survival and reproduction, i.e., that define an optimal release location.

A redistribution strategy for a particular agent will be based on the total number of agents collected and the number of suitable, potential release locations identified. The trade-off between number of releases and release size, discussed previously, may be partially resolved after an analysis of establishment data from the field insectary site. In general, making many releases of smaller numbers of agents may be a sound general release strategy because the establishment of individual releases is not as critical as it is with initial field insectary releases. However, the costs of initiating and subsequently monitoring each release must also be considered before deciding how many releases to make.

## References

Bellows, T. S., Jr. 1993. Shipping natural enemies. Pages 43-51 *in* R. G. Van Driesche and T. S. Bellows, Jr., eds. Proceedings: steps in classical arthropod biological control. Thomas Say Publications in Entomology. Entomol. Soc. Am., Lanham, MD.

Briese, D. T., W. J. Pettit, and A. D. Walker. 1996. Multiplying cages: a strategy for the rapid redistribution of agents with slow rates of increase. Pages 243-47 *in* V. C. Moran and J. H. Hoffmann, eds. Proc. IX Int. Symp. Biol. Contr. Weeds, 19-26 January 1996, Stellenbosch, South Africa, Univ. of Cape Town.

Clark, S. E., R. G. Van Driesche, N. Sturdevant, J. Elkinton, and J. P. Buonaccorsi. 2001. Effects of site characteristics and release history on establishment of *Agapeta zoegana* (Lepidoptera: Cochylidae) and *Cyphocleonus achates* (Coleoptera: Curculionidae), root-feeding herbivores of spotted knapweed, *Centaurea maculosa*. Biol. Control 22: 122-30.

Coombs, E. M., G. L. Piper, and B. Villegas. 1999. Why do weed biocontrol agents fail? Pages 24-25 *in* D. Isaacson and M. H. Brookes, eds. Weed Biocontrol: Extended Abstracts from the 1997 Interagency Noxious-Weed Symposium. USDA Forest Service Forest Health Technology Enterprise Team, Publ. FHTET-98-12. Morgantown, WV.

Coulson, J. R. 1992. Documentation of classical biological control introductions. Crop Prot. 11: 195-205.

Ehler, L. E. 1998. Invasion biology and biological control. Biol. Control 13: 127-33.

Grevstad, F. S. 1999a. Experimental invasions using biological control introductions: the influence of release size on the chance of population establishment. Biol. Invasions 1: 313-23.

Grevstad, F. S. 1999b. Factors influencing the chance of population establishment: implications for release strategies in biocontrol. Ecol. Appl. 9: 1439-47.

Hansen, R. W. 1995. Simple techniques for distributing *Aphthona* flea beetles from established field insectary sites. Page 19 *in* Proc. 1995 Leafy Spurge Symp. North Dakota State Univ., Fargo.

Hansen, R. W. 1999. Managing releases of weed biocontrol agents. Pages 22-23 *in* D. Isaacson and M. H. Brookes, eds. Weed Biocontrol: Extended Abstracts from the 1997 Interagency Noxious Weed Symposium. USDA Forest Service Forest Health Technology Enterprise Team. Publ. FHTET-98-12. Morgantown, WV.

Harley, K. L. S., and I. W. Forno. 1992. Biological Control of Weeds: A Handbook for Practitioners and Students. Inkata Press, Melbourne, Australia.

Ireson, J. E., S. M. Leighton, R. J. Holloway, and W.S. Chatterton. 2000. Establishment and redistribution of *Longitarsus flavicornis* (Stephens) (Coleoptera: Chrysomelidae) for the biological control of ragwort (*Senecio jacobaea* L.) in Tasmania. Aust. J. Entomol. 39: 42-46.

Memmott, J., S. V. Fowler, and R. L. Hill. 1998. The effect of release size on the probability of establishment of biological control agents: gorse thrips (*Sericothrips staphylinus*) released against gorse (*Ulex europaeus*) in New Zealand. Biocontrol Sci. Technol. 8: 103-15.

Memmott, J., S. V. Fowler, H. M. Harman, and L. M. Hayes. 1996. How best to release a biological control agent. Pages 291-96 *in* V. C. Moran and J. H. Hoffmann, eds. Proc. IX Int. Symp. Biol. Contr. Weeds, 19-26 January 1996, Stellenbosch, South Africa. Univ. of Cape Town.

Nowierski, R. M., Z. Zeng, D. Schroeder, A. Gassmann, B. C. Fitzgerald, and M. Cristofaro. 2002. Habitat associations of *Euphorbia* and *Aphthona* species from Europe: development of predictive models for natural enemy release with ordination analysis. Biol. Control 23: 1-17.

Pedigo, L. P., and G. D. Buntin. 1994. Handbook of Sampling Methods for Arthropods in Agriculture. CRC Press, Boca Raton, FL.

Pitcairn, M. J. 1999. Monitoring goals in weed biocontrol. Pages 35-38 *in* D. Isaacson and M. H. Brookes, eds. Weed Biocontrol: Extended Abstracts from the 1997 Interagency Noxious Weed Symposium. USDA Forest Service Forest Health Technology Enterprise Team. Publ. FHTET-98-12. Morgantown, WV.

Shea, K., and H. P. Possingham. 2000. Optimal release strategies for biological control agents: an application of stochastic dynamic programming to population management. J. Appl. Ecol. 37: 77-86.

Southwood, T. R. E. 1978. Ecological Methods, with Particular Reference to the Study of Insect Populations. Chapman and Hall, London, UK.

Van Driesche, R. G., and T. S. Bellows, Jr. 1996. Biological Control. Chapman and Hall, New York, NY.

Ziolkowski, H. W., and R. D. Richard. 2000. A passive sorting device for separating *Aphthona* species biological control agents of leafy spurge from larger insects, seeds and plant debris associated with crude insect sweep net collection. Page 146 *in* N.R. Spencer, ed. Proc. X Int. Symp. Biol. Contr. Weeds, 4-14 July 1999, Montana State Univ., Bozeman, MT.

# Post-release Procedures for Biological Control Agents of Aquatic and Wetland Weeds

Ted D. Center and Paul D. Pratt

Invasive plants adversely affect aquatic and wetland ecosystems by disrupting ecosystem processes and reducing their capacity to recover after disturbance. In the absence of natural enemies, these weeds have spread at an alarming rate, displacing native vegetation, altering fire ecology, and negatively affecting the functioning and biological integrity of many vital ecosystems. Long-term, effective management of these plants is critical to the restoration of degraded habitats and can contribute to the sustainability of water supplies for both agricultural and urban uses as well as preserving biodiversity in natural areas.

Effective integrated weed management programs require a thorough understanding of the biology and ecology of invasive plants and the invaded communities. Information about plant demography, seed bank dynamics, seedling recruitment, plant growth and development, and methods of reproduction help identify vulnerabilities to be exploited in integrated weed management systems. In addition, it is critical that the causes of plant invasion be understood so that they can be alleviated. Research into these areas of weed ecology will improve our ability to select the most efficacious biological control agents and will aid in the development of integrated management of weed infestations, new monitoring and management systems, reduced weed impacts, and effective restoration of infested areas.

Weed biological control often has been approached in a very passive manner, using the "release and pray" approach—releasing the biological control agent and hoping for the best. This is not advisable and the end-user should take a much more active role in implementing and evaluating biological control attempts. This requires specific skills and knowledge of how to rear, release, and manage new biological control agents to maximize the chances for a successful program.

## Acquisition of Stock Material

The classical biological control of a weed entails a multi-step process. Initial phases of a project involve determination of the foreign origin of the weed, faunal inventories of associated plant-feeding arthropods and phytopathogens in the area of origin, and preliminary host ranges of promising agents. Candidates that exhibit a limited host range and seem able to effectively damage the target weed are then transferred to domestic quarantine facilities for final evaluation. The organism is usually colonized while in quarantine to provide experimental subjects needed to support a rather prolonged screening process. This colony may be maintained for several years without infusions of new stock during this evaluation stage. The lengthy regulatory review that follows renders continued rearing logistically impractical, often forcing elimination of the quarantine

71

colony before the agent is released. As a result, supply of the agent may not be available to support releases after permits are finally issued, so new material must be acquired. Even if a quarantine colony has been maintained during the interim, it may still be advisable to acquire fresh stock to minimize founder effects that can lead to inbreeding depression.

Stock material can be acquired from various sources including areas other than the original source of the quarantine colony. This includes alternative areas within the native range of the organism or perhaps other countries where the agent has been introduced for weed control. However, host races of plant-feeding insects may exist within species, so host range data for insects from one area may not be applicable to those from a different area. The stock used to support field release must therefore represent the population evaluated during risk assessment. Sometimes this is not possible, so it becomes necessary to seek alternative sources. In these cases, additional testing must be done to ensure conformity with specificity requirements and new release permits may be required.

## Shipping Agents

The procedures for shipping agents associated with aquatic and wetland weeds are not very different from those for terrestrial species. In fact, many aquatic weed biological control agents are terrestrial or have terrestrial life stages, so handling procedures for terrestrial species are directly applicable. However, packing methods can be quite specialized for other species. Aquatic insects may have to be shipped in water so care must be taken to ensure that dissolved oxygen levels are not depleted during transport. Whenever possible, it is best to hand-carry such material so that conditions can be monitored in transit (however, new security regulations prohibit hand-carrying insects into the United States). Portable battery-powered aerators can be used to maintain aeration. When it is not possible to attend to material while in transit, it should be placed with a minimal amount of water in sealable plastic bags. The bags should be inflated with oxygen as they are being sealed, then packed tightly and shipped.

Generally, it is best not to include plant material in the parcels. Plants respire but do not photosynthesize in the dark, so oxygen can be quickly depleted while carbon dioxide increases in airtight containers. Plant material can release toxic volatiles that reach lethal concentrations in sealed air spaces and can kill the insects. Also, the accumulation of frass on leaf surfaces being consumed by the agents can spread pathogens from one individual to the entire group. In some cases, though, the inclusion of plant material can protect delicate insects. This may be especially appropriate with endophagous stages such as *Hydrellia* fly larvae in hydrilla (*Hydrilla verticillata*).

Insects associated with wet environments generally require higher humidity than terrestrial species, but too much humidity can be problematic. The trick is to provide enough humidity to prevent desiccation while minimizing condensation. The inclusion of plant material can help in this regard by increasing humidity levels without adding free water. Moist excelsior (stranded wood fibers manufactured from aspen trees for use as protective packaging and filling material) may work equally well. The excelsior is thoroughly soaked then drained to eliminate free water before being loosely packed into a shipping tube (not sealed in plastic). The insects may be placed directly into the tube within the excelsior. The open tangle of wood strands provides structure for them to cling to as well as protected nooks and crannies. This system has worked well for the shipment of semiaquatic weevils (e.g., *Neochetina* spp.). Depending on the insect, it might be appropriate to starve them for a day prior to shipment so as to preclude frass production during transit.

Plastic Petri dishes lined with moistened filter paper, the excess water drained off, may also be useful shipping containers. The insects are placed in the Petri dish before sealing the edges with wax film (e.g., Parafilm®). Several dishes may be stacked in a mailing tube. This has worked well for the shipment of adult waterhyacinth moths (e.g., *Niphograpta albiguttalis*) and their eggs. A single pair of moths is placed in each dish along with a prepared leaf (see species account in Section II) to provide an oviposition substrate. These may be held a day or two before shipping to allow the female to oviposit into the leaf material. The adults can be shipped along with the leaves or removed to send only the egg-laden leaves. Salvinia weevils (*Cyrtobagous salviniae*) can also be shipped in plastic Petri dishes sandwiched between layers of moistened paper toweling. However, in this case, the dishes are not sealed, so that air exchange can occur.

Dampened artificial sponges are useful for shipping small Lepidoptera pupae. The sponges used generally have a pattern of diamond-shaped depressions on the surface that conform to the size of the pupae. Several naked pupae are laid on the sponge in these depressions. A second sponge is positioned over the first so that the surface depressions are aligned. They are taped together, then shipped in a suitable container. Care should be taken to avoid shipping older pupae that are likely to emerge in transit, because adults will not survive between the sponges.

As with terrestrial species, parcels of aquatic species must remain cool (but not cold) during shipping. This can be accomplished by using an insulated container containing cold packs. The cold packs should be secured so that they do not crush the consignment. The container should be carefully sealed and packed in a sealed, corrugated cardboard carton or other suitable box. A copy of export and import permits (PPQ Form 599) should be affixed conspicuously to the outside of the box to allow for easy inspection, along with telephone numbers of recipient quarantine facilities and any specific instructions (e.g., live insects—keep cool, open only in an authorized quarantine, etc.). As always, the fastest delivery method is best, by courier if possible but if not, by express delivery. Parcels should never be shipped via surface mail.

All parcels containing exotic agents for biological control of weeds must be delivered to an authorized quarantine facility. Most will have pre-established contacts with -U.S. Department of Homeland Security inspectors at ports of entry, who will cooperate to expedite the transfer of the parcel. It is always wise to alert the inspectors that the package is en route and request that they call a responsible party when it arrives. Preferably, quarantine personnel should pick up the parcel at the agricultural inspection office, but if this is not possible, an approved courier service can deliver the parcel to the appropriate facility. Under no circumstances should the seals on the package be broken before entry into quarantine.

## Quarantine Procedures

After the parcel is received into quarantine, unwanted organisms (parasitoids, pathogens, weed seeds, etc.) that may have accompanied the consignment must be eliminated. The parcel is usually taken into a receiving room, normally the most secure area in quarantine, where it is opened. The insects are removed from the packing material, which is then sterilized and destroyed. Any included plant material should also be destroyed, although it may be held for a while to extract the organisms. The original stock should be retained in the receiving area on fresh plant material until they reproduce, and then be preserved as voucher specimens or otherwise destroyed to eliminate unemerged parasitoids. Taxonomic specialists should confirm the identification of the organisms in the shipment and a sample should be checked for pathogens by qualified personnel. Only progeny of the original stock should be allowed out of the receiving area. Eggs may be surface sterilized with a weak (10%) bleach solution prior to removal; however, this does not eliminate pathogens that can pass from the female into the egg (e.g., microsporidia). Hence, it is wise to check the progeny for pathogens prior to their removal from the receiving area. Normally, about 10% of the individuals are examined, but field infection rates can be quite low (<5%), so examination of a higher percentage might be required. After completion of these procedures, the insects may be moved into general quarantine for limited rearing, or permitted agents may be released from quarantine for direct field release or for colonization elsewhere.

The precautionary principle applies when dealing with disease organisms—never risk releasing diseased agents; they cannot be recalled. Two options are available when pathogens are found: either destroy the entire lot or eliminate the disease. The former option does not guarantee that new stock will not also be infected; the shipper must take additional precautions to avoid resending infected organisms. Eliminating the disease, however, can be labor intensive and time consuming, depending on the disease organism. Some pathogens can be eliminated by selectively purging from the colony individuals that appear sick or show symptoms of disease and keeping the others isolated from one another as much as possible. Other disease organisms, particularly those that can be transovarially transmitted, are best eliminated by developing disease-free breeding lines. This involves isolating individual pairs of adults, allowing them to reproduce, isolating the progeny, then sacrificing the parents and checking them for disease. If no disease is found, the progeny can be released or used to initiate a new colony. If disease organisms are present in either parent, then their progeny should be destroyed and any containers or plant material that they had been in contact with should be sterilized and, if disposable, destroyed.

Hygiene is extremely important. Instruments used to handle one breeding line should be cleaned and sterilized before being used to handle another. Table surfaces and containers should be cleaned and disinfected after handling each group. Dead insects should not be allowed to accumulate in cages or rearing areas. Numbers of insects in cages should be carefully controlled to avoid crowding and fecal contamination of food plants. If a disease materializes in a colony after it has been removed from the receiving area, it is difficult to know whether it originated from abroad or from a local source, perhaps on food plants brought in from the field. Plant material brought in from the outside should therefore be carefully cleaned and disinfected. Qualified insect pathologists should be consulted for advice on how best to handle a particular disease problem. To repeat, the precautionary principle should apply—never release diseased agents.

## Field Colonization

Only small numbers of biological control agents are usually available after their first release from quarantine. Oftentimes quantities are insufficient for a release program so numbers must be increased either though a rearing program or by establishing nursery sites. Artificial diet may offer one rearing option, but this has drawbacks. Diets are not available for most agents and the development of a new diet can take quite a long time. Furthermore, prolonged rearing on artificial media can produce diet-adapted strains that may not fare well under field conditions.

Generally, if brief rearing is needed, it is simpler to use plant material. Again, however, care must be taken to avoid producing agents that are maladapted to field conditions. The hydrilla fly (*Hydrellia* spp.), for example, was originally reared for many generations in gallon jars containing hydrilla. The flies produced from these colonies (which had also been reared for many generations in quarantine) seemed sluggish and not inclined to fly. When released in the field they established marginal populations, but with difficulty and only after many attempts, whereas fresh insect material from overseas established readily. This loss of vigor was ascribed to inbreeding depression due to excessive laboratory colonization. Rearing should therefore be done under conditions that are as natural as possible, on whole plants in greenhouses or screenhouses, and over as few generations as possible.

Most aquatic plants can be easily grown in tanks, aquaria, or even jars. Floating aquatic macrophytes, such as waterhyacinth (*Eichhornia crassipes*), waterlettuce (*Pistia stratiotes*), or giant salvinia (*Salvinia molesta*), are easily produced in pools filled with water and fertilized with commercially available, high-nitrogen, water-soluble fertilizers. Insects can be released directly on the cultivated plants. Fresh plant material should be added periodically to replace older, declining plants, which should be removed and may be transplanted at field sites as a means of release. Cages erected over the pools may be required to prevent the agents from dispersing and to exclude pests or predators.

If the pools are deep enough, some agents (e.g., the adult weevils *Neochetina* spp., *Neohydronomous affinis*, and *Cyrtobagous salviniae*) can be easily extracted by forcing the plants below the water then collecting the adults as they surface. Otherwise, Berlese funnel extraction or hand collecting may be required. Sometimes extraction from the host material is not necessary, inasmuch as infested plants can be moved directly to the field. The flea beetle (*Agasicles hygrophila*) has been successfully released in this way on alligatorweed (*Alteranthera philoxeroides*), as was the waterlettuce weevil (*N. affinis*). Some agents are more difficult. Moths, in particular, because the adults are relatively fragile, require more care. After several unsuccessful attempts, the waterhyacinth moth was finally established by releasing eggs. Females were allowed to oviposit in stripped leaves. The egg-laden leaves were held until they were ready to hatch and were then tucked into the central portion of a plant within a partially unfurled leaf. The emerging larvae readily transferred to the field plants.

The hydrilla leaf-mining flies (*Hydrellia pakistanae* and *H. balciunasi*) were originally reared in water-filled gallon jars that contained the leafy terminal portions of hydrilla stems. The amount of hydrilla in each jar proved to be critical: enough was needed to support several hundred larvae but not so much that light was unable to penetrate throughout. If the plant material was packed too tightly so that the bulk of it was unable to photosynthesize, dissolved oxygen levels decreased and the plant material rotted. It was also important to hold the jars in areas with high levels of natural sunlight so as to maximize photosynthesis and maintain dissolved oxygen levels. The colonies were initiated with adults placed in a sleeve cage with leafy sprigs of hydrilla. The adults were fed a mixture of sucrose (7 g), water (10 ml), and yeast hydrolysate (4 g), which was spotted onto the caps of pill vials placed in the bottom of the box. The hydrilla was placed horizontally in a thin layer of water contained in a shallow dish, such that the leaves on one side of the sprigs were exposed. The adult flies laid copious numbers of eggs on the exposed leaves. The egg-laden sprigs were removed and replaced after one or two days and then released in the field or placed in rearing jars. As the eggs hatched, the larvae readily moved to the hydrilla in the jar or at the field site.

The best strategy for increasing the numbers of agents available for release entails the use of nursery sites or field insectaries. This involves selecting sites that best match the agent's native habitat, then making a concerted effort to establish founder colonies. After the populations build up, agents can be collected from the site and redistributed to other areas. Regular monitoring of nursery sites will determine the abundance of the agent and the optimal timing for collection of a specific life stage. Care should be taken to avoid overharvesting.

Conditions can sometimes be manipulated to increase the likelihood that the agent will successfully colonize or to induce more rapid build-up of populations. For example, reproduction in some insects is related to the nutritive value of the plant tissue (e.g., *C. salviniae*). Tissue nitrogen levels of floating plants can be increased by fertilizing portions of the mats, thereby facilitating population build-up. Other species may need flushes of new plant growth [e.g., the weevil (*Oxyops vitiosa*) on melaleuca (*Melaleuca quinquenervia*)], which can be induced by pruning or mowing. Cages have rarely been used for agents of aquatic and wetland plants because releases are often in public areas subject to vandalism. One exception, however, was the abovementioned hydrilla flies. Open releases on topped-out hydrilla beds at field sites proved ineffective. Flying predators, especially dragonflies and damselflies, consumed so many of the adult flies that they failed to establish. Various cages were tried to resolve the predation problem. Low-profile floating cages produced the best results. These consisted of flat, screened boxes mounted over topped-out hydrilla (see Center et al. 1997 for design details). Egg-laden sprigs from the abovementioned sleeve cages or plants infested with larvae from the rearing jars were placed in the cage among the existing hydrilla. The cages enclosed a space that extended only a few centimeters above the water surface. They were too shallow for flying predators to maneuver in, but allowed plenty of space for the minute flies. They also kept the emerging flies together, thus maximizing opportunities for mating and reproduction.

Agents can be collected from nursery sites using a variety of methods. Waterhyacinth weevils often aggregate in furled young leaves just beginning to open. The weevils are easily collected by selectively choosing plants bearing leaves at this stage, unrolling them, and picking out the adults by hand. Collection may be facilitated by inserting a funnel through the lid of a receiving container and dropping the weevils through the funnel into the container as they are collected. The receiving container can be provisioned with leaves or dampened excelsior. The funnel can be temporarily plugged using a neoprene stopper to prevent the weevils from escaping. The lid is replaced with an intact one after collecting is done and the container then serves to transport the weevils. Care is taken to guard against excessive heat.

Waterhyacinth weevils can also be collected at night by sweep netting the uppermost leaves of waterhyacinth mats. This is not effective during the daylight hours after the weevils have moved to the lower portions of the plants. Sweeping, which is difficult from the shore, requires traversing the plant mat in an airboat or on water shoes. (Water shoes resemble water skis but consist of blocks of buoyant foam, usually covered in a hardened shell, with ski bindings attached. They are awkward to use and can be dangerous, so sweep netting often is not realistic.) Alligatorweed flea beetles are readily collected using sweep nets, as are melaleuca weevils and the melaleuca psyllid (*Boreioglycapsis melaleucae*).

Waterhyacinth weevils, as well as other subaquatic weevils, can also be collected by submerging the plant material. This can be done directly at the site, or the plant material can be collected in bulk in large plastic bags and taken to a suitable location for extraction. The plants are then placed in water-filled tanks and submerged. The weevils can be collected with aquarium dip nets as they abandon the plants and rise to the surface.

Berlese funnels can also be used to extract many sorts of insects, but unless large, custom-built funnels have been constructed, they are not useful for processing large quantities of plant material. These extraction devices progressively dry the plant material in an enclosed container from the top down, thus forcing insects in the plants to gradually move downward, fall through a screen, pass through a funnel, and ultimately drop into a receiving container. Aquatic plants, however, may contain as much as 95% water so they take a long time to dry. As the plants desiccate, the rising moisture-laden air causes condensation that accumulates, then rains down through the plant material, rewetting it and causing the receiving jar to overflow. Hence, when Berlese funnels are used to extract insects from aquatic plants, they often must be vented above and below to create an upward flow of air to dissipate the moisture.

A rather unusual technique was developed for collecting adult hydrilla flies, which rest on the surface of topped-out hydrilla beds. Collecting is done from a boat with one person driving and one or more people collecting. The collectors, leaning over the gunwales of the slowly moving boat, place a thin sheet of polystyrene on the water surface and slowly push it across the hydrilla beds. As the polystyrene disturbs the resting flies, they fly up onto it where they are aspirated with a battery-operated hand vacuum modified for the collection of insects (a BioQuip® Insect Vac).

A similar technique has been used to collect the waterhyacinth moth; an airboat is used to cruise through a waterhyacinth mat just after dusk. A white sheet is spread out on the front deck and either a black light or a mercury vapor light is placed on the sheet. The boat disturbs the moths as it slowly pushes through the plants. The disturbance causes them to become active, fly toward the light, and land on the sheet where they can be trapped or coaxed into collecting containers.

Many of these agents will come to lights at night, but usually not in sufficient quantities to be useful. Waterhyacinth weevils are not always capable of flight, but large numbers sometimes do fly and accumulate at lights. They can be easily collected during these episodes, but this phenomenon is sporadic and not reliable. The waterhyacinth moth also occasionally has been collected at light traps in large numbers.

There are no efficient techniques for collecting some aquatic agents. The alligatorweed stem borers (*Arcola malloi*), for example, are best collected as larvae or pupae by searching for the characteristic wilted stem tips, harvesting them, and splitting them open to search for the insect. Likewise, alligatorweed thrips (*Amynothrips andersoni*) can best be collected by looking for the characteristic distorted leaves and then aspirating them from infested plants.

## Site Selection

Selection of a suitable site is of paramount importance when attempting to establish a new biological control agent. This is especially true when the agent has not yet established anywhere and as nursery sites are developed. Care must be taken to match the characteristics of the habitat and the plants to the requirements of the agent. One of the first and most obvious considerations is climate matching. Insects that originate in tropical areas, for example, may not fare well in temperate areas and vice-versa. The crudest form of climate matching involves latitudinal comparisons: climates at similar latitudes are assumed to be similar. More sophisticated multivariable matching is possible, however, using computer software programs (e.g., CLIMEX) that have been developed for this purpose.

Hydroperiod can also be important, especially for wetland species. Melaleuca weevils, for example, pupate in dry soil so they thrive at dry or seasonally wet sites, but not at permanently wet locations. A weevil (*Bagous affinis*) was introduced into Florida to destroy subterranean hydrilla turions. It cannot access these structures, however, unless systems become dewatered to expose the hydrosoil, such as during prolonged droughts or during reservoir drawdown. Such events were too infrequent in Florida, so the weevil failed to establish.

The predominant plant type or growth stage present can also thwart establishment. Waterhyacinth moths, for example, thrive on open-grown plants that are short in stature with soft-textured, bulbous leaf petioles. They do not do well on tall, dense plants with hardened leaves. Early attempts to establish them on the latter plant type failed, whereas later attempts on the proper plant type succeeded. The alligatorweed moth and flea beetle require plants with thick, hollow stems. They do not do well on plants with thin, solid stems of the type that typically grow in terrestrial habitats. In contrast, the alligatorweed thrips prefers terrestrial plants. One of the waterhyacinth weevil species (*Neochetina bruchi*) seems to require plants with high nitrogen content, whereas the other species (*N. eichhorniae*) seems less sensitive to plant quality. Hydrilla fly populations often fail when thick algal mats grow over beds of hydrilla or when thick mats of topped-out hydrilla trap a surface layer of exceedingly hot water.

Plant genotype or biotype might also be an important consideration. For example, waterlettuce, being widely distributed throughout the tropics, is considered native to South America, Asia, and Africa. It is considered to be a single, monotypic species, despite this broadly disjunct distribution. The waterlettuce moth (*Spodoptera pectinicornis*), which originated from Thailand, was introduced into Florida but failed to establish despite an intensive release program. One hypothesis for this failure was that the donor plants from Thailand were genetically different from the recipient plants in Florida. Genetic fingerprinting techniques are now readily available to facilitate genotypic matching of plants so, whenever possible, it is advisable to confirm that the plants in the adventive region are genetically similar to those from the native range of the candidate agents.

Host plants might differ in other ways that can affect the insect's ability to establish. For example, we now know that at least two melaleuca chemotypes exist in Florida. The differences are manifested in the composition of mixtures of essential oils in the foliage. Melaleuca biological control agents in the laboratory perform differently on these two biotypes (although these differences have not yet been observed in the field). Hence, it is important to consider the possibility that different forms of the host plant may not be equally acceptable to the agents.

The ability of an ecologically naïve biological control agent to colonize a new habitat may be viewed in the context of a "lock and key." The characteristics of the agent must fit with the characteristics of the plant *and* the recipient habitat. All too often, the selection of candidate agents has focused solely on characteristics of the agent (e.g., reproductive rate, size of adults, feeding behavior, etc.) without due consideration to matching these traits with characteristics of the targeted weed or the recipient habitats. For example, if a weed propagates primarily by vegetative means, a seed feeder will not do much good no matter how prolific it is. Likewise, generation times of agents may not synchronize with the growing season of the plant, and agents may consequently emerge when the host is scarce. Species that disperse prior to mating may leave the release site, never find mates, and therefore fail to establish breeding populations. Some species may require a symbiotic association with another organism and may not be able to persist without its mutualistic partner. Defoliators may be ineffective against weeds with large underground storage organs. Successful biological control is an information-intensive process. Success hinges upon knowing as much as possible about the system prior to choosing an agent or formulating a release strategy. Sometimes this requires experience that can only be acquired through trial and error.

## Release Strategies

When a new biological control agent is cleared for release, the available supply may be nominal. Thus, decisions must be made on how best to distribute the few available agents in the field, i.e., whether to allocate them among a few sites and make large releases, several sites using small releases, or take an intermediate approach (see previous chapter). The strategy ultimately chosen may depend on knowledge of the field situation and the characteristics of the agent. For example, if the host plant type seems highly variable and the preferences of the agent are uncertain, then it might be best to spread the available stock over a wide range of plant types to maximize the likelihood of encountering the correct form. On the other hand, if the preferences of the agent are clear, it may be more appropriate to release greater numbers at a few select locations. The longevity and vagility of the agents might also dictate the release strategy. Long-lived, philopatric (home-loving) agents, such as the melaleuca weevil, might easily establish a population from the release of a single fecund female, whereas short-lived, vagile agents might emigrate so rapidly that they never establish a minimum viable population (i.e., the smallest population size that is likely to allow for long-term persistence), as may have occurred in the case of the waterlettuce moth.

How many biological control agents should be released? There is no easy answer to this question and prescribed numbers should be viewed with a healthy measure of skepticism. Care must be taken to distinguish simple counts of individuals (the census population) from the numbers of viable, reproducing individuals (the effective population) in the release stock. If the sex ratio is strongly skewed toward males or if a high percentage of the females are aged and post-reproductive, the chances of establishment will be diminished. Furthermore, even though large numbers may be released, they might possess very little genetic diversity if they have originated from a few progenitors. They may thereby have difficulty adapting to changing conditions even if they do establish, and they could be vulnerable to demographic, environmental, or genetic stochasticity or natural catastrophes. It might be better to release a few genetically diverse individuals rather than large numbers of inbred derelicts. However, small populations face a high risk of extinction so the rule of thumb should be "the more, the better."

Serial releases might overcome the need for large releases when only small numbers of agents are available at any one time. This, of course, requires the availability of an ongoing supply of the agent. Initial small releases are made, later supplemented with periodic infusions of additional stock. The supplemental releases must occur while the individuals from earlier releases are still present so as to augment the existing population. The strategy is to prevent the nascent population from dying out while artificially maintaining a minimal viable population at the site until they become self-sustaining.

Cages can be helpful in this effort, especially if they are designed as modular units that can be adjoined and enlarged as the population grows. Cages can serve several useful functions: they mark the exact release point, which facilitates finding the individuals that have previously been released; they exclude many predators; they retain a minimal viable population within a restricted area by preventing emigration of the agents; and they may favorably modify the enclosed microclimate. There are an equal number of drawbacks: by marking the location, they draw attention to the exact point of a release, possibly inviting vandalism or theft; they prevent the agents from dispersing, thereby risking overpopulation and a resultant population crash; they encourage a rapid build-up of any contained predators and parasitoids; and they adversely alter the enclosed microclimate and light intensity.

As founder populations establish from initial releases, efforts may then be directed toward establishing additional populations at different sites in diverse areas. This is indeed important, inasmuch as it is desirable to minimize the consequences of the loss of a site in the event of some catastrophic event (e.g., a fire, herbicide treatment of the site, drought, etc.). However, the distribution of those populations must be considered. The distributional pattern of releases can be optimized by considering the agent's ability to disperse relative to the range of the weed. Optimal release patterns can be determined by modeling the time it will take for the agent to disseminate throughout the range of the weed (see Pratt et al. 2003). Species that disperse rapidly and prior to mating may depend on nearby populations to find mates and reproduce. In this case, it is important to establish field colonies in a pattern that allows populations to interact. Establishment of this "metapopulation" might be facilitated by making clusters of releases in the same general area (e.g., on the same water body).

## Documenting Releases

All releases of weed biological control agents should be documented, as explained in the chapter on terrestrial insects. Forms are available from the USDA-ARS Biological Control Documentation Center located in the National Agricultural Library. USDA-ARS Form AD-942 is used to document agents collected from foreign sources and USDA-ARS Form AD-943 for agents collected from domestic sources.

## Post-release Monitoring

Various types of monitoring are needed after agents are released. The first involves population viability analysis to determine whether the agent has developed self-sustaining populations and can thereby be considered established. As a rule of thumb, we have found in subtropical areas, such as southern Florida, that when an agent increases in number and persists without assistance for five generations (or an equivalent time period), it will generally persist indefinitely. In more temperate areas where freezing weather is of concern, persistence through at least one winter may be necessary before drawing any conclusions. There are no flawless criteria for making this determination and it depends on one's knowledge of the system.

To determine whether a population has established, it is usually appropriate to mark the point of release to facilitate locating the founder population later. Hand-held GPS receivers are now sufficiently accurate for this purpose, so a physical marker may not be necessary. Marking a release is difficult in aquatic habitats, especially where water levels fluctuate greatly or in situations where plants float from one location to another. In the case of hydrilla and the hydrilla fly, the aforementioned cages were used to mark the release location. When caging was no longer needed, the screen was removed so the unscreened frame marked the release point. Two D-rings attached on opposite sides of each cage allowed cages to rise or fall with water levels while riding on vertical poles that had been driven into the substrate.

Floating plants are even more problematic. Floating square frames (usually 1 m on a side) made from polyvinylchloride (PVC) pipe can be used to mark releases. The frame can be anchored with a concrete block or tied to a nearby stationary object using water-resistant rope with sufficient scope to allow for fluctuating water levels. This works well when the mats are stable, but is not effective when flooding occurs or strong currents flush the plants downstream. Sometimes the plants themselves can be marked using brightly colored surveyor's tape or spray paint.

Knowledge of diffusion rates can be useful in determining the need for and establishing a pattern of additional releases. This requires a second type of monitoring designed to estimate rates of dispersal. Acquisition of these data can be difficult and sometimes involves unjustifiable assumptions. It is generally expected that the agent will spread radially, like a gas, from the original source point outward in a more or less symmetrical pattern. Thus, the most common form of dispersal monitoring incorporates transects that radiate outward from the release point in four cardinal directions. Observers may simply follow the transect while continuously searching for the agent. Alternatively, points may be established at intervals along the transect that are monitored for the presence of the agent. Sentinel plants or passive traps also can be placed at the monitoring points to

81

facilitate detection of the agent. The distributional limits are delineated by the distance from the release point to the furthest transect point at which the agents can be found. The dispersal distance averaged over all transects divided by the time after release (or after establishment) provides the population diffusion rate.

The assumption of radial dispersion can lead to false conclusions, however, and cause one to overlook other patterns of dispersion. Long-distance dispersal may result in satellite colonies well beyond the expected search radius and, over time, agents may expand their range independently of the parent population. The waterhyacinth moth, for example, could not be found anywhere near the exact release point when it first established, but appeared several kilometers away in the same water body. Knowledge of the species' dispersal patterns and jump distances is needed to determine the best approach.

A great deal of concern has arisen in the past few years over potential nontarget effects of introduced plant-feeding insects. It is therefore very important to field-verify laboratory-derived host range predictions. A third monitoring system may be needed for this purpose. Potential nontarget hosts should be identified at release sites and periodically checked for damage. Field plots containing test plants can be established in weed-infested areas then monitored on a regular basis. The biological control agent can be released directly onto potential nontargets, then monitored to determine residence time. Any number of approaches may be useful for this purpose, but it is imperative that the extent of nontarget effects, or the lack thereof, be documented. In fact, conditions on release permits (PPQ Form 526) now require that nontarget impacts be monitored and results reported to the permitting agency within six months of the initial release.

A fourth type of monitoring evaluates the performance of the agent in the field. This actually involves several different approaches in accordance with the different levels of scale that must be considered. The first and most detailed evaluation is usually done at a few selected sites on an extremely local level, the objective being to obtain robust measures of the impact of the agent on the plant population. Ideally, plant populations should be experimentally compared between sites where the agent has been released and control or check sites. Studies at this level might even involve intentional plantings. Baseline data, from which changes can be measured following establishment of biological control agents, are advantageous, so monitoring should begin before releases are made. Experimental controls may also be provided by using exclusion devices such as cages or insecticide treatments. Various parameters of plant performance are usually monitored over time at this stage, and relationships between plant traits are elucidated. Allometric relationships can be established that enable the estimation of variables that require destructive sampling from more easily obtained morphometric measures (e.g., biomass from plant height). This approach is instructive in that it elucidates the effects of the agents on the plants, thereby allowing later simplified recognition of these effects on a broader scale.

Another approach applies information derived from the more detailed evaluation to a more regional scale. After simplified relationships are developed to enable rapid assessments of impacts, the geographical extent of these impacts should be determined. This involves assessments at many sites, which logistically precludes collection of detailed data. One example of how this is applied involves the hydrilla fly (*H. pakistanae*). Hydrilla grown in tanks was artificially stocked with varying numbers of flies, then later harvested.

Detailed data were collected, but the percentage of leaf whorls damaged provided an easily measured indicator of larval intensity. Damage thresholds of about 60 to 70% of the leaf whorls were necessary to reduce plant biomass. However, monitoring during different times of the year at several locations revealed that damage levels rarely exceeded 15% of the leaf whorls. It was determined from this that the flies were not providing adequate control and that a complex of biological control agents would be needed.

The final, least-detailed evaluation approach involves broad regional assessments of the abundance of the weed, usually expressed in terms of total statewide or regional acreage. This involves aerial surveys or the concerted effort of large numbers of participants doing ground-based surveys. Such assessments are usually done by resource management agencies rather than individual research groups. Examples include the annual assessments of waterhyacinth acreage by the Louisiana Department of Wildlife and Fisheries, the periodic aquatic plant surveys done by the Florida Department of Environmental Protection, and the invasive weed surveys done by the South Florida Water Management District. These data, when coupled with information from the previous evaluation phases, can provide a complete picture of the impact of biological control on a landscape level.

## *References*

Center, T. D. 1981. Release and establishment of *Sameodes albiguttalis* for biological control of waterhyacinth. Tech. Rep. A-81-3, U.S. Army Engineers, Waterways Experimental Station, Vicksburg, MS.

Center, T. D. 1984. Dispersal and variation in infestation intensities of waterhyacinth moth *Sameodes albiguttalis* (Lepidoptera: Pyralidae) populations in peninsular Florida. Environ. Entomol. 13: 482-91.

Center, T. D. 1994. Chap. 23. Biological control of weeds: Waterhyacinth and waterlettuce. Pages 481-521 *in* D. Rosen, F. D. Bennett, and J. L. Capinera, eds. Pest Management in the Subtropics: Biological Control—A Florida Perspective. Intercept Publishing Co., Andover, UK.

Center, T. D., and W. C. Durden. 1981. Release and establishment of *Sameodes albiguttalis* for the biological control of waterhyacinth. Environ. Entomol. 10: 75-80.

Center, T. D., F. A. Dray, Jr., G. P. Jubinsky, and M. J. Grodowitz. 1999. Insects and Other Arthropods that Feed on Aquatic and Wetland Plants. USDA Agric. Res. Serv. Tech. Bull. No. 1870, Washington, DC.

Center, T. D., M. J. Grodowitz, A. F. Cofrancesco, G. Jubinsky, E. Snoddy, and J. E. Freedman. 1997. Establishment of *Hydrellia pakistanae* (Diptera: Ephydridae) for the biological control of the submersed aquatic plant *Hydrilla verticillata* (Hydrocharitaceae) in the southeastern United States. Biol. Control 8: 65-73.

Center, T. D., T. K. Van, M. Rayachhetry, G. R. Buckingham, F. A. Dray, S. Wineriter, and M. F. Purcell. 2000. Field colonization of the melaleuca snout beetle (*Oxyops vitiosa*) in south Florida. Biol. Control 19: 112-23.

Coulson, J. R. 1977. Biological control of alligatorweed, 1959-1972. A review and evaluation. USDA Agric. Res. Serv. Tech. Bull. No. 1547, Washington, DC.

Dray, F. A., Jr., and T. D. Center. 2001. Lessons from unsuccessful attempts to establish *Spodoptera pectinicornis* in Florida. Biocontrol Sci. Technol. 11: 301-16.

Dray, F. A., Jr., T. D. Center, D. H. Habeck, C. R. Thompson, A. F. Cofrancesco, and J. K. Balciunas. 1990. Release and establishment in the southeastern United States of *Neohydronomus affinis* (Coleoptera: Curculionidae), an herbivore of waterlettuce. Environ. Entomol. 19: 799-802.

Grodowitz, M. J., T. D. Center, A. F. Cofrancesco, and J. E. Freedman. 1997. Release and establishment of *Hydrellia balciunasi* (Diptera: Ephydridae) for the biological control of the submersed aquatic plant *Hydrilla verticillata* (Hydrocharitaceae) in the United States. Biol. Control 9: 15-23.

Harley, K. L. S., and I. W. Forno. 1992. Biological Control of Weeds. A Handbook for Practitioners and Students. Inkata Press, Melbourne, Australia.

Jacob, H. S., and D. T. Briese. 2003. Improving the selection, testing, and evaluation of weed biological control agents. Proc. Commonwealth Government's Cooperative Research Centre for Australian Weed Management Biological Control of Weeds Symposium and Workshop, 13 September 2002, Univ. of Western Australia, Perth. CRC for Australian Weed Management Tech. Series No. 7.

Julien, M. H., M. W. Griffiths, and J. N. Stanley. 2001. Biological control of water hyacinth. The moths *Niphograpta albiguttalis* and *Xubida infusellus*: biologies, host ranges, and rearing, releasing and monitoring techniques. Australian Centre for Int. Agric. Res. Monogr. No. 79.

Julien, M.H., M.W. Griffiths, and A.D. Wright. 1999. Biological control of water hyacinth. The weevils *Neochetina bruchi* and *N. eichhorniae*: biologies, host ranges, and rearing, releasing and monitoring techniques for biological control of *Eichhornia crassipes*. Australian Centre for Int. Agric. Res. Monogr. No. 60.

Madeira, P. ., R. Hale, T. D. Center, G. R. Buckingham, and M. Purcell. 2001. Whether to release *Oxyops vitiosa* from a second Australian site onto Florida's melaleuca: a molecular approach. Biocontrol 46: 511-28.

Pratt, P. D., D. H. Slone, M. B. Rayamajhi, T. K. Van, and T. D. Center. 2003. Geographic distribution and dispersal rate of *Oxyops vitiosa* (Coleoptera: Curculionidae), a biological control agent of the invasive tree *Melaleuca quinquenervia* in south Florida. Environ. Entomol. 32: 397-406.

Van Driesche, R. G., S. Lyon, B. Blossey, M. Hoddle, and R. Reardon. 2002. Biological Control of Invasive Plants in the Eastern United States. USDA Forest Service Publ. FHTET- 2002-04, Morgantown, WV.

Wheeler, G. S., and T. D. Center. 2001. Impact of the biological control agent *Hydrellia pakistanae* Deonier (Diptera: Ephydridae) on the submersed aquatic weed *Hydrilla verticillata* (L.f.) Royle (Hydrocharitaceae). Biol. Control 21: 168-81.

# Factors that Affect Successful Establishment of Biological Control Agents

Eric M. Coombs

Classical biological control of weeds is an important tool in the arsenal in the fight against the unwanted impacts of invasive plants throughout the world. As biological control becomes more popular, the application of its principles becomes more available to the general public who collectively may lack the understanding and training for effective implementation. With increased successes in biological control of weeds, there comes the other side of the coin in that increased failures may be proportionate (Crawley 1989b). For the most part, biological control is employed as a tool on regional projects where other methods have been economically and ecologically unsuccessful. Therefore, the overall 30% rate of success using biological control is still a very impressive record. Despite the failures of other methods, there seems to be an overly confident anticipation of success when using biological control against recalcitrant weeds at regional scales.

Perhaps the best way to increase success of biological control is to avoid the causes of failure identified by others. It is often difficult to identify and quantify reasons why biological control agents fail to establish. We do not purposefully kill off biological control agents—they are valuable and we want them to control the target weed. A major hurdle in a biological control of weeds program is the introduction and establishment of viable populations of biological control agents in a region, locality, or site. Much has been learned through trial and error. This chapter discusses issues surrounding the successes and failures in agent establishment and concludes with practical considerations for implementation.

Success and failure rates of classical biological control of weeds vary with the scale on which they are measured: site, country, state, county, region, habitat, etc. Rates of success and failure have been analyzed by numerous scientists (Crawley 1989a, Julien 1989, Julien and Griffiths 1998, McFadyen 1998, Moran 1985, Syrett et al. 2000), whereas analyses examining the causes of failures are less prevalent (Beirne 1985, Crawley 1989b, Lawton 1990, Simberloff 1989, Stiling 1990, 1993). One of the inherent dangers in relying on analyses of success and failure rates throughout the world is that not all successes and failures are created equal in either scale or assessment. Successful biological control is a process and not an event. Biological control projects pass through sequential stages, i.e., target selection (McClay 1989, 1995, Peschken and McClay 1995); agent testing and safety (Coulson and Soper 1989, Klingman and Coulson 1983, McEvoy 1996); and introduction, recovery, establishment, redistribution, impact on host, and impact on plant community (Huffaker and Kennett 1959, McEvoy and Rudd 1993). Thus, when examining and comparing successes and failures in biological control, it is important to note both the scale of the observation and the stage of the program that was assessed. Crawley (1989b) gave the following breakdown of causes of failure based on literature review and surveys (a single case may have more than one cause): climate 44%, host incompatibility 33%, predators 22%, competition 12%, parasitoids 11%, and disease 8%.

Comparing the successes and failures by country, order of agent, etc., may be useful for generalizations, but not necessarily for local managers. Projects that appear promising at the national scale may be difficult to implement at local scales because of different environmental conditions and local cultural practices. Researchers should develop an evaluation form to assist personnel in determining the possible causes of why biological control agents fail to establish or control their target weeds. This would be helpful when introducing new agents that are not widely established in the region and for retrospective analyses.

## Establishment Success and Failure—Regional and Local

In this chapter, I have combined collective experiences from the Oregon biological control program and regional observations from a number of my colleagues in order to identify some of the causes of why biological control agents have not established. Comparisons were also made with biological control projects at international, national, state (Julien and Griffiths 1998), and county scales (Coombs 1999).

Maintaining and updating databases are an effective means of collecting and disseminating data. In 2000, a general status survey was conducted of weed biological control projects (number of weed-agent combinations) from experts in 11 western states in the United States (Figs. 1 and 2). As of 2003, a new database was being constructed to include data collected throughout the United States with a Web interface that can be queried by the general public and directly updated by practitioners.

Factors were identified that may have contributed to the failure of specific biological control agents to establish and/or control their host weeds. More than 12,000 Oregon Department of Agriculture biological control records were evaluated to determine the

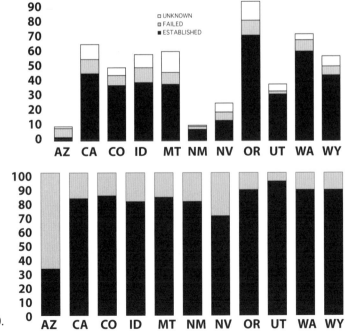

Figure 1. Number and status of biological control projects for 11 western states in 2000.

Figure 2. Percent of biological control projects that are successful (black) and failures (gray) in 11 western states in 2000.

presence and impacts of biological control agents throughout the state. Other data were obtained from the recollections and observations of scientists who had previously worked on projects in the Pacific Northwest. The data are somewhat subjective and proportional to the scale to which they are applied. However, observations from the local scale can prove beneficial at larger scales. The rates of establishment and failure are dynamic and vary each year in relation to the scale of analysis.

## What is Failure?

Failure to establish, as defined by Harris (1991), is when a biological control agent has not been recovered three or more years after release. For example, of the 71 species of biological control agents released in Oregon from 1947 to 2003, 58 species established (82%), seven failed, and six are in the unknown category. Some agents may be recovered for a year or two, but for various reasons fail thereafter. Residual failures are those that remain failures and are unlikely to be reintroduced. Transitional failures are those that later become or are actually found to be established. Verified failures are those that are reported by experienced biological control professionals after three years of monitoring. Unverified failures are often reported by less experienced personnel.

Verified residual failures in Oregon include those that will always remain failures because importation and release will never be attempted again (e.g., *Phrydiuchus spilmani* on Mediterranean sage [*Salvia aethiopis*]), as well as those species that are established in other states, but because of their ineffectiveness have not been reintroduced to Oregon (e.g., *Spurgia esula* and *Hyles euphorbiae* on leafy spurge [*Euphorbia esula*] and *Zeuxidiplosis giardi* on St. Johnswort [*Hypericum perforatum*]). A number of transitional failures were rereleased in Oregon in later years and establishment was obtained (i.e., *Oberea erythrocephala* on leafy spurge, *Ceutorhynchus litura* on Canada thistle [*Cirsium arvense*], and *Pterolonche inspersa* on diffuse knapweed [*Centaurea diffusa*]).

Intermittent failures occur when environmental conditions cause sporadic die-offs of tenuously established populations, e.g., *Microlarinus* spp. on puncturevine (*Tribulus terrestris*) in the northern limits of their range. These failures occur when agents exist at the edge of their climatic zones and are subject to variables such as severe winters, moisture, and drought.

Ecological failures occur through competition when the host has been controlled by another more successful biological control agent (e.g., *Larinus minutus* on diffuse knapweed) to a point below the carrying capacity required by the agent in peril (e.g., *Pterolonche inspersa*). Ecological failure can also happen when an agent occurs in extremely low numbers or when it has a negligible impact on the host (e.g., *Coleophora* spp. moths on Russian thistle [*Salsola tragus*]).

## Measuring Success and Failure

Data on the success and failure of biological control are dependent on accurate documentation, which hinges on the competency of those who make the observations. Biological control agents reported to have failed to establish by less-experienced personnel may later be found established when monitored by professionals. Thus, one of the best ways to increase the success rate is to increase the amount and quality of monitoring. Regional success rates over time may be driven more by information than biology.

Establishment success rates can be increased by using "transfer projects" (importing successful agents) from neighboring regions, states, or counties (Fig. 2). For example, in the western United States, Utah has many transfer projects and therefore the highest success rate (>90%), yet it ranked eighth of 11 western states in the total number of projects. In the U.S. 2003 survey, 24 states reported 100% establishment rates (not including unknowns), but these states generally have fewer projects (<15). States that implement more projects tend to have lower success rates because they often receive releases of new unproven agents. In some cases, failure rates can be decreased if previous unsuccessful projects are reimplemented. However, due to program prioritization and poor agent performance in other areas, there may be little incentive to attempt reintroductions. The International Code of Best Practices (see "International Code of Best Practices for Biological Control of Weeds," this volume) states that biological control practitioners should not introduce ineffective agents; however some agents that perform poorly in one region but may succeed in another remain untried.

Establishment and failure rates may be determined by the number of attempts, number of sites, competency of cooperators, and how these interact with chance events in nature. An interesting example of how statistics can affect success or failure rates involves the knapweed root moth (*P. inspersa*). It was first released in Oregon in 1986 and several times thereafter, was finally recovered in 1997, and now may be only tentatively established. The establishment success rate in the countries where it was released (United States and Canada) is 100%. The success rate in the states (n=4) where it was released is 25%. The success rate in the counties in Oregon (n=8) where it established is 12.5%. Finally, the success rate based on the number of attempts to establish the moth in Oregon (n=24) is 4.2%. It matters how the question is asked and to what scale the data are applied.

When older, successful biological control projects are compared, the establishment rates tend to be similar. For example, the tansy ragwort flea beetle (*Longitarsus jacobaeae*) was first released in the United States in 1969. The success rate for establishment in countries (n=4) is 100%, for states in the United States (n=4) is 100%, for counties in Oregon (n=24) is 79%, and for the total number of releases in Oregon (n=3,199) is 95%.

When should establishment and control success be counted? After three and 10 years, respectively, when biological control agents have had time to adapt to regional conditions. By using those intervals, we will help reduce the number of unknown elements that lead to underestimation of the actual success rates.

## Practical Considerations

We can improve successful establishment by determining the causes of past failures and then avoid repeating them. Measuring failures is not a one-step process or an end. Failure rates depend on the number of attempts, number of sites, and the procedures utilized by cooperators as they interact with the chance events that occur in nature (Crawley 1989a, b, Grevstad 1999). It is important not to blame factors such as climate or agent performance if human error can be identified as a contributor to failure, because this would reduce the likelihood of retrying agents that are successful in other areas. The best of agents can fail when handled poorly; therefore, good husbandry practices are vital.

When considering a site for a biological control release, the entire system must be considered, as well as what happens at the site throughout the year (grazing, flooding, fire, or other disturbances). Picking the best biological control targets does not always guarantee successful rehabilitation of weed-infested environments because the easiest target weeds to control may not be the most problematic. Biological control practitioners should function as members of an Integrated Pest Management (IPM) team to help solve specific weed control problems, which can lead to increased success rates.

The following discussion identifies some of the abiotic, biotic, and procedural aspects that can lead to the failure of biological control agents to establish and control their host weeds (Bierne 1985, Coombs et al. 1999). In most cases, success and failure are products of multiple interacting factors.

## Abiotic

A number of abiotic (nonliving) factors can contribute to the failure of a biological control agent. Climate comprises the variables concerning weather. The degrees and extremes of temperature and precipitation (intensity, duration, season, frequency) can impact the range and periods of activity available to biological control agents. Site characteristics such as soil, slope, aspect, shade, and moisture can all lead to site and microclimatic variations that inhibit some biological control agents. The variability in habitat requirements for the leafy spurge flea beetles (*Aphthona* spp.) is a good example. Elevation gradients that affect temperature and precipitation can limit distribution and abundance of biological control agents. Latitude affects the weather extremes, seasons, and day length at regional scales, which can limit how far north an agent can survive or how far south it can reproduce. Fire frequency and intensity are generally the result of plant community types, weather, and human causes. Many areas where target weeds are abundant are disturbed sites that tend to be more vulnerable to fires that can harm biological control agents in susceptible stages.

## Biotic

Many biotic or living factors interact and impact biological control agents at the community and organism levels.

**Community or synecological interactions** are probably the most complex of variables and are both the bane and inspiration of ecological modelers. Natural enemies are the predators, parasitoids, and pathogens that comprise the old associations that impact the biological control agent in its homeland. Great efforts are made to separate the agent from its natural enemies to allow it to achieve its biological potential in its new habitat in the United States. But native species in the United States can become predators, parasitoids, or pathogens that can reduce the survival and limit the effectiveness of introduced biological control agents. Little can be done to prevent these new associations, but fortunately, most are opportunistic and have minor impact. Host density can determine the numbers, density, and distribution of biological control agents, especially those that are density dependent. Defoliating moth larvae generally fare better in more dense stands of their host. Dense stands of the host plant can affect the microclimate by lowering the temperature within the stand, thus affecting the growth and movement of some species of biological control agents. Competition with other species of biological control agents can reduce the probability of establishing new and perhaps more effective

89

species. This may reduce the overall success of the program because in many cases one species of the suite of agents is often responsible for the majority of the control. The successional state of the plant community may change over time. The community structure and composition may change to the point where conditions no longer favor an effective biological control agent. These changes can affect plant diversity, which determines the abundance and location of important nectar sources, predator habitats, and other invasive plants that may compete with the host weed.

**The organismal or autecological aspects** are those that are inherent to the biological control agent itself. Synchronization of the agent with the host is necessary for the correct timing of oviposition and larval development. Physiology, including unique biotypes, allows some biological control agents to adapt to variations of the host plant, environmental gradients, and can provide resistance to disease and parasitoids. Biological control agents must be able to build up sufficient reserves to survive periods of long inactivity such as hibernation and aestivation. Some species are hygroscopic as larvae and swell when conditions are too moist, which may prevent development or cause mortality. Variability in the fecundity of biological control agents can affect their ability to reproduce. Certain innate capacities of behavior determine the ability and success of an agent to find its host, search for mates, and escape from predators. Sometimes the agents may emigrate or move to a more suitable location, making recovery difficult during initial investigations. Finally, releases of small numbers of agents simply tend to be more susceptible to failure for reasons that include: 1) problems adapting to the local environment due to insufficient genetic diversity; 2) the male-to-female sex ratio may become skewed toward males (reducing reproduction rates); or 3) local changes in the habitat, such as a fire or flood, may destroy the nascent population (Grevstad 1999).

## Procedural

The human or procedural factors are probably the most important yet most overlooked aspects leading to biological control establishment failures. They are also the categories over which we have the greatest control. These categories can be grouped into what is done or not done during the prerelease, release, and post-release periods, and personnel variables. Many of the variables previously discussed certainly interact with the human errors discussed below. Procedural errors are likely to be the most underreported for a variety of reasons: embarrassment, fear of jeopardizing funding, ignorance, lack of prioritization, and in some instances, local politics.

**Prerelease.** Site selection is one of the most important choices we can make at the beginning of a biological control program. Most people assume that if the target weed is present, the agent can be released, but many factors can negatively impact success. Factors that contribute to the demise of a recently released biological control agent include grazing, flooding, proximity to roads, fire regimes, accessibility, pesticide use, and refugia for hibernation and aestivation. It is critical to know what happens to the release site throughout the year in order to understand which variables suppress agent populations. The colony source can determine the fitness of the new agent to a new site. Laboratory-reared agents may be weakened from too many generations in captivity. The wrong agent biotype may have been supplied. In addition, the development of the agent may not be synchronized with the host. This often happens when agents are collected at elevations or latitudes quite different from those of the release site.

One of the most common problems that occurs when untrained personnel are involved is the method of collection. Delicate agents can be damaged during sweep netting by hard debris in the net such as heavy seed heads, straw, grass seed, dirt, and rocks. Sweeping needs to be carefully performed, targeted at the sites where the agents are most easily collected, using the correct type of net, and avoiding too many sweeps per collection period. When using vacuum apparatus including aspirators, care should be taken to avoid too long or strong suction that may swirl and damage agents. Agents that are physically damaged, especially flying species like flies and moths, have little chance of establishing at new sites.

Once insects are collected, the method and timing of shipment becomes very important. The method, duration, humidity, temperature, access to food (if needed), hiding places, season, freedom from predators, and opportunity for mating can all affect the number and quality of agents before they are released. Biological control agents should generally be kept cool and dry (except aquatic species). It is best to provide insects with a substrate on which to crawl and hide in the transport container, and to avoid constant motion. The webbing from spiders, even small ones that do not feed on the agents, can prevent agents from moving about after release. Some moth species should have a single male and female placed together in a vial to avoid pheromone-induced mating disruption that may occur when too many are kept in close confinement.

The amount of time in quarantine will determine the remaining life span of the agents. Agents shipped late in the development cycle can be less healthy and fecund. Some weevils become very shiny because of the loss of scales; moths also become shiny and often have ragged wings. The quality or general health of the agents affects their fecundity, depending on, for example, the source (laboratory-reared), condition due to parasitoids and pathogens, and maturity (life span remaining). The quantity variables such as the number released, sex ratios, and genetic diversity, can push populations below a recovery threshold level for establishment or cause local extinction altogether. If agents are released before mating, the sex ratio becomes very important in order to provide breeding opportunities before and after release.

**Release**. Once biological control agents are ready for release, a number of factors can be manipulated to reduce the probability of failure. The method of the release (open versus caged, speed of release, available plant canopy, and escape routes) can influence survival and reproduction. Many practitioners wrestle with the choice of caged versus open releases. Cages are beneficial when few agents are released in order to improve chances for mating and oviposition on site. However, cages can also limit available resources and prevent escape from enclosed predators. Releasing agents by dumping them out into the air may cause flying species to disperse too rapidly to maintain contact with each other. It is best to make releases near the ground and allow insects to crawl up the plants and disperse normally. To avoid competition, releases of new species of agents should not be made in proximity to nursery sites of established species on the same host.

Occasionally the wrong agents are collected and shipped to cooperators because they were misidentified or may have included unapproved species. This may occur when similar species of weevils, flea beetles, and seed head flies are collected. Agents may also be released on the wrong host; for example, host misidentification has occurred with thistles because of local names and confusion with native species.

Timing of agent release can be an important factor. Temperature, time of day, season, and weather during or shortly after release can affect the activity and survival of biological control agents. It is generally best to release agents in the early part of the day so they have the opportunity to behave normally. Releases during or before storms may lead to failure because agents can be blown away from the site or damaged by precipitation. The duration and density of agents during confinement can reduce their fitness before release. Minimize the amount of time agents spend in confinement. Too many agents per container may cause severe stress when they crawl over each other, keeping them in constant motion. Also, the buildup of feces in the container may lead to increased susceptibility to disease.

The life stage released can affect susceptibility to mortality. Transplanting infested host material may occasionally work, but the plants are often already under stress. Maintaining adequate moisture for the transplants and controlling competition with other plants is often difficult.

Releases must be documented to provide records for future monitoring to determine establishment and impact. Poor recordkeeping, not poor agent performance, can be a source of false failures.

**Post-release.** After biological control agents are released, procedural elements still can affect their status. Changes in land ownership can lead to denial of access to monitor and collect the agents. Site management can contribute to failure. Changes in land use such as grazing, cultivation, logging, or construction can cause biological control agents to fail to establish or control the target weed. Many releases are initially made on public lands where ownership and management practices are stable. Vandalism occasionally can disrupt establishment when cages, host plants, and the biological control agents themselves are destroyed. Detection of agent populations after release may be difficult due to low numbers of agents. When inexperienced field staff or cooperators are unable to find a biological control agent that is actually present, it may be recorded as a failure.

**Personnel.** A difficult issue to control is the personnel who become involved in the biological control process after releases are made. Problems that arise from these variables can lead to failures and false failures. Inadequacy and ignorance of untrained, uninterested personnel can result in site trampling, overharvesting, or premature harvesting of recently established populations. Therefore, cooperators should be provided with sufficient information and training. Continuity problems occur with high turnover rates of personnel (especially summer staff), which increase the overall lack of experience and can lead to loss of data from the institutional memory. Changes in political leadership and funding can shift the prioritization of projects before completion, perhaps leaving no opportunity for follow-up studies and monitoring.

## Summary

Identifying and avoiding common errors will help land management personnel increase the probability of successfully establishing biological control agents of weeds. The causes of failure are grouped into three general categories: abiotic, biotic, and procedural. Procedural errors are frequently underreported and may be a major source of failure in biological control. Most reported biological control failures have been attributed either to climatic or biological factors. Each failure should be individually appraised based on the relationship between the natural enemy and its environment (nonliving and living) and the procedures that were used. Factors in one category can be influenced by those in other categories; therefore failures can be attributed to multiple causes. Successes and failures are mostly reported by large-scale political units and rarely by geographical area and ecosystem classification. Any stage of the biological control process can fail, especially as the number of sites targeted increases. During the past decade, more government agencies and other interested parties have elected to incorporate biological control into their integrated weed management programs. Because more personnel lacking expertise and experience are becoming involved with biological control, the number of local failures reported has increased. By improving the techniques of collecting, handling, and releasing biological control agents, establishment success rates can be improved. Fortunately, once natural enemies become abundant in several areas, failures at local sites becomes less important because additional releases can be easily and economically made. However, the ease of obtaining additional release organisms should not serve as an excuse for carelessness. Adequate training and provision of information to secondary users (those who receive biological control agents after they have already been established locally) will help improve the establishment and success rates of weed biological control.

Successful biological control advances in degrees, and we are generally quick to claim success on the easy parts while neglecting more weighty matters such as impact to the target plant population, the surrounding community, the impact to ecosystem processes, or potential nontarget impacts. The causes for failure in biological control are legion, but we shall probably find the road to success paved with well-defined ecological protocols for research, implementation, monitoring, and evaluation.

## Acknowledgment

I thank Shon S. Schooler and Janet K. Clark for reviewing and making suggested improvements on this chapter.

## References

Bierne, B. P. 1985. Avoidable obstacles to colonization in classical biological control of insects. Can. J. Zool. 63: 743-47.

Coombs, E. M. 1999. Status of weed biocontrol in the Northwest. Pages 26-33 *in* M. Brooks and D. I. Isaacson, eds. Weed Biocontrol: Extended Abstracts from the 1997 Interagency Noxious-Weed Symposium. USDA Forest Service Pub. FHTET-98-12. Morgantown, WV.

Coombs, E. M., P. B. McEvoy, G. L. Piper, and B. Villegas. 1999. Why do weed biocontrol agents fail to establish or control their host weeds locally? Pages 24-25 *in* M. Brooks and D. I. Isaacson, eds. Weed Biocontrol: Extended Abstracts from the 1997 Interagency Noxious-Weed Symposium. USDA Forest Service Pub. FHTET-98-12. Morgantown, WV.

Coulson, J. R., and R. S. Soper. 1989. Protocols for the introduction of biological control agents in the U.S. Pages 1-35 *in* R. P. Kahn, ed. Plant Protection and Quarantine, Vol. 3: Special Topics. CRC Press, Inc., Boca Raton, FL.

Crawley, M. J. 1989a. Chance and timing in biological invasions. Pages 407-23 *in* J. A. Drake, H. A. Mooney, F. di Castri, R. H. Groves, F. J. Kruger, M. Rejmanek, and M. Williamson, eds. Biological Invasions: A Global Perspective. John Wiley, NY.

Crawley, M. J. 1989b. The success and failures of weed biocontrol using insects. Biocontrol News Info. 10: 213-23.

Grevstad, F. S. 1999. Factors influencing the chance of population establishment: implications for release strategies in biocontrol. Ecol. Appl. 9: 1439-47.

93

Harris, P. 1991. Classical biocontrol of weeds: Its definition, selection of effective agents, and administrative-political problems. Can. Entomol. 123: 827-49.

Huffaker, C. B., and C. E. Kennett. 1959. A ten-year study of vegetational changes associated with biological control of Klamath weed. J. Range Manage. 12: 69-82.

Julien, M. H. 1989. Biological control of weeds worldwide: trends, rates of success and the future. Biocontrol News Info. 10: 299-306.

Julien, M. H., and M. W. Griffiths. 1998. Biological Control of Weeds. A World Catalogue of Agents and their Target Weeds. CABI Publishing, CAB International, Wallingford, UK.

Klingman, D. L., and J. R. Coulson. 1983. Guidelines for introducing foreign organisms into the United States for the biological control of weeds. Bull. Entomol. Soc. Am. 29: 55-61.

Lawton, J. H. 1990. Biological control of plants: a review of generalisations, rules, and principles using insects as agents. Pages 3-17 *in* C. Bassett, L. J. Whitehouse, and J. A. Zabkiewicz, eds. Alternatives to Chemical Control of Weeds. Proc. Int. Conf. Rotorua, New Zealand, July 1989. Ministry of Forestry, FRI Bull. 155.

McClay, A. S. 1989. Selection of suitable target weeds for classical biological control in Alberta. Alberta Environmental Centre, Vegreville. AECV89-R1.

McClay, A. S. 1995. Beyond "before and after": experimental design and evaluation in classical weed biocontrol. Pages 213-19 *in* E. S. Delfosse and R. R. Scott, eds. Proc. VIII Int. Symp. Biol. Contr. Weeds, 2-7 February 1992, Lincoln Univ., Canterbury, New Zealand. DSIR/CSIRO, Melbourne, Australia.

McEvoy, P. 1996. Host specificity and biological pest control. BioScience 46: 401-5.

McEvoy, P. B., and N. T. Rudd. 1993. Effects of vegetation disturbances on insect biological control of tansy ragwort *Senecio jacobaea.* Ecol. Appl. 3: 682-98.

McFadyen, R. E. C. 1998. Biological control of weeds. Annu. Rev. Entomol. 43: 369-93.

Moran, C. 1985. The Silwood International Project on the biological control of weeds. Pages 65-68 *in* E. S. Delfosse, ed. Proc. VI Int. Symp. Biol. Control Weeds, 19-25 August, Vancouver, Canada. Agric. Canada, Ottawa.

Peschken, D. P., and A. S. McClay. 1995. Picking the target: A revision of McClay's scoring system to determine suitability of a weed for classical biological control. Pages 137-43 *in* E. S. Delfosse and R. R. Scott, eds. Proc. VIII Int. Symp. Biol. Contr. Weeds, 2-7 February 1992, Lincoln Univ., Canterbury, New Zealand. DSIR/CSIRO, Melbourne, Australia.

Simberloff, D. 1989. Which insect introductions succeed and which fail? Pages 61-75 *in* J. A. Drake, H. A. Mooney, F. di Castri, R. H. Groves, F. J. Kruger, M. Rejmanek, and M. Williamson, eds. Biological Invasions: A Global Perspective. John Wiley, NY.

Stiling, P. 1990. Calculating establishment rates of parasitoids in classical biological control. Am. Entomol. 36: 225-30.

Stiling, P. 1993. Why do natural enemies fail in classical biological control programs? Am. Entomol. 39: 31-37.

Syrett, P., D. T. Briese, and J. H. Hoffmann. 2000. Success in biological control of terrestrial weeds by arthropods. Pages 189-230 *in* G. Gurr and S. Wratten, eds. Biological Control: Measures of Success. Kluwer Academic Publishers, The Netherlands.

# Monitoring in Weed Biological Control Programs

Bernd Blossey

## Introduction

Biological weed control programs attempt to reduce populations of invasive weeds to levels where their continued presence is no longer of economic or ecological concern. As such, releases of biological control agents constitute ecological experiments on a grand scale that may have impacts at local, regional, and continental scales. Widespread impacts from local releases are seen as benefits by proponents of biological control programs and as potential ecological disasters by those critical of the technology. Despite an apparently excellent safety record (Crawley 1989, McFadyen 1998), skepticism concerning the need for and safety and effectiveness of insect introductions for weed control remains high among the general public, administrators, and scientists (Hager and McCoy 1998, Louda et al. 1997, Sagoff 1999). Some may argue that the increased emphasis on nontarget effects of biological control in North America has focused on two nonspecific weed control agents, *Rhinocyllus conicus* and *Cactoblastis cactorum*. Both species were released despite their documented broad host range (Blossey et al. 2001a, b, Pemberton 2000). However, the increased scrutiny of weed biological control programs may also be a function of increasing numbers of programs initiated over the last two decades (Julien and Griffiths 1998, McEvoy and Coombs 1999). In addition to an increase in the number of target plants, the number of control agents released per plant has also increased substantially (McEvoy and Coombs 1999). Unfortunately, the increased activity in biological control has not resulted in an increase in the knowledge of how control agents contribute to suppression of weed targets, nor how they are integrated into new food webs (Blossey 1999). In fact, recent reviews of biological control programs found that few evaluations provided objective measures of the decrease in weed abundance (Crawley 1989, McFadyen 1998), and when monitoring programs were implemented they generally focused on presence and spread of control agents (McClay 1992). This lack of data on the effects of control organisms on target host plants and their associated fauna and flora is a severe handicap for improving the scientific basis of biological weed control. In addition, weed biological control is unable to meet the increased public demands for data on impacts of invasive plants, effectiveness in weed suppression, economic and ecological benefits of control using herbivores, and lack of nontarget effects.

While many scientists consider the invasion of nonindigenous plants a primary threat to the integrity and function of ecosystems (Drake et al. 1989, Mack et al. 2000, Westbrooks 1998), there is little quantitative or experimental evidence for ecosystem impacts of invasive plant species. Justifications for control are often based on potential, but not presently realized, recognized, or quantified, negative impacts and the need to prevent future degradation (Blossey et al. 2001a). Given the poor state of monitoring in biological control, it is ironic that a recently completed review by the National Academies (NRC 2002) on predicting effects of invasions of nonindigenous plants and plant pests concluded that "there is no uniform agreement among investigators in this field on how to judge the severity of an impact, partly because scientists have not been able to carry out comprehensive and carefully designed studies of invasions. *Probably the best published data on establishment and impact of introduced species come from work on biological control,* which is particularly relevant for assessing impacts in forest, range, and agricultural ecosystems because most biological control efforts on both islands and mainlands are in these ecosystems" (emphasis added). This lack of scientific certainty about impacts has led to harsh criticism of attitudes toward nonindigenous species and control efforts (Hager and McCoy 1998, Sagoff 1999, Slobotkin 2001), culminating in a statement that "if a species is not clearly a medical or agricultural pest, let's learn to love it!" (Slobotkin 2001). While such harsh and pointed statements do not represent a majority opinion, they clearly expose a fundamental weakness in our ability to assess impacts of introduced species on individual species or ecosystem processes. The lack of evidence does not necessarily imply lack of impacts (Blossey et al. 2001a) and the impacts of invasive plant species may be greatly underestimated, yet we lack clear and convincing evidence due to lack of long-term data. The most important question becomes whether lack of scientific certainty about negative impacts should result in postponing control measures.

Any serious evaluation of the impacts of invasive plants on native species and ecosystems, as well as an assessment of how biological control programs affect the species composition and functioning of ecosystems, has to rely on long-term data about species abundances and ecosystem function at various spatial scales (Blossey 1999). Biological control, once implemented, cannot be discontinued if nontarget effects materialize after control agents are established. This sets it apart from other weed control options. While potential negative impacts of herbicides on applicators, ecosystem function, and on nontarget species are widely recognized, ecosystem impacts of traditional techniques (mechanical, physical, chemical) to control invasive plants are rarely considered. Many people assume that local control efforts have localized impacts and can be discontinued if unwanted side effects occur. Control at one site does not affect populations of the target at other sites, and landowners are able to control management options. But once biological control options are explored, a unique set of rules and regulations are in effect and an integral part of any petition to introduce biological control agents is a review of evidence for the negative impacts of the target weed (Knutson and Coulson 1997, USDA-APHIS 1998). While direct evaluation of ecosystem impacts of invasive plants is not the responsibility of scientists researching biological control, the burden of proof for negative ecosystem impacts is placed onto those proposing to utilize biological control (McEvoy and Coombs 1999). The question is whether long-term monitoring programs are in place in the United States that could provide such crucial information to guide management programs.

## The Status of Environmental Monitoring in the United States

Sustaining the quality of the nation's ecological and natural resources requires effective management of those resources (Olsen et al. 1999). Effective resource management relies on accurate, timely, and complete information on the extent, condition, and productivity of those resources. To obtain this information, federal, state, and local agencies have established ecological and natural resource monitoring programs (Olsen et al. 1999). Monitoring programs associated with agriculture have an extensive history; one of the most sophisticated is the National Resource Inventory (NRI) administered every five years through the Natural Resource Conservation Service (NRCS). This inventory collects data on clean water, prime farmland, wetlands, wildlife habitat, and environmental effects of agriculture using permanent sampling locations established across the United States (Nusser and Goebel 1997). Lately the NRI relies primarily on remote sensing techniques (Nusser and Goebel 1997) and it does not collect information about presence or abundance of any plant or animal species (other than crop species). Therefore, the long-term monitoring activities of the NRI have not been able to provide sufficient or relevant data for environmental assessments or effective natural resource management (Goebel 1998). Most state departments (agriculture, natural resources, natural heritage, and fish and game) conduct censuses (often of game species) and all endangered species are periodically censused. Other long-term data sets include the North American Breeding Bird Survey and the Christmas Bird Count. In general, however, many formal censuses of animal species are regional in scale and short-term in duration.

Overall, the status of long-term monitoring programs of biological resources in the United States continues to be in serious disarray, and information about biological diversity or individual species is patchy, at best. Vague objectives, a piecemeal selection of species chosen for monitoring, and poor linkages between monitoring projects and the decision-making process plague such programs (Woodward et al. 1999). In addition, lack of consistent standards for data collection across agencies makes it difficult to increase statistical power by combining data (Woodward et al. 1999). Two recent reports of the National Academies evaluated long-term monitoring programs; the first evaluated and proposed a system of bioindicators that may be used to assess the status of the nation's biological resources (NRC 2000); the second provided an overview and critique of monitoring capabilities to assess potential effects of transgenic plants on North American ecosystems (NRC 2002). The bottom line of all current reviews is that there is no system in place to even begin a coordinated monitoring—although the need for the development of such a program has been recognized for quite some time (Olsen et al. 1999).

An important and sophisticated example of the power of long-term monitoring to assess and manage populations is the annual waterfowl surveys administered through the U.S. Fish and Wildlife Service to set hunting seasons and bag limits. The surveys were established through the Migratory Bird Treaty Act in 1918 (Anderson and Heny 1972) in response to rapidly declining waterfowl populations. Aerial surveys, banding, reporting of harvests, computer models, and weather indicators are used to estimate annual populations (Shaeffer 1998, Shaeffer and Malecki 1996); this estimate in turn is used to establish annual hunting regulations. Similar sophisticated assessments of how invasive plants or biological control agents may affect native and introduced organisms are needed to move from perception to data. Assessments of the nation's biological resources will become more sophisticated once decades of data have been accumulated. However, at present, biological weed control workers cannot rely on established monitoring systems to provide data on populations of native and introduced organisms and their ecological interactions. While there is hope for future improvements, the current situation requires that biological control scientists and practitioners engage in long-term monitoring to satisfy both scientific and societal concerns. Without at least minimal monitoring providing data on the success and safety of biological control as a management technique, the future of biological weed control may be jeopardized.

## Long-term Monitoring Benefits Biological Weed Control

Although biological control practitioners clearly identified the need for long-term follow-up work decades ago (Huffaker and Kennett 1959, Schroeder 1983), little progress has been made in collecting quantitative data on the effect of biological control agents on target plant performance or in documenting the response of associated plant and animal communities (Blossey 1999, McClay 1992). Logistical difficulties, long duration, and resistance by sponsors to fund monitoring appear to be the main stumbling blocks (McFadyen 1998). Most biological control scientists and practitioners also lack the time, training, and skills to conduct monitoring independently. However, sponsors, practitioners, and weed biological control scientists all need to recognize that this lack of follow-up work has stymied improvements in the state-of-the-art of weed biological control. Our ability to select successful control agents has not progressed much over the past decades (Crawley 1989). Trial-and-error approaches still dominate release strategies (introduction of single versus multiple agents, small or large releases, root-feeders or leaf-feeders, etc.) (Lawton 1990, McEvoy and Coombs 1999, McFadyen 1998). Too much of the emphasis in biological control programs has been on finding, screening, releasing, and distributing control organisms (Blossey 1999, McEvoy and Coombs 1999) and a better balance of these activities with follow-up work is required in the future. Development of long-term monitoring programs is one of the most important tools for improving the scientific basis of biological control. A better understanding of successes (and failures!) should over time lead to better selection and release strategies and replace the current "lottery model" (Malecki et al. 1993, McEvoy and Coombs 1999). The responsibility for developing monitoring protocols rests with biological control practitioners familiar with the target plant and its response to control agents.

Monitoring protocols should be completed and implemented before control agents are actually field released. Simple before-and-after pictures or anecdotal evidence from visual observations will not be sufficient. Monitoring programs should also include multiple taxa to increase knowledge about the effect of biological control agents on native biota and ecosystem processes. Well-executed long-term monitoring programs offer exciting opportunities to merge basic ecological research with applied work using teams of investigators. However, to allow widespread adoption of protocols and participation by nonacademic personnel, monitoring protocols should balance scientific sophistication with ease of application (Blossey and Skinner 2000). Incorporating long-term monitoring will certainly increase costs associated with biological control programs, but these costs may be offset (at least in part) by savings of resources currently wasted on unsuccessful agents. Credibility for biological control will significantly increase among ecologists and with the public if we can demonstrate convincingly that release of biological control agents is beneficial to native organisms and not a potential or realized ecological disaster (Blossey et al. 2001a, Hager and McCoy 1998, Louda et al. 1997, Simberloff and Stiling 1996).

## Development of Monitoring Programs

The real challenges are in determining what factors (or organismal interactions) drive population fluctuations or changes in ecosystem function. Natural ecosystems are notoriously complex (although invaded systems may have lost much of their original complexity), and the prevalence of indirect interactions makes it difficult to predict the response of even well-understood systems to environmental change or perturbations (Polis and Strong 1996, Yodzis 1988). Large oscillations in populations of birds, insects, and mammals can be associated with the North Atlantic Oscillation and the El Niño Southern Oscillation (Mysterud et al. 2001, Sillett et al. 2000). Arguing with confidence that conditions are better, worse, or just different due to changes associated with spread of introduced species (for example, herbivorous insects released as biological control agents) is impossible unless such impacts can reliably be distinguished from natural oscillations or plant succession. Lag effects between the initial perturbation and the measurable impact that is often associated with habitat fragmentation or invasion of nonindigenous species make the detection and mitigation of impacts even more challenging (Byers 2002, Parker et al. 1999). Laboratory and small-scale field experiments cannot adequately replicate interactions that occur in the field; the only way to capture the full range of ecological effects of the release of biological control agents is by observing actual ecosystems.

The inherent difficulty in predicting indirect interactions, cumulative or synergistic effects (Attayde and Hansson 2001) makes the use of long-term monitoring essential in our assessment of environmental effects associated with biological weed control. However, it is not immediately clear which organisms or processes should be monitored when implementing biological control programs. Long-term monitoring programs are expensive and labor-intensive, require standardization of monitoring units, and data collected must be verified for accuracy. Providing sufficient financial and taxonomic expertise to monitor more than a fraction of the biota will be extremely challenging. The development of ecological indicators as proposed by the National Research Council (NRC 2000) may offer a cost-effective alternative in the future. (For a discussion and development of a framework of indicators, see NRC 2000.) At a minimum, assessments should include measures of control agent abundance and impact on the host plant and host-plant population, and performance measures for associated plant communities at release and control sites. These measures would capture many of the direct effects associated with release of biological control agents. However, indirect effects are prevalent in ecological communities and many species and trophic levels are linked through such indirect interactions. For example, invertebrate and vertebrate predators may respond with a population explosion to an increase in control agent abundance; however, it is entirely unclear which predator species (e.g., spiders, ladybird beetles, ground beetles, wasps, mice) may respond and how changes in populations of predators may affect their original prey and the associated food webs. What type of biological inventories should accompany the assessment of control agent and target weed populations will be case specific—monitoring for effects of biological control targeting rangeland weeds will be different from monitoring in forests or wetlands—yet the overall goals are the same. It is at the discretion of those implementing biological control programs to select, justify, and defend the most appropriate metrics and the scope of the monitoring. The long-term nature of such monitoring programs will require collaboration with colleagues in other disciplines.

The ultimate goal of the release of biological control agents is the restoration of invaded areas. Uninvaded reference sites may serve as examples for restoration goals, yet even better would be long-term documentation of communities before the invasion of nonindigenous species (Blossey 1999). This may be a difficult proposition given the current poor state of biological inventories in the United States. However, clear specification of goals (Malecki et al. 1993) before implementing biological control programs will help in later evaluations of programs.

## Guidelines for Development of Monitoring Protocols

While this chapter is not intended to present a standardized "toolkit" for monitoring in weed biological control programs, a number of specific guidelines apply:

• *General considerations*

Monitoring in biological control programs should deliver information about establishment and performance of biological control agents and how their feeding affects target plant individuals and populations. In addition, monitoring should deliver information about presence or absence of *direct* nontarget effects (e.g., Do the biological control agents remain host-specific?) and how associated plant communities respond to feeding of the biological control agents on the target weed (e.g., Can biological control be used to restore invaded communities?).

1.  Develop monitoring protocols early, before release of biological control agents.
2.  Begin monitoring before control agents are released. Time series (the standardized collection of data over long time periods) become more powerful in separating true effects from "noise" the longer data are collected. This long-term duration requires that monitoring sites be available for extended periods of time, often a decade or more.
3.  Monitor at multiple release and control sites (where no purposeful biological control agent introductions have occurred). This requires careful consideration of habitat and climatic influences among sites.
4.  Standardize procedures so the same measures can be used across the range of biological control releases.
5.  Monitoring will need to be conducted for many years. Repeat visits are necessary to assess changes at release sites over time. Carefully consider nondestructive sampling (such as permanent quadrats) before implementing destructive (often much more time-consuming but also more accurate) measures.
6.  Balance scientific sophistication with the need for technology transfer. Monitoring will need to be conducted at multiple sites (potentially hundreds) across climatic

regions. Assessments need to be conducted by personnel with very different backgrounds and taxonomic expertise.

7. Publish early monitoring protocols on websites. Refine as needed. Make sure that those participating in this activity are trained. Conduct workshops to introduce concepts, methods, and organisms to potential participants. This will help standardize procedures and implementation of monitoring programs.

8. Monitoring protocols should be designed to capture large changes in plant performance and plant populations. As populations of control agents grow, severe impacts such as defoliation or target plant population collapses may follow, sometimes over large areas. Anticipate declines in target weed populations to very low levels. A change of 5% in seed output may be statistically significant, but it is unlikely to affect a weed population. In successful control programs, reductions to less than 5% of the original weed abundance can be common (McEvoy et al.1991).

9. Develop procedures that are able to assess impacts over various temporal scales. There may be no immediate impact in the year(s) following insect releases, yet effective monitoring protocols should have the ability to demonstrate establishment (for example, through presence of control agents or of their feeding). Monitoring needs to continue even if there is no immediate impact and after weed populations decline to very low levels. Often weed populations bounce back followed by recolonization of control agents and renewed suppression of the weed. True restoration of plant and animal communities may require many years. It is extremely important to document the duration of this process and changes in abundances of various organisms over time.

10. Develop procedures that can assess changes at various spatial scales. On-the-ground monitoring can capture small-scale changes in small plots or at an individual site (for example on a square meter basis). However, the analysis of impacts of successful biological control programs also requires landscape-scale measures. Remote sensing may offer exciting opportunities in the future; the current state of the technology does not allow for monitoring (of plant species) at small enough scales. While large populations can be detected using appropriate imagery, the required resolution to detect individuals or small populations covering areas as small as a square meter is generally not available. Aerial surveys using low-flying aircraft offer a possible alternative but evaluations are time-consuming. In addition, remote sensing is unable to assess populations of biological control agents. However, a combination of on-the-ground monitoring and remote sensing can be powerful in capturing large-scale impacts and processes.

## • Details of monitoring protocols

At a minimum, each protocol should include the following:

11. Careful documentation of control agent releases (numbers, dates, species, stages, locations, and sources).

12. All monitoring sites should be geo-referenced (GPS, or UTM coordinates) to allow easier incorporation in future evaluations and remote sensing evaluations.

13. Use photo-documentation at regular intervals to visually capture changes at monitoring sites. Pictures should be taken at the same time each year and from the same location.

101

14. Measure control agent abundances (adults, eggs, feeding marks). Timed counts are useful to standardize search efficiency across observers. Timing of field visits is important to capture control agents when they are visible. Difficulties remain if activity periods do not overlap or when species are cryptic or nocturnal.

15. Measure target plant performance. Good measures that are sensitive to herbivory include height, number of stems, and reproductive output. The best time to assess impact of herbivore feeding is before plants senesce but after most of the growing season is over. This may not be the same time as appropriate for measuring control agent abundance. Thus, it may be necessary to visit monitoring sites two or more times over the season, especially with agents that may have two or more generations per year.

16. Document presence or absence of control agent feeding, oviposition, or larval development on plants other than the target weed. When control agents are observed feeding or ovipositing on other plant species, quantify attack—do not rely on presence/absence data! Contact local authorities to verify nontarget impact.

17. Measure plant communities at monitoring sites. Plant diversity, composition, abundance, and performance may change as the target weed is suppressed. Monitoring protocols should be able to capture such changes (for example, in reproductive effort, height, cover, or species composition).

18. Measure other biological activity at monitoring sites. This may include assessments of predators as well as (secondary) presence and abundance of other herbivores (insects and pathogens). Populations of biological control agents as well as target weed populations may be greatly affected by these organisms. Predation has been implicated in failures of many biological control programs (Lawton 1990); monitoring protocols should enable the documentation of such effects.

19. Monitoring protocols need to capture fluctuations in abiotic conditions (flooding in wetlands, for example) that may greatly influence herbivores and their host plants. While these cannot necessarily be measured continuously, measures should occur at the time of monitoring.

20. Visit release sites during the off-season to better understand or capture seasonal changes that may affect the biological control agents.

### • Refinement and extension of monitoring protocols

Monitoring protocols established to assess impacts of biological weed control agents should ideally be embedded in a network of monitoring activity to assess the status of natural resources and organisms across North America. As discussed previously, such a network is currently not available. However, weed biological control programs offer exciting opportunities for ecological and applied work. Areawide or nationwide successful programs offer opportunities for participation by many different audiences. Biological control practitioners and scientists should recognize this opportunity and encourage work with very different audiences.

102

21. Encourage more detailed ecological studies at select sites. Implement factorial designs releasing single and multiple agents at both release and control sites (no purposeful insect releases). These experiments can provide in-depth evaluations of indirect and food web effects.

22. In successful programs, not all release sites can be monitored with the same detailed protocols. Develop "quick assessment procedures" such as visual scoring of defoliation status, control agent populations, extent of flowering, etc., that can be accomplished with minimal time commitment at each field site. Make sure that those participating in this activity are trained and use the same scoring system. Do not consider such quick assessments substitutes for long-term monitoring programs. These less detailed studies may become valuable once observations can be pooled and analyzed for large-scale effects.
23. Develop analysis guidelines for participants, particularly for those without an academic background. Graphical representation of trends over time is a powerful way to present data.
24. Establish data depositories and attempt meta-analyses. Combining monitoring data of many observers is powerful for evaluating regional or continental programs.
25. Publish monitoring results in both academic and nonacademic formats.

## Summary and Outlook

The current status of environmental monitoring in the United States is seriously deficient to a point that informed decisions about management of natural resources are nearly impossible. Biological control scientists and practitioners cannot rely on established monitoring programs to assess changes in species abundances or ecosystem processes that may be associated with the release of biological weed control agents. At the same time, it is impossible for scientists and society to request comprehensive monitoring for biological control programs without dramatically increasing the available funding for this work. However, it is the responsibility of biological control programs to achieve a new balance for expenditures associated with study, detection, importation, and follow-up monitoring.

The responsibility for developing standardized monitoring protocols rests with biological control practitioners. The inability to discontinue biological control after control agents are established distinguishes biological control from more traditional weed control methods. In the absence of data on ecosystem impacts of an invasive plant, the potential risks associated with the introduction of biological control agents may appear high compared to a "no action" scenario.

Future support (financial and otherwise) for biological control programs will be linked to the ability of biological control scientists and practitioners to comply with growing public and scientific demands to deliver data on impact of invasive plants and for quantitative measures of safety and effectiveness of control organisms. Comprehensive monitoring, public involvement, and timely publication of such results assure the public that concerns are being addressed seriously, and provides reassurance that biological control scientists are being careful to keep risks low. Without long-term data, our claims that biological control of weeds is effective and safe will always be looked at with suspicion. Financial and logistical challenges in establishing and maintaining sophisticated long-term monitoring programs may appear daunting. However, willingness to begin implementation, even of minimal monitoring programs as outlined above, will benefit biological weed control directly by improving its scientific basis and its public acceptance as a sophisticated (but not risk-free) measure for control of invasive plant species.

# References

Anderson, D. R., and C. J. Heny. 1972. Population ecology of the mallard. I. Review of previous studies and the distribution and migration from breeding areas. USDI Fish Wildlife Serv., Washington, DC.

Attayde, J. L., and L. A. Hansson. 2001. Press perturbation experiments and the indeterminacy of ecological interactions: effects of taxonomic resolution and experimental duration. Oikos 92: 235-44.

Blossey, B. 1999. Before, during, and after: the need for long-term monitoring in invasive plant species management. Biol. Invasions 1: 301-11.

Blossey, B., and L. C. Skinner. 2000. Design and importance of post-release monitoring. Pages 693-706 in N.A. Spencer, ed. Proc. X Int. Symp. Biol. Contr. Weeds, 4-10 July 1999, Bozeman, MT.

Blossey, B., L. C. Skinner, and J. Taylor. 2001a. Impact and management of purple loosestrife (Lythrum salicaria) in North America. Biodiversity Conserv. 10: 1787-1807.

Blossey, B., R. Casagrande, L. Tewksbury, D. A. Landis, R. N. Wiedenmann, and D. R. Ellis. 2001b. Nontarget feeding of leaf-beetles introduced to control purple loosestrife (Lythrum salicaria L.). Nat. Areas J. 21: 368-77.

Byers, J. E. 2002. Impact of non-indigenous species on natives enhanced by anthropogenic alteration of selection regimes. Oikos 97: 449-58.

Crawley, M. J. 1989. The successes and failures of weed biocontrol using insects. Biocontrol News Info. 10: 213-23.

Drake, J. A., H. A. Mooney, F. di Castri, R. H. Groves, F. J. Kruger, M. Rejmanek, and M. Williamson. 1989. Biological Invasions: A Global Perspective. J. Wiley and Sons, NY.

Goebel, J. J. 1998. The National Resources Inventory and its Role in U.S. Agriculture. Agricultural Statistics 2000. In Proc. Conf. Agric. Statistics, Natl. Agric. Statistics Inst., 6 February 2001. Available online: http://www.statlab.iastate.edu/survey/nri#Goebel.

Hager, H. A., and K. D. McCoy. 1998. The implications of accepting untested hypotheses: a review of the effects of purple loosestrife (Lythrum salicaria) in North America. Biodiversity Conserv. 7: 1069-79.

Huffaker, C. B., and C. E. Kennett. 1959. A ten-year study of vegetational changes associated with biological control of Klamath weed. J. Range Manage. 12: 69-82.

Julien, M. H., and M. W. Griffiths. 1998. Biological Control of Weeds. A World Catalogue of Agents and Their Target Weeds, 4th ed. CABI Publishing, Wallingford, UK.

Knutson, L., and J. R. Coulson. 1997. Procedures and policies in the USA regarding precautions in the introduction of classical biological control agents. Bull. OEPP 27: 133-42.

Lawton, J. H. 1990. Biological control of plants: A review of generalisations, rules and principles using insects as agents. Pages 3-17 in C. Basset, L. J. Whitehouse, and J. A. Zabkiewicz, eds. Alternatives to Chemical Controls of Weeds. FRI Bulletin 155, Rotorua, New Zealand.

Louda, S. M., D. Kendall, J. Connor, and D. Simberloff. 1997. Ecological effects of an insect introduced for the biological control of weeds. Science 277: 1088-90.

Mack, R. N., D. Simberloff, W. M. Lonsdale, H. Evans, M. Clout, and F. A. Bazzaz. 2000. Biotic invasions: Causes, epidemiology, global consequences, and control. Ecol. Appl. 10: 689-710.

Malecki, R. A., B. Blossey, S. D. Hight, D. Schroeder, L. T. Kok, and J. R. Coulson. 1993. Biological control of purple loosestrife. BioScience 43: 680-86.

McClay, A. S. 1992. Beyond "before and after": experimental design and evaluation in classical weed biocontrol. Pages 213-19 in E. S. Delfosse and R. R. Scott, eds. Proc. VIII Int. Symp. Biol. Contr. Weeds, 2-7 February 1992, Lincoln Univ., Canterbury, New Zealand. DSIR/CSIRO, Melbourne, Australia.

McEvoy, P. B., and E. M. Coombs. 1999. Biological control of plant invaders: regional patterns, field experiments, and structured population models. Ecol. Appl. 9: 387-401.

McEvoy, P., C. Cox, and E. Coombs. 1991. Successful biological control of ragwort, *Senecio jacobaea*, by introduced insects in Oregon. Ecol. Appl. 1: 430-42.

McFadyen, R. E. C. 1998. Biological control of weeds. Annu. Rev. Entomol. 43: 369-93.

Mysterud, A., N. C. Stenseth, N. G. Yoccoz, R. Langvatn, and G. Steinheim. 2001. Nonlinear effects of large-scale climatic variability on wild and domestic herbivores. Nature 420: 1096-99.

National Research Council (NRC). 2000. Ecological indicators for the nation. National Academy Press, Washington, DC.

National Research Council (NRC). 2002. Environmental impacts associated with commercialization of transgenic crops: issues and approaches to monitoring. National Academy Press, Washington, DC.

Nusser, S. M. and J. J. Goebel. 1997. The national resources inventory: a long-term multi-resource monitoring programme. Environ. Ecol. Stat. 4: 181-204.

Olsen, A. R., J. Sedransk, D. Edwards, C. A. Gotway, W. Liggett, S. Rathbun, K. H. Reckhow, and L. J. Young. 1999. Statistical issues for monitoring ecological and natural resources in the United States. Environ. Monitor. Assess. 54: 1-45.

Parker, I., D. Simberloff, M. Lonsdale, K. Goodell, M. Wonham, P. Kareiva, M. Williamson, B. von Holle, P. Moyle, J. E. Byers, and L. Goldwasser. 1999. Impact: toward a framework for understanding the ecological effects of invaders. Biol. Invasions 1: 3-19.

Pemberton, R. W. 2000. Predictable risk to native plants in weed biocontrol. Oecologia 125: 489-94.

Polis, G. A., and D. R. Strong. 1996. Food web complexity and community dynamics. Am. Nat. 147: 813-46.

Sagoff, M. 1999. What's wrong with alien species? Rep. Inst. Philosophy and Public Policy 19: 16-23.

Schroeder, D. 1983. Biological control of weeds. Pages 41-78 *in* W. E. Fletcher, ed. Recent Advances in Weed Research. Commonwealth Agricultural Bureau, Farnham, UK.

Shaeffer, S. E. 1998. Recruitment models for mallards in eastern North America. AUK 115: 988-97.

Shaeffer, S. E., and R. A. Malecki. 1996. Predicting breeding success of Atlantic population Canada geese from meteorological variables. J. Wildlife Manage. 60: 882-90.

Sillett, T. S., R. T. Holmes, and T. W. Sherry. 2000. Impacts of global climate cycle on population dynamics of a migratory songbird. Science 288: 2040-42.

Simberloff, D., and P. Stiling. 1996. How risky is biological control? Ecology 77: 1965-74.

Slobotkin, L. B. 2001. The good, the bad and the reified. Evol. Ecol. Res. 3: 1-13.

USDA-APHIS 1998. Reviewer's Manual for the Technical Advisory Group for Biological Control Agents of Weeds: Guidelines for Evaluating the Safety of Candidate Biological Control Agents. USDA-APHIS-PPQ, Marketing and Regulatory Programs. 03/98-01.

Westbrooks, R. 1998. Invasive plants, changing the landscape of America: fact book. Federal Interagency Committee for the Management of Noxious and Exotic Weeds (FICMNEW), Washington, DC.

Woodward, A., K. J. Jenkins, and E. G. Schreiner. 1999. The role of ecological theory in long-term ecological monitoring: a report on a workshop. Natural Areas J. 19: 223-33.

Yodzis, P. 1988. The indeterminacy of ecological interactions. Ecology 69: 508-15.

# Nontarget Impacts of Biological Control Agents

Eric M. Coombs, Shon S. Schooler, and Peter B. McEvoy

A common question concerning biological control is, "What will the biocontrol agents eat when the weed is gone?" In this chapter, we will briefly focus on some of the nontarget effects of biological control.

The nontarget risks inherent to weed biological control are similar to those of the pharmaceutical industry. Many of our best medicines have risks or side effects associated with their use. Practitioners of biological control strive to prescribe specific natural enemies on targeted weeds to accomplish a natural and self-regulating form of pest control, but realize that this is not without risks and that those risks should be predictable and acceptable. The risks of biological control can be reduced through critical review and revision of safety, testing, and implementation protocols. Practitioners are strongly urged to implement the safety and ethical protocols of the recently developed biological control International Code of Best Practices for Classical Biological Control of Weeds (see "International Code of Best Practices for Classical Biological Control of Weeds." this volume).

Any effect that a weed biological control agent exerts on an organism other than the intended target weed can be considered a nontarget impact. Nontarget impacts can be positive or negative. Besides the reduction of the target weed, positive effects include the restoration and function of native flora and fauna. Negative effects disrupt plant and animal community functions and abundance.

Several nonsanctioned exotic species that are known to attack some native plant species have been utilized for biological control of weeds. Because they have not been approved by the Technical Advisory Group for Biological Control Agents of Weeds (TAG) and United States Department of Agriculture Animal Plant Health Inspection Service Plant Protection and Quarantine USDA-APHIS-PPQ, we do not consider them as classical biological control agents and therefore they are not included in this book. Examples include *Cassida rubiginosa* and *Larinus planus*. Several states have imported one or both of these natural enemies for control of Canada thistle (*Cirsium arvense*) even though they were not approved. Biological control practitioners should strive to prevent the introduction into new ecosystems of natural enemies that are known to have unacceptable negative impacts as well as those that are not approved for release. Once a biological control agent is established, its impact is essentially irreversible, so careful prerelease screening for host specificity and effectiveness is essential.

It is generally expected that biological control practitioners will monitor their actions and do all they can to prevent negative nontarget impacts. Many of the nontarget impacts that have been observed were reported by biological control researchers and ecologists studying plant systems related to the target weeds. Post-release monitoring of the biological control agents and the plant communities in which they work is very important, but often the weakest link in the documentation process.

One of the best-documented examples of a biological control agent impacting nontarget plants is *Rhinocyllus conicus,* a seed head weevil introduced to control weedy thistles in the genus *Carduus.* In the 1980s, the weevil was observed attacking native thistles in the genus *Cirsium* (Turner et al. 1987), and recent work documents the quantitative impact on some native species in the central United States (Louda et al. 1997). When this insect was introduced into the United States in 1969, there was only limited government regulation of biological control agents and scientists were primarily concerned about the risk to agricultural plants. The prevailing attitude was that "the only good thistle is a dead thistle." This attitude has changed as we have learned to appreciate the functional role that all native plants—even thistles—play in diverse ecosystems. Current safety protocols would undoubtedly disapprove introducing *R. conicus* as a weed biological control agent today.

Through the years and with changing attitudes toward the environment, scientists' and government regulators' concerns have expanded to include nontarget effects to animal and plant communities with particular attention paid to threatened and endangered (T&E) species. The U.S. Fish and Wildlife Service has played a more prominent role in recent years to ensure that T&E species have been considered before new biological control agents are released in the United States.

A number of native plants in North America are congeners (plants in the same genus) of weeds targeted for biological control, e.g., *Cirsium, Euphorbia, Hypericum,* and *Senecio.* Included among these are sensitive and T&E species. It is within this group that nontarget impacts cause the greatest concern. As noted in the earlier chapter on host specificity testing, the most vulnerable nontarget plants are generally tested first. The plants at highest risk of nontarget impacts are congeners, then those in the same tribe, and then family. Plants targeted for biological control in the future that have native congeners (such as *Cynoglossum, Hieracium, Lepidium,* and *Potentilla),* therefore, will undergo comprehensive and meticulous testing to avoid nontarget impacts.

Risk can be computed as a function of: 1) the probability of the nontarget plant being exposed to the agent; and 2) the amount of potential damage should the nontarget plant be attacked (Harris and McEvoy 1995). Climate, season, elevation, ecological zone, and other factors affect which species are exposed to biological control agents and the degree of exposure. The economic and environmental risks of not using biological control also should be factored into a weed control program.

When biological control agents have a major impact on changing the structure and function of a plant community and its fauna, meeting host specificity requirements may not be a sufficient guarantee of an agent's harmlessness. For example, the control of saltcedar (*Tamarix ramosissima*) may reduce nesting sites for the threatened Southwestern willow flycatcher (*Empidonax traillii extimus*) (DeLoach et al. 1999). However, it is anticipated that the biological control of saltcedar will eventually allow for the return of native trees that will provide natural nesting sites.

One of the concerns identified by scientists is that of genetic changes and host shifts in the biological control agents. The evolutionary process through genetic change and natural selection may allow biological control agents now considered safe to cause

107

negative impacts in the future. Novel associations of an agent with a suite of new potential hosts, especially when there are no native competitors, may set the stage for adaptive radiation. This same process may work against biological control agents over time. Native parasitoids, predators, and pathogens may attack the agents and diminish both weed control and nontarget impacts. In another possible scenario, the pressure of competitive exclusion may enforce host specificity in some biological control agents.

## Biological Control Agent Species of Concern

The following examples (listed by weed and then agent) describe biological control agents that are known or strongly suspected to cause negative nontarget impacts. Additional nontarget impacts, if any, are listed under each agent in Section II.

### Musk and other thistles: *Rhinocyllus conicus*

As mentioned earlier in this chapter, the most serious nontarget problem currently is the thistle seed head weevil *Rhinocyllus conicus*. This weevil was first introduced into the United States in 1969, primarily to control musk thistle (*Carduus nutans*). Other strains of the weevil were later introduced to control milk thistle (*Silybum marianum*) and Italian thistle (*Carduus pycnocephalus*) (Goeden 1995a, b). Since then, it has been actively spread throughout many states infested with *Carduus* thistles with varying degrees of success. During the initial host specificity testing, the weevil was shown to feed on some native *Cirsium* thistles, but with a stronger preference for *Carduus* thistles (Zwölfer and Harris 1984). In the mid-1980s, *R. conicus* was found to attack several native *Cirsium* thistles in California and other states (Turner et al. 1987). In 1989, the Oregon Department of Agriculture discontinued redistribution of the weevil because of concern for nontarget impacts on native thistles. Other state departments of agriculture and universities have been following suit since the startling observations documented by ecologist Svata Louda and others. They documented *R. conicus* colonizing native thistles at their study sites and measured the associated decline of thistle seeds and the native thistle fly *Paracantha culta* (Louda 2000, Louda et al. 1997). It is expected that the list of 22 native thistles (about a quarter of our native thistle species) attacked by this weevil will continue to grow (Pemberton 2000). *R. conicus* can readily use the ubiquitous Canada thistle (*Cirsium arvense*) as a "host highway" to disperse throughout North America. At greatest risk are those species that flower early in the season. Practitioners should not use *R. conicus* as a biological control agent. The USDA-APHIS-PPQ has denied permits for the interstate shipment of *R. conicus* since 10 June 1997, and revoked all active permits as of 30 August 2000.

### *Opuntia* cacti: *Cactoblastis cactorum*

The pyralid moth *Cactoblastis cactorum* was introduced from South America into Australia in 1926 and later to other countries to control weedy *Opuntia* cacti. Spectacular results were achieved in Australia. Because the moth has not been approved or intentionally introduced into the United States as a classical biological control agent, it is not covered in Section II of this book. However, the moth may have been accidentally transported to the United States on ornamental cacti shipped from infested areas in the Caribbean. Because the moth is oligophagous (it feeds within one genus), the endangered semaphore cactus (*Opuntia spinosissima*) and several other species in Florida are at risk (Johnson and Stiling 1998, Pemberton 1995). Although this insect has been used in other regions of the world as a biological control agent, it is not suitable for all regions. Its accidental introduction into the United States is similar to the introduction of any other plant pest. If this moth should spread west, it may attack numerous native *Opuntia* species. This moth should not be used as a biological control agent in the United States.

## Purple loosestrife: *Galerucella calmariensis* and *G. pusilla*

This duo of leaf-skeletonizing beetles (Chrysomelidae) was introduced in 1992 to control the invasive wetland plant purple loosestrife (*Lythrum salicaria*). Host specificity tests indicated that the beetles could not complete their life cycle on any of the other plant species tested (Blossey et al. 1994). However, the testing did indicate that they would feed—but not reproduce—on the introduced ornamental shrub crepe myrtle (*Lagerstroemia indica*). Concern about nontarget impacts from spillover feeding led the California Department of Food and Agriculture (CDFA) to deny permission to release the beetles in California until the potential impact was further studied. Additional field studies in Oregon found that nontarget damage approached zero at 50 m from the heavily attacked loosestrife infestation (Schooler et al. 2003). Also the distributions of the two plant species exhibited minimal overlap. The limited potential demonstrated for nontarget impact encouraged the CDFA to allow release of the beetles within California. During this study, *Galerucella* were also observed to attack hyssop loosestrife (*Lythrum hyssopifolia*), another exotic species. Other potential impacts are being studied.

The initial host specificity studies also predicted that the beetles would feed—but not reproduce—on two native species, swamp loosestrife (*Decodon verticillus*) and winged loosestrife (*Lythrum alatum*) (Blossey et al. 2001). Similar to the example above, this prediction was confirmed as temporary feeding with low potential impact to either native plant species. Population-level effects have not been reported.

## Puncturevine: *Microlarinus* species

The puncturevine (*Tribulus terrestris*) weevils *Microlarinus lareynii* and *M. lypriformis* were introduced into California in 1961 and have spread throughout much of the southwestern United States. They were observed feeding on California caltrop (*Kallstroemia californica*) and Arizona poppy (*K. grandiflora*) in low numbers (Turner 1985). Published evidence of population-level impacts by these weevils on the native *Kallstroemia* species has not been found.

## St. Johnswort (Klamath weed): *Chrysolina quadrigemina* and others

The leaf beetle *Chrysolina quadrigemina* was introduced in 1946 to control St. Johnswort or Klamath weed (*Hypericum perforatum*). The spectacular results in California (Huffaker and Kennett 1959) were later duplicated in the Pacific Northwest. In 1985, researchers reported that the beetle was feeding on the native *Hypericum concinnum* and an ornamental ground cover, *H. calycinum*. *C. quadrigemina* is widespread throughout the western United States. No population-level impacts have been reported on native species. The root-boring beetle *Agrilus hyperici* and the bud gall midge *Zeuxidiplosis giardi* have also been reported to cause minor impacts on native *Hypericum* species in California (Andres 1985), but no long-term impacts have been reported.

**109**

## Tansy ragwort: *Tyria jacobaeae*

The cinnabar moth (*Tyria jacobaeae*) was introduced into the Pacific Northwest in 1959 to control the toxic range weed tansy ragwort (*Senecio jacobaea*). During the 1980s, it was found to attack the native triangle leaf groundsel (*Senecio triangularis*) (Diehl and McEvoy 1990), the exotic weed common groundsel (*S. vulgaris*), and the introduced ornamental dusty miller (*S. bicolor*). Although no population-level declines have been shown in the nontarget species, the moth can cause severe defoliation and is a cause for concern. In later studies, stream bank butterweed (*Packera pseudaurea*) (=*S. pseudaureus*) was also attacked by cinnabar moth larvae in the Cascade Mountains of Oregon (Fuller 2002). This case study illustrates an important principle: although the physiological host range established in laboratory tests may be broad, the ecological host range expressed in the field can be narrow due to biogeographical, phenological, habitat, behavioral, and physiological barriers to host use. The moth should not be introduced into new areas of the United States where it may encounter *Senecio* and *Packera* species for which it was not tested.

## Documenting Nontarget Impacts

When a nontarget impact is observed, the first objective is to identify the attacking agent and the affected plant. Some biological control agents have congeners that are serious pests, e.g., the gorse spider mite (*Tetranychus lintearius*) may be misidentified as a pest because it looks similar to the two-spotted spider mite (*T. urticae*) and strawberry mite (*T. turkestani*). During outbreak conditions in Bandon, Oregon, in 1997, gorse spider mites created a short-term alarm as they left behind extensive webbing on grass, ornamental plants, and even children's toys as the colonies moved en masse in search of more gorse. In another case, the Canada thistle stem gall fly (*Urophora cardui*), which looks similar to the walnut huskfly (*Rhagoletis completa*), was suspected of attacking walnuts in Corvallis, Oregon. The thistle flies were observed after they emerged in a walnut storage facility. It was later determined that galls from infested Canada thistle adjacent to the orchard had been accidentally collected in the fall when walnuts were vacuumed from the ground. The galls and walnuts are similar in shape and size and not easily sorted. Therefore, it is important to lay the blame on the proper organism and for the right reason.

The second objective is to document damage levels to individual nontarget plants. Through testing or observation, it should be determined whether a biological control agent is tasting versus feeding on a nontarget plant. During an outbreak period when the agent population is peaking, the degree of damage can be misleading when many agents may sample a plant and move on. Because of the sheer number of agents, substantial cumulative damage may occur. A case like this was observed during the 1980s in Oregon with the cinnabar moth on tansy ragwort. Hordes of larvae moved en masse from the decimated host plants to the ornamental dusty miller, which they subsequently stripped as well. Thousands of larvae drowned in swimming pools, invaded homes, and caused considerable alarm in neighborhoods. However, the resulting biological control of tansy ragwort caused an associated decline in cinnabar moth abundance and this type of nontarget impact is now uncommon.

The third objective is to determine effects on the nontarget plant population (as opposed to effects on individual plants). Based on the abundance of the nontarget species, monitoring can help determine whether the agent's feeding is opportunistic or preferential. If a nontarget species is attacked when the density of the plant and biological control agent are low, then the nontarget impact could be a serious problem and should be reported to authorities (state department of agriculture or university biology department). It is important to determine whether the individual plant or the abundance and function of its population in the plant community are affected. Some nontarget impacts may occur late enough following the reproductive period of the plant that there may not be a population-level effect.

The behavior of the female biological control agent can be the determining factor for host specificity by preferentially laying eggs on the target plant even though the larvae may be able to complete their life cycle on related species. Host specificity tests conducted showed that the larvae of *Agonopterix ulicis,* a gorse shoot tip moth, fed on several native lupines in laboratory no-choice tests, but in the field females only laid eggs on gorse. In this case, the degree of host specificity is regulated by the behavior of the female, not the larvae, which generally stay put. Additional ecological and behavioral testing must be completed to demonstrate that the moth is safe enough beyond its physiological host range before it is released in the Pacific Northwest.

The fourth objective is to determine whether nontarget impacts alter the function of plant and animal communities. Changes in nutrient cycling, species diversity and abundance, erosion, fire frequency, etc., may be beneficial or harmful. At some sites in Montana, populations of white-footed deer mice (*Peromyscus maniculatus*) have increased substantially in areas where they feed on the abundant seed gall flies (*Urophora* spp.) on spotted knapweed (*Centaurea stoebe* ssp. *micranthos*) (Pearson et al. 2000). As an example of another possible community effect, published evidence indicates that when the root-boring moth *Agapeta zoegana* attacks spotted knapweed, the growth of a desired forage grass, Idaho fescue (*Festuca idahoensis*), may be inhibited by physiological changes induced in the knapweed's root chemistry (Callaway et al. 1999). Biological control agents may displace more effective ones. For example, the leaf beetle *Chrysolina quadrigemina* may have inhibited the establishment of *Agrilus hyperici* on St. Johnswort in Australia by causing boom-bust oscillations of the weed (Briese 1977). More attention to prerelease ecological testing may help reduce "revenge effects" (McEvoy and Coombs 2000).

Anticipated nontarget impacts of candidate biological control agents are studied during the prerelease host specificity testing phase. These no-choice tests identify the physiological host range of nontarget plants on which the candidate agent can develop. During the choice testing phase, in which candidate agents are observed in a cage with the target plant as well as likely nontarget plants, behavioral responses of the candidate agent to nontarget species can be identified. The organism-level testing is often conducted in controlled circumstances in the laboratory or in cages, whereas the community-level testing is often conducted in the field after release. It is generally expected that the ecological host range will be narrower than the physiological host range. However, if significant nontarget impacts are documented in one region, then efforts to control the errant agent and prevent it from being released into new regions may be warranted.

## Conclusion

Nontarget impacts of some classical biological control agents have been documented, while for others the documentation may be forthcoming. The majority of negative nontarget impacts identified have occurred with older (pre-TAG) biological control programs with agents that would not likely be released under the present safety protocols.

No biological control agent is risk-free. An unrealistic expectation of complete safety may cause undue delays that could limit the accrual of benefits of using biological control in pest management and ecological restoration. Considering the options of allowing invasive plants to spread unchecked or applying herbicides as a temporary solution is a sobering comparison of risk. The best method to reduce biological control risk is through improved host specificity testing and pre- and post-release monitoring for ecological and economic nontarget effects. We should be cautious to prevent the introduction and use of natural enemies (approved or not) that are known to have substantial negative impacts elsewhere. Poor decisions today may lead to increased regulations that further restrict the practice of biological control of weeds.

### *Acknowledgment*

We thank Svata Louda for her initial review of this chapter.

### *References*

Andres, L. A. 1985. Interaction of *Chrysolina quadrigemina* and *Hypericum* spp. in California. Pages 235-39 *in* E. S. Delfosse, ed. Proc. VI Int. Symp. Biol. Contr. Weeds, 19-25 August 1984, Vancouver, Canada. Agric. Canada, Ottawa.

Blossey, B., D. Schroeder, S. D. Hight, and R. A. Malecki. 1994. Host specificity and environmental impact of two leaf beetles (*Galerucella calmariensis* and *G. pusilla*) for biological control of purple loosestrife (*Lythrum salicaria*). Weed Sci. 42: 134-40.

Blossey, B., R. Casagrande, L. Tewksbury, D. A. Landis, R. N. Wiedenmann, and D. R. Ellis. 2001. Nontarget feeding of leaf-beetles introduced to control purple loosestrife (*Lythrum salicaria* L.). Nat. Areas J. 21: 368-77.

Briese, D. T. 1977. Biological control of St. John's wort: past, present, and future. Plant Prot. Q. 12: 73-80.

Callaway, R. M., T. H. DeLuca, and W. M. Belliveau. 1999. Biological control herbivores may increase competitive ability of the noxious weed *Centaurea maculosa*. Ecology 80: 1196-1201.

DeLoach, C. J., R. I. Carruthers, J. E. Lovich, T. L. Dudley, and S. D. Smith. 1999. Ecological interactions in the biological control of saltcedar (*Tamarix* spp.) in the United States: Toward a new understanding. Pages 819-73 *in* N.R. Spencer, ed. Proc. X Int. Symp. Biol. Contr. Weeds, 4-14 July 1999, Montana State Univ. Bozeman, MT.

Diehl, J., and P. B. McEvoy. 1990. Impact of the cinnabar moth (*Tyria jacobaeae*) on *Senecio triangularis*, a nontarget native plant in Oregon. Pages 119-26 *in* E. S. Delfosse, ed. Proc. VII Int. Symp. Biol. Contr. Weeds, 6-11 March 1988, Rome, Italy. Ist. Sper. Patol. Veg. (MAF).

Fuller, J. 2002. Assessing the safety of weed biological control: a case study of the cinnabar moth *Tyria jacobaeae*. MS Thesis. Oregon State Univ., Corvallis.

Goeden, R.D. 1995a. Italian thistle. Pages 242-44 *in* J. R. Nechols, L. A. Andres, J. W. Beardsley, R. D. Goeden, and C. G. Jackson, eds. Biological Control in the Western United States: Accomplishments and Benefits of Regional Research Project W-84. Univ. Calif. Div. Agric. Nat. Res. Pub. No. 3361. Oakland, CA.

Goeden, R. D. 1995b. Milk thistle. Pages 245-47 *in* J. R. Nechols, L. A. Andres, J. W. Beardsley, R. D. Goeden, and C. G. Jackson, eds. Biological Control in the Western United States: Accomplishments and Benefits of Regional Research Project W-84. Univ. Calif. Div. Agric. Nat. Res. Pub. No. 3361. Oakland, CA.

Harris, P., and P. McEvoy. 1995. The predictability of insect host plant utilization from feeding tests and suggested improvements for screening weed biological control agents. Pages 125-31 *in* E. S. Delfosse and R. R. Scott, eds. Proc. VIII Int. Symp. Biol. Contr. Weeds, 2-7 February 1992, Lincoln Univ., Canterbury, New Zealand. DSIR/CSIRO, Melbourne.

Huffaker, C. B., and C. E. Kennett. 1959. A ten-year study of vegetational changes associated with biological control of Klamath weed. J. Range Manage. 12: 69-82.

Johnson, D. M., and P. D. Stiling. 1998. Host specificity of *Cactoblastis cactorum* Berg., an exotic *Opuntia*-feeding moth in Florida. Florida Entomol. 25: 743-48.

Louda, S. M. 2000. Negative ecological effect of the musk thistle biological control agent *Rhinocyllus conicus*. Pages 167-94 *in* P. A. Follett and J. J. Duan, eds. Nontarget Effects of Biological Control. Kluwer Academic Publishers, Boston, MA.

Louda, S. M., D. Kendall, J. Conner, and D. Simberloff. 1997. Ecological effects of an insect introduced for the biological control of weeds. Science 277: 1088-90.

McEvoy, P. B., and E. M. Coombs. 2000. Why things bite back: unintended consequences of biological weed control. Pages 167-94 *in* P. A. Follett and J. J. Duan, eds. Nontarget Effects of Biological Control. Kluwer Academic Publishers, Boston, MA.

Pearson, D. E., K. S. McKelvey, and L. F. Ruggiero. 2000. Nontarget effects of an introduced biological agent on deer mouse ecology. Oecologia 122: 121-28.

Pemberton, R. W. 1995. *Cactoblastis cactorum* (Lepidoptera: Pyralidae) in the United States: An immigrant biocontrol agent or an introduction of the nursery industry? Am. Entomol. 41: 230-32.

Pemberton, R. W. 2000. Predictable risk to native plants in weed biological control. Oecologia 125: 489-94.

Schooler, S. S., E. M. Coombs, and P. B. McEvoy. 2003. Nontarget effects on crepe myrtle by *Galerucella pusilla* and *G. calmariensis* (Chrysomelidae), used for biocontrol of purple loosestrife. Weed Sci. 51: 449-55.

Turner, C. E. 1985. Conflicting interests and biological control of weeds. Pages 203-25 *in* E. S. Delfosse, ed. Proc. VI Int. Symp. Biol. Contr. Weeds, 19-25 August 1984, Vancouver, Canada. Agric. Canada, Ottawa.

Turner, C. E., R. W. Pemberton, and S. S. Rosenthal. 1987. Host utilization of native *Cirsium* thistles (Asteraceae) by the introduced weevil *Rhinocyllus conicus* (Coleoptera: Curculionidae) in California. Environ. Entomol. 16: 111-15.

Zwölfer, H. and P. Harris. 1984. Biology and host specificity of *Rhinocyllus conicus* (Froel.) (Col., Curculionidae), a successful agent for biological control of the thistle, *Carduus nutans* L. Z. Agnew. Entomol. 97: 36-62.

# Integrated Weed Management and Biological Control

Gary L. Piper

Biological control, as discussed in this book, refers to the purposeful importation, conservation, or augmentation of various natural enemies such as invertebrate and vertebrate animals and plant pathogens to reduce vigor, occurrence, and reproduction of invasive plants. Of these procedural approaches, importation, also known as "classical" or "inoculative" biological control, has undoubtedly received the greatest attention from weed control specialists (Andres 1982, Masters and Sheley 2001). This type of biological control involves obtaining an undesirable plant's natural enemies from within its native range and, after subjecting the organisms to rigorous testing procedures to confirm their safety, reuniting them with the plant in its naturalized areas. For over a century, biological control has been employed worldwide and is viewed by many individuals as being an economical, effective, and environmentally acceptable way to limit intrusive plant species. When it is highly successful, people frequently tend to utilize this approach to the exclusion of all other weed suppressive measures. This type of thinking should be avoided because biological control, like most other weed control practices, has limitations and is by itself often unable to reduce undesirable plant populations to nondamaging levels. But despite any inherent shortcomings, the practice of biological control will remain a valued component of many present and future weed management efforts (Sheley and Petroff 1999).

In recent years, a structured, more insightful approach for dealing with problematic plants has emerged and readily gained acceptance by weed scientists, agricultural commodity producers, and others. This approach is termed "integrated weed management" (IWM) (Sheley and Petroff 1999). It involves the deliberate selection, integration, and application of effective, environmentally safe, and sociologically acceptable practices to suppress invasive weed populations. The goal of IWM is optimization of production or protection of an ecosystem through the concerted use of scientific knowledge, preventive tactics, monitoring procedures, and skillful application of control practices (CIPM 2002). IWM is rather unusual in that it is founded upon ecological principles and incorporates different methodologies in developing ecosystem management strategies that are sustainable, economical, and protective of public and environmental health (Masters and Sheley 2001). It organizes multiple yet mutually supportive techniques of plant population reduction in a way that maximizes merits and minimizes limitations of the component methods (CNAP 2000).

The focus of IWM is not on the short-term elimination of offensive plant species. The approach considers the natural or agricultural ecosystems in which weed infestations occur and stresses the manipulation of these invaded systems in ways that lead to sustained levels of plant suppression. This is important because only by studying and understanding weed-ecosystem dynamics can a land manager then sensibly develop methods to impact undesired vegetation and restore long-term productivity and integrity of the afflicted land or water resource (Heady and Child 1994). If the underlying causes of plant invasions are not adequately addressed, IWM efforts will most likely be woefully inadequate or fail miserably.

Invasive weeds, be they natives or nonnatives, are those species that demonstrate an ability to rapidly reproduce and disperse in an altered ecosystem (DiTomaso 2000, Masters and Sheley 2001). They can be conveniently classified as "potential new invaders," "new invaders," and "established invaders." A potential new invader is a weedy plant that has not yet been recorded in at a site but a strong possibility of its imminent invasion exists. New invaders are weeds whose population densities and distributions are such that all seed, turion, fragment, tuber, or nutlet production can be prevented at a site by the application of appropriate management methods. On the other hand, those weeds whose population levels and distributions are such that all propagule production cannot possibly be prevented at a site are considered to be established invaders. An understanding of these differences is of utmost importance because management responses will usually differ based upon the actual or potential threat that these weeds pose to a particular area.

It is necessary to determine whether eradication, suppression, or containment is the management outcome sought (CNAP 2000). IWM characteristically involves a manipulation of the ecosystem so that weeds, especially established invaders, are kept at noninjurious levels without ever being completely eliminated (Masters and Sheley 2001). Populations of well-established weeds should be managed through the use of suppression and containment measures to prevent their movement into areas not yet infested. Eradication is an inappropriate objective for an established weed infestation (DiTomaso 2000, Sheley and Petroff 1999). In the case of new invaders, rapid elimination of small populations, if practicable, and not long-term management might be the preferred objective. Early treatment is required to halt propagule production and deter all off-site spread of new invader species. IWM can often successfully reduce new invader abundance over time to the point where eradication sometimes becomes possible. The desirability and feasibility of eradication is heavily dependent upon the likelihood of site reinvasion by the weed in the future.

## Construction of an Integrated Management Program

Many of the problems encountered by persons during the formulation and implementation of weed management plans are not technical, but rather social in nature (Sheley and Petroff 1999). Public involvement during the inception of an IWM program is highly desirable, especially where management activities involve county-, state- or federally owned properties (CIPM 2002). In an effort to minimize social conflicts, opportunities must be provided for special interest groups and concerned citizens to comment upon and participate in coordinated weed management planning. This is particularly true for large-scale weed suppression programs where public understanding, acceptance, and support of the effort are indispensable to a successful outcome. The public usually supports programs they understand and opposes those they do not. A community awareness program should begin during early IWM project planning stages and continue throughout the implementation and evaluation phases. All available and

115

effective information dissemination methods—including traditional mass media outlets, Internet Web sites, personal contacts, public meetings, and educational workshops— should be used to reach and solicit input from an affected and often politically- connected citizenry (Masters and Sheley 2001).

Most IWM programs proceed in a stepwise manner over an extended period of time. Because development of a specific IWM plan depends on the invasive status of the target species and infestation density, the resource to be protected, environmental and human health concerns, and financial, labor, and other constraints, it is not possible to provide definitive procedures to be followed in every potential management situation. However, there are certain general program guidelines that are applicable to most management efforts. The reader should consult the References for detailed information on IWM plan assembly and for examples of specific weed management plans.

### • *Information acquisition*

An essential first step in developing an IWM plan is to generate and maintain an information gathering system. Information acquisition is necessary to delineate the problem area, determine damage and management action levels, select and time the most effective and least environmentally intrusive management options, and monitor and evaluate the program once implemented. The process, including surveying, monitoring, record-keeping, and evaluation, represents an ongoing activity in any IWM program.

### • *Surveys*

Systematically conducted surveys provide information about undesirable species composition, distribution, dominance, and impacts within the designated management unit (DiTomaso 2000). Various survey methods can be used depending on the level of accuracy required and budgetary limitations. A list of weeds to be surveyed must be compiled and their legal and invasive status determined. Survey personnel need to be able to accurately recognize designated weeds or know where to seek assistance when unfamiliar species are encountered. It is important to prioritize areas subject to repetitive disturbance and intense human activity during the surveys, as these sites favor weed colonization. Emphasis should first be placed on locating and inventorying potential and new invader sites, followed by an assessment of sites occupied by established invader species. Seasonal surveys should be conducted both prior to and after the implementation of treatment measures. All infestations should be mapped, and maps should be updated annually once management has begun (CIPM 2002). Maps are important because they present survey results in an easily understandable form and allow resource managers to document year-to-year changes in weed incidence (Sheley and Petroff 1999).

### • *Monitoring*

The purpose of monitoring (i.e., time-repeated inspection of weed-infested or management areas) is to compile and record site-specific information upon which decisions about management options for invasive species are to be based (Sheley and Petroff 1999). Monitoring provides weed management personnel with baseline data, reference points for all management decisions, a mechanism to evaluate the effectiveness of short- and long-term management effort effectiveness, and an opportunity to assess any environmental damage resulting from the management methods employed. Each monitoring system must be site-specific because the level of effort must reflect IWM program objectives and the constraints of monitoring personnel time and skill levels. Furthermore, a surveillance system is valuable only if accurate records are maintained. The format and process by which monitoring personnel maintain records should be standardized to achieve continuity during long-term management program efforts (CNAP 2000).

## • *Damage and action thresholds*

An assessment of weed populations in a management system is worthwhile only if infestation levels can be meaningfully related to potential damage. Basing management decisions on damage and "action thresholds" is extremely important in IWM. Thresholds provide weed managers with definitive ways of determining whether and when weed populations warrant manipulative actions.

A damage threshold is the plant population density or growth stage that results in intolerable injury to the affected resource and/or its occupant biota. It is based on: 1) an assessment of the weed's potential to establish and spread; 2) determination of the extent of a geographic area susceptible to invasion; and 3) evaluation of the affected resource area's management objective(s). To the extent that a weed is capable of damaging a resource, a damage potential exists. The damage potential is a subjective or objective (where quantitative data exists) estimate of the potential for damage based upon aspects of weed biology and ecology, propagule transport potential, and habitat susceptibility.

The action threshold is determined relative to the damage threshold. The action threshold is the plant abundance level at which suppressive techniques must be implemented to prevent the population from reaching the damage threshold. An action threshold is established to guide the selection and implementation of weed management treatments.

In the case of new invader weed species, the damage and action threshold are considered to be one plant of that species in the area of concern. This determination is based on the potential of any new noxious plant to adversely affect natural resources and increase management costs as it spreads. The low damage and action thresholds indicate immediate action is needed to prevent reproduction. An established invader would have higher damage and action thresholds because management goals emphasize prevention of reproduction and spread.

Thresholds are most useful for decision-making when infestations of well-established weeds require annual management. In such situations, establishment of realistic damage and action thresholds depends on species-specific, site-specific, phenological, economic, and other relevant factors. For an established invader, occasional readjustment of established damage and action thresholds is usually required because management strategies and objectives change in response to weed population reductions or increases. Threshold evaluation and modification should occur continuously in any IWM program.

**117**

## Weed Management Methods and Use

In a management program, various measures to reduce plant populations are put into place only when monitoring indicates thresholds have been reached. The focus of integrated management is the well-orchestrated use of diverse methods to effectively limit troublesome plants. No single method can or ever should be relied on to combat established invaders or deter new invasive species (DiTomaso 2000). Many factors must be considered when deciding which management options would be most effective and practicable in a specific situation. Selecting these measures must be based on a detailed knowledge of available options, the nature of the resource to be protected, and weed biology and ecology (CIPM 2002). An understanding of these factors enables the management practitioner to identify maximum vulnerability periods in the life cycles of weeds and then apply appropriate, timely tactics to achieve optimum population suppression. Increased attention should be given to any methods that disrupt weed propagule production, dispersal, survival, and minimize habitat disruption to inhibit further undesirable plant establishment.

### • *Prevention*

Preventive measures, along with one or more intervention methods, should be directed against all invasive species. Prevention is the process of forestalling contamination of an uninfested area by certain plant species known to be serious pests. It includes early detection procedures (surveys) and methods taken to alleviate the conditions that cause and foster the spread of undesirable vegetation (CNAP 2000). A preventive strategy must always precede and accompany any weed management effort. It is always more environmentally desirable and cost effective to prevent weed infestations than to have to subsequently manage them (CIPM 2002).

The success of preventive measures varies with the weed species involved, its means of dissemination, and the amount of effort expended. Specific preventive practices include: 1) enacting and/or enforcing federal and state laws restricting weed movement; 2) minimizing weed invasion potential in pasture and rangeland through improved grazing management practices; 3) promptly revegetating disturbed soils with desirable plant species to resist weed invasion; 4) curtailing weed development in wastelands or along transportation and utility line corridors; 5) inspecting and/or cleaning machinery, vehicles, and watercraft prior to their movement from weed-infested to uninfested sites; 6) managing livestock and wildlife feed and manure to prevent seed dissemination; 7) restricting weed propagule transport in fill-dirt, gravel, mulch, hay bales, and in irrigation water delivery systems; and 8) enhancing educational awareness among resource managers and users concerning the identification of both potential new and new invader species and the harmful impacts associated with their spread.

### • *Mechanical/physical, cultural, and chemical controls*

When weeds must be managed, the selection of a treatment method should be based upon its relative effectiveness and environmental impacts, and its cost and the costs of feasible alternatives (CNAP 2000). Methods suitable for the control of unwanted plant populations can be categorized as mechanical/physical, cultural, chemical, and biological

(DiTomaso 2000, Sheley and Petroff 1999). Not all methods may be applicable or available in every management situation. Mechanical/physical controls are used to uproot, bury, or impede plant growth and development, and include hand removal, hoeing, cutting, tillage, mowing, crushing, smothering, flooding, blading or bulldozing, chaining or cabling, dredging, scalding, flaming, and controlled burning. Cultural techniques decrease the suitability of the environment for weed entry, establishment, and proliferation. Specific methods include soil fertility enhancement, site revegetation, and improved livestock grazing management practices (CNAP 2000, Heady and Child 1994). Revegetation with site-desired, competitive plant species is the best method for suppressing weed infestations, but remains the method most neglected or inadequately addressed by many weed managers (Sheley and Petroff 1999). Chemical management of weeds employs various selective and nonselective herbicides. Herbicides are important tools in many IWM programs and their use requires careful application to minimize disruption of the ecosystem into which they are introduced (Newman et al. 1998). Suppressive techniques in an IWM program should be used in a mutually supportive or synergistic manner. Each technique selected must not impair the efficacies of other management methods. The References provide additional sources for further discussion of the advantages and disadvantages inherent to these particular management methods.

## Integration of Biological Control

The focus of biological control is on the long-term regulation of weed populations rather than the immediate elimination of offending individuals (DiTomaso 2000). If effective biological control agents exist for problem weeds and if the agents do not yet occur in an infested area of sufficient size, every attempt should be made to introduce them. If they are present at a site but their overall numbers are low, additional agents should be obtained and released at appropriate times to augment resident populations. Once they are established, every effort should be made to conserve biological control agents by avoiding disruptive practices (e.g., improperly timed herbicide applications, grazing or mowing operations) that may interfere with their development and survival (Andres 1982, Newman et al. 1998).

Biological control organisms are best suited for use against established invader weeds and to a lesser extent against new invaders. Their use is never of value if eradication is the desired outcome of a management effort. Established invader species typically occupy extensive acreages of low-value and often inaccessible lands. Some weeds may expand their populations at rates faster than those of their biological control agents. When dealing with established invader populations, the management approach may need to be more involved and take longer to implement. In such situations it is frequently advantageous to employ biological control agents along with other suppression methods to prevent plant movement into unoccupied areas. The selection of a management method is based upon plant location within the infestation area. The first priority should be management of outlier populations, followed by plant suppression on the perimeter and then in core infestation zones, respectively. Outlier and perimeter populations can be reduced in size immediately by mechanical/physical and chemical methods while biological control agents such as insects and plant pathogens become established and more effectively contribute to suppression of the core population. Forced grazing by livestock could also

119

be used in the core area to enhance the impacts realized by bioagent deployment (DiTomaso 2000). For example, intensive, short-duration sheep grazing has been used in combination with *Aphthona* spp. root-feeding beetles to successfully reduce leafy spurge (*Euphorbia esula*) populations (Sheley and Petroff 1999). Spurge populations are grazed in the spring as flowering begins and again in late summer or early fall during spurge's regrowth phase. By following this schedule, grazing occurs before adult beetles emerge in midsummer and after larvae have moved into roots and soil, thus effectively curtailing seed production without adversely harming bioagent survivability.

Biological control agents require living target plants on which to develop and reproduce. Management methods that hinder normal plant growth and development may not be compatible with biological control. Timing is a fundamental consideration in the successful integration of nonbiological and biological methods. For example, herbicides should only be used when and where they will not impede development of agents (Messersmith and Adkins 1995). In some instances, reduced herbicide application rates may stress surviving weeds just enough to make them more vulnerable to attack by the biological control agent. Mowing, cutting, or grazing, depending on when the activity is conducted, can either increase or decrease agent performance. Burning, if improperly timed, could destroy biological control agents overwintering within stems or seed heads but might have little effect on root-infesting natural enemies. Enhancement of the native and/or introduced plant community in the management area can frequently aid performance of a natural enemy by reducing opportunities for weed reinvasion (Heady and Child 1994, Sheley and Petroff 1999). Once biological control agents have reduced invasive plant densities, the agents' positive impacts on the plant community can be maintained or intensified through the establishment of diverse, site-adapted plant species whose presence improves the community's competitiveness, durability, and resistance to weed recurrence. Regardless of the methods selected for a management plan, each one should be carefully scrutinized to determine its compatibility or incompatibility with biological control organisms.

## Conclusion

IWM is an inherently dynamic framework designed to address problematic plant species (Sheley and Petroff 1999). Once an IWM plan has been implemented, project accomplishments (impacts of control) must be regularly reviewed and evaluated. Information gathering and interpretation through time enables weed management specialists to detect and respond to changes that may occur in the ecosystem. Evaluation of monitoring results enters the management system at all levels and provides updated information for subsequent rounds of decision- making necessary to modify and improve the system. Continued implementation of IWM programs must remain an important priority for all entities charged with the protection of natural, agricultural, and built environments from degradation by invasive species.

## References

Andres, L. A. 1982. Integrating weed biological control agents into a pest-management program. Weed Sci. 30 (Suppl.): 25-30.

Center for Invasive Plant Management (CIPM). 2002. Guidelines for Coordinated Management of Noxious Weeds: Development of Weed Management Areas. http://weedcenter.org/management/guidelines/tableofcontents.html

Colorado Natural Areas Program (CNAP). 2000. Creating an Integrated Weed Management Plan: A Handbook for Owners and Managers of Lands with Natural Values. Caring for the Land Series Vol. IV. Colorado Natural Areas Program, Colorado State Parks, Colorado Department of Natural Resources, and Division of Plant Industry, Colorado Department of Agriculture. Denver, CO. http://parks.state.co.us/cnap/IWM_handbook/IWM_index.htm

DiTomaso, J. M. 2000. Invasive weeds in rangelands: Species, impacts, and management. Weed Sci. 48: 255-65.

Heady, H. F., and R. D. Child. 1994. Rangeland Ecology and Management. Westview Press, Inc., Boulder, CO.

Masters, R. A., and R. L. Sheley. 2001. Principles and practices for managing rangeland invasive plants. J. Range Manage. 54: 502-17.

Messersmith, C. G., and S. W. Adkins. 1995. Integrating weed-feeding insects and herbicides for weed control. Weed Technol. 9: 199-208.

Newman, R. M., D. C. Thompson, and D. B. Richman. 1998. Conservation strategies for the biological control of weeds. Pages 371-96 in P. Barbosa, ed. Conservation Biological Control. Academic Press, San Diego, CA.

Sheley, R. L., and J. K. Petroff, eds. 1999. Biology and Management of Noxious Rangeland Weeds. Oregon State Univ. Press, Corvallis.

# Economic Benefits of Biological Control

Eric M. Coombs, Hans Radtke, and Thomas Nordblom

Classical biological control of weeds has a very good safety record, but a somewhat scanty track record of documented economic benefits. Reported benefit-cost ratios of biological control programs from around the world vary from 0.99:1 to 112:1, depending upon how successful each program was and which factors were considered in the analysis (Nordblom 2003, Syrett et al. 2000).

It has been suggested that economic analyses depend upon the accurate measurement of biological control projects by ecological standards through the quantification of agent-host population interactions (Syrett et al. 2000). Benefits from biological control also accrue through decreased costs for the same level of weed control by other means or through increased revenue from improved lands (Anderson et al. 2003). A benefit-cost ratio is determined by comparing the benefits (decreased losses and increased revenue) to the costs (losses caused by the target weed and the costs of developing and implementing a biological control project) that accrue over time, adjusted for inflation.

Nonmarket benefits that are often overlooked include decreased environmental contamination by herbicides, decreased health risks, increased biodiversity, reduced application costs, etc. Some benefits may be lost by controlling a targeted weed, for example, decreasing revenue from honeybees that collect nectar from yellow starthistle (*Centaurea solstitialis*) and purple loosestrife (*Lythrum salicaria*), targeting plants used in the nursery trade such as purple loosestrife, decreasing wild fruit sales (some blackberries), or losing native plants when nontarget effects occur.

Conservative analyses of benefit-cost ratios may indicate benefits substantial enough to warrant investing in a biological control program based solely on market factors (McEvoy 1996). It has been suggested that both direct and indirect costs should be included in economic analyses. Assigning costs to various social and environmental parameters can be difficult and can lead to over- or under-estimation of economic evaluations. Future benefits can be measured by surveys that determine the public's willingness to pay for decreased control costs, reduced environmental impacts, and increased productivity through biological control of invasive species.

Most of the funding allocated for biological control projects is expended during the foreign exploration, host specificity testing, and introduction phases, with little appropriated for long-term studies of efficacy or economic benefits (McEvoy and Coombs 2000). However, the desire to reduce economic losses often has been the driving force that justified the initial expenses of many biological control projects.

Initially, rangelands were the primary benefactors of biological control projects. Early targets included tansy ragwort (*Senecio jacobaea*), St. Johnswort (*Hypericum perforatum*), Mediterranean sage (*Salvia aethiopis*), leafy spurge (*Euphorbia esula*), and knapweeds (*Centaurea* spp.), for example. The benefits from rangeland projects are increased forage and decreased animal losses from competitive and toxic weeds—justifications for the expense of biological control. In recent years, invasive plants such as melaleuca (*Melaleuca quinquenervia*), waterhyacinth (*Eichhornia crassipes*), and purple loosestrife have been targeted to reduce their environmental impacts and to protect forests, waterways, and urban green spaces. Support and funding for such projects demonstrate the public's willingness to pay for biological control programs that will improve the overall aesthetics of refuges, parklands, and waterways.

Biological control projects should be evaluated in advance to estimate benefit-cost ratios and internal rates of return under variable subjective scenarios and to periodically document the actual effects (Syrett et al. 2000). A thorough economic analysis in advance can help justify the high initial expenses of a biological control program, which may cost $250,000 to $750,000 and take three to five years to implement. Such analyses have uncertainties. Biological control projects often span the careers of a number of investigators, which further compounds the problem of documentation and post-project analysis. Another problem is the lag time for benefits to accrue; for example, it took nearly 20 years for the tansy ragwort biological control project to become economically and ecologically successful in western Oregon (Coombs et al. 1996).

While economic justifications can create expectations of higher returns on low-value lands, the total acres of low-value lands may total a high value when aggregated at regional scales. A small benefit per acre applied to very large acreages can yield a large return, much as large grocery stores make their profits on high volume to compensate for lower markup per item. When managing invasive species at such a large scale, changes in cultural practices (grazing, logging, etc.) and biological control may be the only practical solutions for the problem. Some control projects may work at the local scale within the confines of a fence (e.g., with sheep and goats), but such programs may not be widely accepted and may not work at the regional level (Sheley and Rinella 2000). Integrated, ecologically based management of invasive weeds will provide the solutions for large-scale economic problems.

The actual value of biological control is difficult to measure, especially the socioeconomic and environmental impacts of weed infestations that are prevented. Biological control is different from chemical control, for example, in that the products (the agents and knowledge of how to use them) are not consumed and their use does not diminish their supply (Tisdell et al. 1984). The knowledge of using biological control is a public asset—difficult for the private sector to supply. Private companies are unlikely to develop biological control programs because of the difficulty in recouping their developmental costs. Therefore, we rely on government agencies to provide the initial services for the public benefit. Land managers may not want to front the cost of bringing in a new agent, when they only need to wait for them to spread from a neighboring property—for free.

In a post-project economic analysis of the tansy ragwort biological control program in western Oregon, an 83% internal rate of return and a conservative benefit-cost ratio of 13:1 were reported (Coombs et al. 1996). On successful long-term projects, benefits can occur as steady-stream returns, e.g., $5 million/year (mostly through increased livestock and forage production) for the Oregon ragwort project, where annual agency investment is now less than $10,000. In this case, the accruing annual benefits of a single biological control project have offset the yearly operating costs of the entire noxious weed control program. Where it is feasible, therefore, it is economically justifiable to expeditiously implement locally successful biological projects throughout an entire region to reduce annual losses in the shortest time possible. By actively redistributing ragwort biological control agents, the Oregon Department of Agriculture accomplished a successful regional project five to 10 years sooner than by the natural spread of the insects, averting $25 to $50 million in losses to agriculture. Successful projects can also provide economic and environmental benefits by reducing the number of new infestations at sites distant from core infestations (Isaacson et al. 1996).

Another example of the economic benefit of biological control of weeds is with waterhyacinth. In Louisiana in the mid-1970s, waterhyacinth increased from a normal overwintering infestation of 80,939 to 485,633 ha by the end of the growing season (Center et al. 2002, Cofrancesco et al. 1985). During the mid- to late 1970s, three biological control agents were released in the state: two weevils (*Neochetina eichhorniae* and *N. bruchi*) and a moth (*Niphograpta albiguttalis*). Due to successful control by these insects, infestations remained at approximately 40,469 to 80,939 ha, never attaining the prior level. Since then, the plant's ability to re-expand has continually decreased. The success of biological control has yielded a very conservative steady-stream benefit of $40 million annually, based on annually treating the same 323,755 ha of waterhyacinth with herbicides at $23 per ha. Benefits to aquatic transportation, recreation, and wildlife production were not measured, nor were the indirect ecological benefits to the aquatic environments, but they are no doubt enormous.

Across southern Australia, 400 populations of the crown weevil *Mogulones larvatus* were established between 1993 and 2000 against the pasture weed Paterson's curse (*Echium plantagineum*), which has become problematic on more than 23 million ha since its introduction as an ornamental in the mid-1800s (Nordblom et al. 2002, Nordblom 2003). A recent analysis that accounts for the slow geographic spread of the biological control agent has concluded that an additional 410 strategically placed releases could fill the gaps more quickly. Assuming restored pasture productivity is valued at a modest A$12 per dry sheep equivalent (DSE), a high discount rate of 10%, a cost of A$2,000 for each new release, and a base project cost of A$1,000,000, such a "stop-gap" program is expected to result in a benefit-cost ratio on the order of 15:1.

## Realistic Expectations

Global establishment rates of biological control agents vary from 30 to 90% and successful biological control projects comprise about 25% of the attempts (McFadyen 1998). A partially successful biological control project, i.e., a project that reduces weed infestations by variable percentages over large areas, can still provide a positive benefit-cost ratio even though the degree of weed control may be less than desired. If biological control reduced the top 12 weeds in Oregon by 30%, the direct annual losses to agricultural production and jobs could be decreased by $20 million. Where biological control is successful, long-term ecological land management is necessary to prevent a treadmill of replacement weeds that could create new economic losses. It has been suggested that concentration of efforts on a single weed also can lead to land management failures and increase the treadmill effect (McEvoy 1996). Unreported, partially effective projects also can promote the biological control treadmill and increase the probability of risk to nontarget species by going through the expense of introducing additional and perhaps unnecessary biological control agents.

Because the economics of biological control are so closely tied with ecology, failures linked to ecological causes (such as climate, parasitism, disease, etc.) will result in an associated economic loss (the costs of the biological control program and the continued impacts of the weed). There may also be a tendency to blame the tool when a biological control project is too slow or fails to control a weed. Land management practices may cause biological control failure, such as when livestock are allowed to graze seed heads and flower buds needed by developing biological control agents.

Some targeted weeds may be controlled by one agent, while another (such as leafy spurge) may require a series of agents that work in varied microhabitats; therefore, improved predictive models are needed (McEvoy 1996). The models may help determine whether the expense for a suite of biological control agents suited to a weed that infests varied habitats is worth the risk.

Some biological control projects are becoming locally successful [for example, leafy spurge, diffuse knapweed (*C. diffusa*), purple loosestrife, and Mediterranean sage] and in time may become regional successes. When a project lacks sufficient scale to cover the initial expense of the biological control program, intervention may be required to increase redistribution or seek additional biological control agents better adapted to recalcitrant infestations.

Because biological control projects are independent of each other and in some cases are area-specific, broad generalizations about economic and ecological efficacy may create unrealistic expectations among those who support and fund local projects. There is a great need for ecologists and economists to work together and fill the pre-project monitoring and evaluation gaps in U.S. biological control programs. Ecologists can develop models to predict the efficacy of biological control projects and where and under what conditions they are likely to succeed. Economists can then evaluate the probable costs and benefits to determine which projects are likely to be good investments before committing funds. This will help us better predict and evaluate the benefits and costs of biological control programs to ensure efficiency and efficacy in our nationwide weed management.

The biological control of weeds community could well use the following as an excellent example of an interdisciplinary, integrated pest management program: the positive consequences and profitability of a large multinational project for biological control of the cassava mealybug were studied in Africa (Zeddies et al. 2001). Long and arduous effort, combining the talents of entomologists, ecologists, chemists, sociologists, and agronomists under the auspices of national and international research agencies, has brought significant benefits to the producers and consumers of cassava, the staple food of the continent. The vision that this long-term effort would more than justify the investment of public funds stands to the credit of those who supported the work in spite of many competing demands and the usual political necessity for short-term results.

# References

Anderson, G. L., E. S. Delfosse, N. R. Spencer, C. W. Prosser, and R. D. Richard. 2003. Lessons in developing successful invasive weed control programs. J. Range Manage. 56: 2-12.

Center, T. D., M. P. Hill, H. Cordo, and M. H. Julien. 2002. Waterhyacinth. Pages 41-64 *in* R. Van Driesche, S. Lyon, B. Blossey, M. Hoddle, and R. Reardon, eds. Biological Control of Invasive Plants in the Eastern United States. USDA Forest Serv. Pub. FHTET-2002-04. Morgantown, WV.

Cofrancesco, A. F., R. M. Stewart, and D. R. Sanders. 1985. The impact of *Neochetina eichhorniae* (Coleoptera: Curculionidae) on water hyacinth in Louisiana. Pages 525-35 *in* E. S. Delfosse, ed. Proc. VI Int. Symp. Biol. Contr. Weeds, 19-25 August 1984, Vancouver, Canada. Agric. Canada, Ottawa.

Coombs E. M., H. Radtke, D. L. Isaacson, and S. P. Snyder. 1996. Economic and regional benefits from the biological control of tansy ragwort, *Senecio jacobaea* (Asteraceae), in Oregon. Pages 489-94 *in* V. C. Moran and J. H. Hoffmann, eds. Proc. IX Int. Symp. Biol. Contr. Weeds, 19-26 January 1996, Stellenbosch, South Africa. Univ. of Cape Town.

Isaacson D. L., D. B. Sharratt, and E. M. Coombs. 1996. Integrating biological control into management of invasive weed species. Pages 27-31 *in* V. C. Moran and J. H. Hoffmann, eds. Proc. IX Int. Symp. Biol. Contr. Weeds, 19-26 January 1996, Stellenbosch, South Africa. Univ. of Cape Town.

McEvoy, P. B. 1996. Host specificity and biological pest control. BioScience 46: 401-5.

McEvoy, P. B., and E.M. Coombs. 2000. Why things bite back: unintended consequences of biological weed control. Pages 167-94 *in* P. A. Follett and J. J. Duan, eds. Nontarget Effects of Biological Control. Kluwer Academic Publishers, Boston, MA.

McFadyen, R. E. C. 1998. Biological control of weeds. Annu. Rev. Entomol. 43: 369-93.

Nordblom, T. 2003. Putting biological reality into economic assessments of biocontrol. Pages 75-85 *in* H. Spafford-Jacob and D. Briese, eds. Improving the Selection, Testing, and Evaluation of Weed Biocontrol Agents. Proc. Biol. Contr. Weeds Symp. and Workshop, Univ. West. Australia, Perth, 13 Sept. 2002. CRC for Australian Weed Management, Tech. Series No. 7. Available at http://www.weeds.crc.org/au/publications/technical_series.html.

Nordblom, T., M. Smyth, A. Swirepik, A. Sheppard, and D. Briese. 2002. Spatial economics of biological control: investing in new releases of insects for earlier limitation of Paterson's curse in Australia. Agric. Econ. 27: 403-24.

Sheley, R. L., and M. J. Rinella. 2000. Incorporating biological control into ecologically based weed management. Pages 211-28 *in* E. Wajnberg, J. K. Scott, and P. C. Quimby, eds. Evaluating Indirect Ecological Effects of Biological Control. CABI Publishing, Oxford, UK.

Syrett, P., D. T. Briese, and J. H. Hoffmann. 2000. Success in biological control of terrestrial weeds by arthropods. Pages 189-230 *in* G. Gurr and S. Wratten, eds. Biological Control: Measures of Success. Kluwer Academic Publishers, Netherlands.

Tisdell, C.A., B.A. Auld, and K.M. Menz. 1984. On assessing the value of biological control of weeds. Prot. Ecol. 6: 169-79.

Zeddies, J., R. P. Schaab, P. Neuenschwander, and H. R. Herren. 2001. Economics of biological control of cassava mealybug in Africa. Agric. Econ. 24: 209-19.

# The Role of Private Industry in Biological Control of Weeds

Noah H. Poritz, Aubrey Mayfield, and Jerry Johnson

Since the mid-1980s, commercial producers of agents for the biological control of weeds have become an increasingly important source of supply in the United States. Each year these businesses provide a progressively more expansive selection of USDA-approved weed-feeding agents attacking a wide variety of target weeds. Prior to commercialization of biological weed control, the primary source of weed-feeding agents was the research scientists actively involved in biological control. Because of time and financial constraints, these scientists were often forced to limit their insect redistribution activities and numerous biological control agents approved for field release had no ongoing redistribution activity. Given the magnitude of the exotic weed problem and the limited number of biological control agents available for redistribution, there was an unsatisfied demand for many of these organisms. The private businesses that emerged to fill that void were predominately owner operated. These businesses now provide a ready supply of formerly scarce or unavailable biological control agents. Today, every level of the weed management community relies on the commercial availability of biological control of weeds agents. Farmers, ranchers, county weed control personnel, extension agents, state and federal land managers, and others all use the services offered by private biological control businesses.

## Quality, Convenience, and Customer Service

The success of the private biological control agent industry depends on its ability to provide: 1) a dependable, affordable supply of weed-feeding agents; 2) acceptable quality of organisms; and 3) a satisfactory level of customer service. The cost-savings and convenience of receiving one's insects from a dependable commercial supplier cannot be underestimated. Outlined in other chapters of this book are the numerous time-consuming and physically demanding steps that culminate in the successful redistribution and field release of a biological control agent. In most cases, ranchers, weed control professionals, public land managers, and others would rather leave those many steps to a commercial supplier. With the simplicity of a phone call, postcard, or e-mail, one can receive healthy, vigorous biological control agents on a date appropriate to the agent's life history and convenient to one's work schedule. This results in spending less time collecting and caring for insects and having more time for other weed-management activities. It is often far less expensive to purchase agents from a commercial source than to collect for oneself. Customers typically receive their biological control agents by overnight delivery with transit times less than 24 hours. Many successful field insectaries and redistribution programs have been established with commercially supplied insects.

One of the most unusual and satisfying aspects of providing biological control agents commercially comes from working closely with customers. Frequently, customers have little or no entomological or biological control experience. Commercial suppliers are often their clients' first connection in learning about the specifics of biological weed control. Clients may have read a newspaper article illustrating the benefits of biological control or perhaps they witnessed firsthand the impact of a specific agent on a target weed. To be successful, the industry must offer enough information about the benefits and limitations of weed biological control to clients so they can decide whether to utilize this tool. Often—and in many different ways—the biological control industry reiterates the earlier chapters presented in this book for the benefit of the end user.

With few exceptions, all commercial suppliers of biotic agents collect their products from field populations on their target weeds. These entrepreneurs are therefore relying on the subsequent generations of the initial USDA-approved field releases of each agent. The quality of the product depends upon the supplier's ability to: 1) accurately identify the agent in the field; 2) know the appropriate collection time in the agent's life history; 3) properly handle the organism during collection and transport; and 4) quickly provide the live agents to the customer. Clearly, anyone—public or private—involved in the redistribution of biological control agents must concentrate on these same quality-control fundamentals to provide a satisfactory level of biological control.

## Added Value

Private suppliers of biological control organisms provide a vital link between the scientific research community and the public. Reprints of scientific studies, extension bulletins, and symposia proceedings can often be readily obtained from these suppliers. Because of their ongoing commitment to biological weed control, these businesses provide, often at no charge, written, verbal, and visual educational information on nonchemical weed management to their customers.

Collectively, these businesses often are active in local, state, and international weed-management and biological control organizations. They participate in scientific meetings and some have ongoing biological control research programs to support the needs of their customers.

The demand for biological control agents has increased in the last several decades. Concomitant with this heightened demand has been a corresponding increase in the number of commercial ventures supplying organisms for biological weed control. Like all businesses, the support and expertise one receives varies depending upon the company. Some commercial suppliers of biological weed control agents have many years of hands-on field experience working with a variety of natural enemies and target weeds in various geographic locations. Some of their personnel are highly trained in the field of biological weed control and have worked for state and federal biological control of weeds programs. A few hold advanced degrees specializing in the field of biological control. Often these businesses provide assistance in completing the required state and federal permit applications and, in many instances, have approved permits in place for their customers' convenience. In all cases, it is worthwhile to ask questions and evaluate whether you are satisfied with the quality of agents and the level of service and support for which you are paying.

These companies also support their customers' needs in the areas of: biological control education and training, and agent identification; weed and insect field surveys and mapping; providing insect collection equipment, methods, and supplies; field release of biological control agents; and post-release evaluation reports. While these services may benefit some customers, the vast majority of consumers desire only to receive as many healthy, accurately identified biological control agents as possible for the least cost. From this consumer view, the biological control agent supply industry differs little from most other mail-order businesses.

## Future Prospects

From the biological control agent supplier's perspective, the industry is filled with uncertainty. Some of these uncertainties are beyond control. Unpredictable climatic conditions can affect a producer's ability to collect adequate numbers of agents. In some areas, commercial and residential development have reduced weed populations to the point that agents have become quite scarce. Ironically, the overwhelming effectiveness of certain biological control agents upon their target weed has had the same effect!

Regulatory and political uncertainties also influence this industry. As mentioned earlier, all commercial suppliers rely upon agents introduced by the USDA. For the industry to grow, it is imperative that businesses have access to newly introduced species. Clearly, the private sector must operate in partnership with the public agencies and scientists. All parties are worthy of recognition and support and should not be regarded as competition to be met with suspicion. Historically, as newly established biological control agents have expanded their populations, businesses and public agency personnel have been careful to nurture these agents, thereby ensuring sustainability. Over-collection has not proven to be an issue.

The emerging industry has made great contributions to the biological control of weeds in recent years. Millions of weed-feeding agents have been commercially redistributed to attack our country's worst rangeland, pasture, and aquatic weeds. The thousands of satisfied customers of this industry are the best testimony to the importance of commercial biological weed control. These clients frequently remark about the important role that commercial suppliers occupy in implementing their weed management objectives.

The private biological control industry is unique; it is a combination of science, hard fieldwork, and customer service. Without question, this industry has made, and will continue to make, a meaningful and lasting contribution toward biological control of weeds in the United States.

### References

Association of Natural Biocontrol Producers. 2003. http://www.anbp.org.

Biological Control of Weeds. 2003. http://www.bio-control.com.

Hunter, C. D. 1997. Suppliers of Beneficial Organisms in North America. California Environmental Protection Agency, Dept. of Pesticide Regulation, Environmental Monitoring and Pest Management Branch. http://www.cdpr.ca.gov/docs/ipminov/bensuppl.htm.

# International Code of Best Practices for Classical Biological Control of Weeds

Joe K. Balciunas and Eric M. Coombs

## The Need for a Code

The subdiscipline of classical biological control of weeds has an enviable long-term record for being safe, environmentally appropriate, and effective. This is the result of the care and diligence of the practitioners involved. Initially, there were only a handful of practitioners in each country, who knew each other and who exercised a communal oversight over the projects and releases made in their country. But as the popularity of classical biological control of weeds increased, so did the number of scientists. Thousands of persons around the globe are now involved in various aspects of weed biological control. Their training, experience, and tolerance for risk vary greatly. At the same time, the practice of biological control is facing increased scrutiny from ecologists and other scientists, as well as from administrators and the general public. Some of their concerns about the efficacy and safety of previous biological control projects have merit. Rather than trying to discredit research that points out the negative impacts and other shortcomings of past projects, we should strive to make new and current projects as rigorous and safe as possible. In order to avoid "revenge effects" (McEvoy and Coombs 2000) and charges that we are releasing agents "willy-nilly" (Strong 1997), everyone in our discipline needs to practice the craft at a consistently high standard. If biological control of weeds is conducted on a trial-and-error basis, soon our errors will be on trial.

Standards are needed so that actions that clearly lie well outside of current, accepted science cannot be construed to represent "normal" biological control practices. Biological control practitioners frequently spend large parts of their careers trying to manage a few weeds, and this can lead to their adopting an attitude of controlling the target weed at any cost. However, they should remember that their primary duty is to protect and enhance this valuable weed-management tool. If practitioners of biological control of weeds do not regulate themselves, they very likely will face additional statutory regulations and possibly total bans.

The Technical Advisory Group for Biological Control Agents of Weeds (TAG) committee, federal and state statutory regulations, as well as the model regulations published by the United Nations Food and Agriculture Organization (FAO) (1996), provide some oversight and guidance to weed biological control practitioners. However, this guidance is primarily concerned with the importation and initial release of a biological control agent. The redistribution and other post-release aspects of classical weed biological control are far less regulated, but this should not imply that they are free of risk, and individuals and organizations involved in redistributing agents should exercise appropriate caution.

---

**Table 1. Twelve guidelines make up the International Code of Best Practices for classical biological control of weeds.**

1. Ensure the target weed's potential impact justifies the risk of releasing nonendemic agents.
2. Obtain multi-agency approval for the target weed.
3. Select agents with potential to control the target weed.
4. Release safe and approved agents.
5. Ensure only the intended agent is released.
6. Use appropriate protocols for release and documentation.
7. Monitor impact on the target.
8. Stop releases of ineffective agents, or when control is achieved.
9. Monitor impacts on potential nontarget species.
10. Encourage assessment of changes in plant and animal communities.
11. Monitor interaction among agents.
12. Communicate results to the public.

---

Because standards covering all aspects of classical biological control of weeds were lacking, in 1998 the senior author formulated 10 guidelines for a Preliminary Code of Best Practices and presented it as a topic for discussion at several meetings attended by weed biological control scientists. Most colleagues agreed that such standards were desirable. In July 1999, he presented a Proposed Code of Best Practices to delegates at the X International Symposium on Biological Control of Weeds that was held in Bozeman, Montana. This led to a group of more than two dozen colleagues meeting in a breakout session to refine and enhance the Proposed Code. The resultant Code of Best Practices, consisting of 12 guidelines (Table 1), was overwhelmingly adopted by the symposium delegates on 9 July 1999 with only two dissenting votes (Balciunas 2000).

We discuss each of the 12 guidelines and provide examples of their application and value. The Code covers all aspects of classical weed biological control. Activities that precede the first establishment of an approved agent are designated as Introductory. Subsequent activities, usually to increase the geographic range of the agent and to document its impact, are denoted as Redistribution. While the applicability of any one guideline will vary with one's job duties, practitioners should try to follow the intent of all the guidelines. When an agency lacks sufficient resources to fulfill some of the tasks (e.g., monitoring impacts on the target and nontargets) that the Code recommends, we encourage developing partnerships with colleagues in other agencies and organizations that might assist in the necessary research and documentation.

### • Guideline 1: Ensure the target weed's potential impact justifies the risk of releasing nonendemic agents

Biological control is a powerful tool, but like other management approaches, it has its limitations and in addition a unique risk—to permanently alter an ecosystem. Those involved in selecting new weed targets should ensure that before significant resources are committed to a new biological control approach, costs and risks are understood by all stakeholders (see Guideline 2).

Those involved in redistribution releases should ensure—on a site-by-site basis—that control of the target weed is necessary and that other control technologies (e.g., hand-pulling, herbicides) are not more appropriate or effective. Avoid releasing agents just because they are available.

### • Guideline 2: Obtain multi-agency approval for the target weed

A plant that one group considers a serious weed may have beneficial and important uses (e.g., fodder, honey production, medicine) for other sections of society. Before initiating a project on a new target for a region, stakeholders, government agencies, environmental groups, and other nongovernmental organizations should be consulted and a consensus reached. A proposed test plant list, including a section on potential conflicts of interest, should be submitted to the TAG.

Those making redistribution releases should ensure that major landholders in the area and the state's department of agriculture concur on the negative impacts of the target weed and the need for releasing new weed biological control agents.

### • Guideline 3: Select agents with potential to control the target weed

The release of effective agents that control the target weed can provide substantial economic and environmental benefits, thereby enhancing the reputation of biological control. On the other hand, releasing ineffective agents can damage biological control's reputation by contributing to the perception that it is only sporadically effective, thereby making the public and administrators wary of committing funds to additional biological control projects. It is also possible that an ineffective agent, especially if it becomes very abundant, may cause unpredicted impacts on nontarget organisms (e.g., Pearson et al. 2000). Efficacy is the flip side of the coin of safety, and scientists seeking and testing new potential agents should be as concerned about the agent's efficacy as about its host specificity. For nonlethal agents, prerelease assessments of potential impacts should be performed (Balciunas 2004a).

While no law presently prohibits the redistribution of ineffective agents, we should refrain from further releases of organisms that have consistently failed to demonstrate a significant impact on their target weed. Releases that do not result in control of the target are a waste of resources, may interfere with more effective agents, and damage the reputation of biological control.

### • Guideline 4: Release safe and approved agents

In the United States (and most other countries), agencies and individuals involved in introducing new biological control agents for weeds must present convincing data supporting the safety of the proposed agent, as well as comply with a lengthy list of other statutory requirements for initial releases.

Agencies and individuals redistributing established agents should verify that they are complying with applicable federal and state regulations. Transporting agents across state lines requires a permit from USDA-APHIS-PPQ and possibly from state authorities. When redistributing agents, only approved agents should be considered. Nonendemic natural enemies that have somehow arrived into a new region should not be used as biological control agents. After release of an approved agent, evidence of nontarget impacts may become apparent—especially for agents approved decades ago when host range testing was far less intensive (Louda et al. 2003, Turner et al. 1987). Continued release of such agents is inadvisable.

### • Guideline 5: Ensure only the intended agent is released.

It is standard operating procedure for quarantine scientists to ensure that agents from overseas are free of parasitoids and pathogens before they are released, and that the release shipment does not contain any sibling species or contaminating organisms. Vigilance is required, since release of nonapproved natural enemies has occurred recently (Balciunas and Villegas 1999).

Those redistributing approved agents should follow similar procedures. Infested weed material, such as seed heads containing overwintering agents, should not be distributed directly into the field. This could result in the introduction of unapproved organisms, including parasitoids and pathogens, that might reduce the impact of natural enemies already present at the site. Also, the seeds in the plant material might introduce new weed strains or lead to hybrid weeds that are even more troublesome. Instead, the plants collected from the field should be held until the agents emerge, and the agents then should be collected and redistributed.

### • Guideline 6: Use appropriate protocols for release and documentation

Initial releases of a new agent are usually well documented, but those involved should stay abreast of their colleagues' research so as not to overlook recording relevant data. Data might include: the geographic source of the agents being released, the number of generations spent in the quarantine lab by the agents, or the sex ratio of the agents being released.

Documentation of redistribution releases has been more sporadic and variable. Data for every redistribution release should be standardized to ensure that relevant core data are recorded then transmitted to appropriate state and federal officials. Also, prerelease surveys at the intended release site should become universal to ensure that the agent (or other agents already controlling the target weed) is not already present.

### • Guideline 7: Monitor impact on the target

Most current and future projects that plan to release new agents to control a weed include protocols for monitoring the agent's impact on the target. However, since the monitoring might easily require a decade or more, scientists and administrators should ensure that adequate resources for such long-term monitoring are sustained.

Individuals and agencies involved in redistributing previously approved agents have a vital role in this long-term monitoring. They should form partnerships with appropriate scientists to document the impact (or lack thereof) of each agent on the target weed in their area. Until monitoring protocols are in place, redistribution of agents whose impacts have never been documented may not be advisable.

**133**

### • *Guideline 8: Stop releases of ineffective agents, or when control is achieved*

After evidence accumulates that an agent is ineffective, further releases should cease. Releases of ineffective agents consume valuable time and resources, increase risk to nontargets, may reduce impact of more effective agents, and promote the perception that weed biological control is ineffective. Likewise, once a weed is controlled, further releases at that site should be curtailed to prevent introduction of parasitoids, pathogens, or competitors that might compromise the control.

### • *Guideline 9: Monitor impacts on potential nontarget species*

The possible impact of agents on nontarget species should be a concern for all involved in weed biological control. Plants closely related to the target weed (i.e., in the same genus or subfamily) should be vigilantly monitored for feeding or damage caused by the biological control agent. The results, including lack of damage to closely related nontargets, should be documented and published (see Guideline 12).

### • *Guideline 10: Encourage assessment of changes in plant and animal communities*

Assessing ecosystem impacts, especially when agents start to exert widespread control on the target, is highly desirable. The benefits of North America's first successful weed biological control project were well documented by Huffaker and Kennett (1959). Likewise, the study that demonstrated that the control of tansy ragwort in Oregon was saving $5 million annually (Coombs et al. 1996) has not only benefited that state, but also increased interest in biological control of weeds in other states. Documenting the landscape-level impacts is likely to require considerable assistance; therefore, efforts should be made to enlist ecologists, economists, Geographic Information Systems (GIS) specialists, and others who are willing to cooperate in this important task.

### • *Guideline 11: Monitor interaction among agents*

Many weed targets have had an array of agents released against them. However, the interactions among these agents are seldom investigated. Releases of additional agents could interfere with established agents and lead to less control. Everyone involved in weed biological control should encourage more research into this overlooked area of investigation.

### • *Guideline 12: Communicate results to the public*

Weed biological control workers may feel that the public does not fully appreciate the benefits and safety record of this subdiscipline. Biological control practitioners should strive to keep the public informed of new projects, the safety protocols that are followed, the ecological and economic results, and successes. Practitioners also must be forthright about limitations of biological control and the nontarget impacts that have accompanied a few older projects.

## Conclusions

Biological control can be a powerful tool to help manage some widespread, invasive weeds. While its safety record is unsurpassed, it is not without risks. Everyone involved in weed biological control is encouraged to use the latest information to help maintain and enhance this record of safety and effectiveness. Risky or ill-advised actions in an attempt to control a particular weed might result in prohibitions and constraints on future projects. The International Code of Best Practices was designed to provide practitioners with internationally agreed-upon guidelines to help select the actions that reduce risk and enhance positive impact.

These guidelines go beyond the traditional concerns for host specificity and now also emphasize the effectiveness of agents. The Code also mandates the monitoring of the consequences of releases, including redistribution releases, and encourages forming partnerships to ensure adequate monitoring as called for by Guidelines 7 through 11. The resources required for monitoring may also mean that we will need to make fewer releases or undertake fewer projects.

The Code attempts to ensure that lessons learned from the shortcomings of past projects, as well as insights from current research in other fields, become incorporated into current and future projects. Adherence to the Code will help ensure that classical biological control remains a viable and desirable option for managing invasive plants (Balciunas 2004b).

## Acknowledgments

The inappropriate attacks by some colleagues on the meticulous research of Svata Louda galvanized the senior author to search for some form of self-regulation that would allow outside observers to distinguish good biological control from the bad and ill-conceived. His subsequent conversation with Jennifer Marohasy about her experiences with Queensland Sugar Grower's Code of Best Practices inspired this approach. We thank Peter McEvoy, Ernest "Del" Delfosse, and others who have assisted in popularizing the Code, and applaud all our colleagues who have adhered to its spirit and intent.

## References

Balciunas, J. K. 2000. Code of best practices for biological control of weeds. Pages 435-36 *in* N. R. Spencer, ed. Proc. X Int. Symp. Biol. Contr. Weeds, 4-14 July 1999. Montana State Univ., Bozeman, MT.

Balciunas, J. 2004a. Are mono-specific agents necessarily safe? The need for pre-release assessment of probable impact of candidate biocontrol agents, with some examples. Pages 252-57 *in* J. M. Cullen, D. T. Briese, W. M. Lonsdale, L. Morin, and J. K. Scott, eds. Proc. XI Int. Symp. Biol. Contr. Weeds, April 27 – May 2, 2003, Australian Nat. Univ., Canberra, ACT, CSIRO Publ., Melbourne, Australia.

Balciunas, J. 2004b. Four years of "Code of Best Practices": is biocontrol of weeds less risky, and receiving greater acceptance? Pages 258-60 *in* J. M. Cullen, D. T. Briese, W. M. Lonsdale, L. Morin, and J. K. Scott, eds. Proc. XI Symp. Biol. Contr. Weeds. April 27 – May 2, 2003, Australian Nat. Univ., Canberra, ACT, CSIRO Publ., Melbourne, Australia.

Balciunas, J., and B. Villegas. 1999. Two new seed head flies attack yellow starthistle. Calif. Agric. 53: 8-11.

**135**

Coombs, E. M., H. Radtke, D. L. Isaacson, and S. Snyder. 1996. Economic and regional benefits from the biological control of tansy ragwort, *Senecio jacobaea*, in Oregon. Pages 489-94 *in* V. C. Moran and J. H. Hoffman, eds., Proc. IX Int. Symp. Biol. Contr. Weeds, 19-26 January 1996, Stellenbosch, South Africa. Univ. of Cape Town.

Huffaker, C. B., and C. E. Kennett. 1959. A ten-year study of vegetational changes associated with biological control of Klamath weed. J. Range Manage. 12: 69-82.

Louda, S. M., R. W. Pemberton, M. T. Johnson, and P. A. Follett. 2003. Nontarget effects—the Achilles' heel of biological control? Retrospective analyses to reduce risk associated with biocontrol introductions. Annu. Rev. Entomol. 48: 365-96.

McEvoy, P., and E. M. Coombs. 2000. Why things bite back: unintended consequences of biological weed control. Pages 167-94 *in* P. A. Follett and J. J. Duan, eds. Nontarget Effects of Biological Control. Kluwer Academic Publishers, Boston, MA.

Pearson, D. E., K. S. McKelvey, and L. F. Ruggiero. 2000. Non-target effects of an introduced biological control agent on deer mouse ecology. Oecologia 122: 121-28.

Strong, D. R. 1997. Fear no weevil? Science 277: 1058-59.

Turner, C. E., R. W. Pemberton, and S. S. Rosenthal. 1987. Host utilization of native *Cirsium* thistles (Asteraceae) by the introduced weevil *Rhinocyllus conicus* (Coleoptera: Curculionidae) in California. Environ. Entomol. 18: 111-15.

U.N. Food and Agriculture Organization. 1996. Code of conduct for the import and release of exotic biological control agents. Pages 1-12 *in* International Standards for Phytosanitary Measures, No. 3: Food and Agriculture Organization of the United Nations. Rome, Italy.

*Section II*

# Target Plants
# and the Biological Control Agents

# Alligatorweed

*Alternanthera philoxeroides*

Amaranth family—Amaranthaceae

## G. R. Buckingham

**Additional common names:** None widely accepted.

**Native range:** Eastern coast of South America from Argentina to Venezuela.

**Entry into the United States:** Alligatorweed is believed to have been introduced in ships' ballast prior to 1897.

## Biology

**Life duration/habit:** Alligatorweed is an herbaceous, aquatic perennial rooted in the bottom or along the shoreline of waterways. The plants are emersed and often form dense floating mats of elongated hollow stems. The tips of the stems may ascend 10 to 50 cm above the water. If a waterway dries, alligatorweed changes to a terrestrial form with smaller, tougher leaves and narrower stems.

**Reproduction:** Reproduction is mostly vegetative from stems, but viable seeds are produced in South America.

**Roots:** Roots develop from nodes closely spaced along the stem. Roots are short and narrow in water but somewhat longer and thicker in terrestrial habitats.

**Stems and leaves:** Stems of floating plants are usually hollow, to 0.8 cm thick. The dark green, stalkless leaves are opposite. Aquatic leaves are smooth and somewhat waxy; terrestrial leaves have hairs and are thinner. The leaves are linear-elliptic, 2 to 9+ cm long, 0.5 to 2.0 cm wide, with pointed to rounded tips.

**Flowers:** There are 6 to 20 flowers in short, head-like spikes. There is only one spike per node, either axillary or terminal, on peduncles 1.5 to 7.5 cm long. Flowers have five white petaloid sepals, five stamens, and a single style. The flowers can be female, male, or perfect even on the same flower spike.

Alligatorweed flower heads. (Photo credit: G. Buckingham, USDA-ARS)

139

**Fruits and seeds:** Seed production has been reported in the United States but it is rare and seeds are believed to be nonviable in the field. The smooth, lens-shaped seed is inside a flattened, indehiscent, hard, one-seeded fruit.

## Infestations

**Worst infested states:** Alabama, Florida, Georgia, Louisiana, Mississippi, North Carolina, South Carolina, and Texas. California has small infestations.

**Habitat:** Alligatorweed grows in flooded ditches, in shallow water, and along banks of all types of waterways. The terrestrial form grows mostly in ditches and dry beds of waterways, but it can be a minor pest of lawns, pastures, and some crops. Alligatorweed populations are most abundant in USDA Plant Hardiness Zone 8a or warmer with some incursion into Zone 7 and even Zone 6b in southwestern Kentucky and northwestern Tennessee.

Alligatorweed infestation. (Photo credit: A. Cofrancesco, U.S. Army Corps of Engineers)

**Impacts:** Alligatorweed disrupts water flow causing increased sedimentation and occasional flooding. Heavy mats break loose and float against bridges and dams, and also reduce oxygen levels beneath them by shading submersed plants. Alligatorweed can outcompete native vegetation. The mats also hinder boat traffic and shore fishing.

## Comments

Alligatorweed is an opportunistic weed that is favored by periods of low water and drought. The terrestrial stems colonize the dry waterway and outgrow other aquatic species when the water returns. In permanent waterbodies, the dense mats often dominate native vegetation along shore. Biological control agents have reduced the importance of alligatorweed in the warmer parts of its range where chemical controls are usually applied only when controlling another weed species. Biologies of the biocontrol agents reported here are mostly from laboratory rather than field studies.

# Agasicles hygrophila

G. R. Buckingham

**Common name:** Alligatorweed flea beetle.

**Type of agent:** Insect: Beetle (Coleoptera: Chrysomelidae).

**Native distribution:** South America, from Argentina to Brazil.

**Original source:** Argentina.

## Biology

**Generations per year:** Four to six.

**Overwintering stage:** Adult.

**Egg stage:** Eggs are pale yellow, elongate, and 1.25 mm long. Eggs are glued in clusters of 12 to 54 in two rows flat against the leaf, usually on the underside. Eggs hatch in four to five days.

**Larval stage:** Neonates are yellow progressing to black as they age. The three instars develop in eight days. Feeding larvae often leave the opposite epidermis, creating a window in the leaf. With heavy feeding, entire leaves and even upper portions of stems are eaten. A mature larva bores into the hollow stem and plugs the hole with masticated stem tissue. It then anchors itself to the stem wall.

**Pupal stage:** After one to two days, a cream-colored pupa is formed, which gradually darkens as the new adult forms. The new adult emerges after about five days and chews its way out of the stem.

**Adult stage:** Adults are about 5 mm long, black, with two yellow stripes on each elytron. The premating and preoviposition period is about six days. Females live about 48 days and deposit an average of 1,127 eggs during their lifetime, one cluster per day. The total life cycle, from egg to egg, is about 25 days. Adults eat small holes in the leaves and eventually entire leaves and stem tissue. When disturbed, they jump "like a flea"; during the heat of the day, they often fly.

## Effect

**Destructive stages:** Larval and adult.

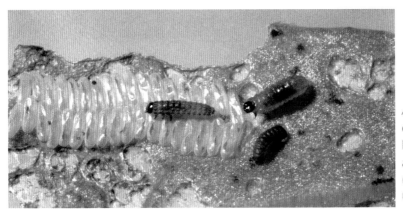

A. hygrophila eggs and larvae on alligatorweed. (Photo credit: USDA-ARS)

141

Adult *A. hygrophila* and leaf damage. (Photo credit: G. Buckingham, USDA-ARS)

**Additional plant species attacked:** None.

**Site of attack:** Primarily leaves, but also upper stems during heavy attack.

**Impact on the host:** Defoliation is spectacular during heavy attack. Regrowth is also attacked, leading to eventual submergence of the yellow, defoliated floating mat and clearing of the waterway.

**Nontarget effects:** There have been no confirmed reports of this flea beetle attacking any plant but alligatorweed after 38 years of establishment.

## Releases

**First introduced into the United States:** 1964, California and South Carolina.

**Established in:** Alabama, Florida, Georgia, Louisiana, Mississippi, Texas, and possibly South Carolina.

**Habitat:** The flea beetle only oviposits on aquatic alligatorweed, although minor adult feeding can occur on recently terrestrial alligatorweed. The flea beetle survives winter in USDA Plant Hardiness Zone 8b and warmer, but extends during summer into Zone 8a and along the Mississippi Valley into Zone 7b. Hot, dry summers are believed to reduce beetle populations.

**Availability:** Adults can be collected in early spring in Florida and in portions of Gulf Coast states where the mean January temperature is 11° C or warmer. During summer and early autumn, adults should be available at sites throughout their range.

**Stages to transfer:** All life stages are relatively easy to transfer.

**Redistribution:** Adults and larvae can be swept from the plants with a heavy sweep net. At the same time, many broken stems with eggs are also collected. Pupae can be located by finding the larval entrance holes and collecting the infested stems; stems with heavily damaged leaves and no larvae will usually have pupae in them. Adults and larvae can be fed cut stems for several days and bouquets of stems in water for longer periods. Releases onto alligatorweed mats are best made in early morning or early evening to avoid midday sun and heat. Adults can be shipped in cardboard mailing tubes with moist wood excelsior (long, thin, curled wood shavings) and small alligatorweed bouquets.

Ditch infested with alligatorweed (1965). (Photo credit: J. Lotz, Florida Department of Agriculture)

Ditch cleared of alligatorweed by *A. hygrophila* (1968). (Photo credit: J. Lotz, Florida Department of Agriculture)

## Comments

Successful control of alligatorweed by the flea beetle in the warmer parts of the plant's range stimulated later biocontrol programs on waterhyacinth and other aquatic weeds. Flea beetles have been collected during spring in Florida for many years by the U.S. Army Corps of Engineers (COE), Jacksonville District, and shipped to colder regions. COE officials should be contacted for advice by anyone planning large-scale collections and releases.

## *Amynothrips andersoni*

G. R. Buckingham

**Common name:** Alligatorweed thrips.

**Type of agent:** Insect: Thrips (Thysanoptera: Thripidae).

**Native distribution:** South America, from Argentina to Brazil.

**Original source:** Argentina.

143

## Biology

**Generations per year:** Four to five in Argentina with no reproductive diapause.

**Overwintering stages:** Adults, with some eggs and larvae.

**Egg stage:** Eggs are amber colored, elongated oval, and 0.44 mm long. They are glued onto hairs in the nodes of the apical leaves.

**Larval stage:** The two larval stages take eight to 13 days. The first-stage larva is pale amber but the second-stage is bright orange to red. Larvae feed on meristematic tissue and in leaf curls along the edges of the expanding new leaves. Larvae are gregarious.

**Pupal stage:** The orange to red prepupae and pupae are found on the plant in leaf curls. They appear similar to larvae and are mobile but sluggish.

**Adult stage:** Adults are black, elongate, and 2 mm long with fringed wings. Wing length varies; long-winged forms that can fly are uncommon. The mean life cycle, from egg to egg, is 28 days. Females

*A. andersoni* larva and eggs beneath the leaf margin. (Photo credit: J. Lotz, Florida Department of Agriculture)

*A. andersoni* adult. (Photo credit: J. Lotz, Florida Department of Agriculture)

deposited a mean of 201 eggs with a four-day preoviposition period. Unmated females produce males, but fertilized females produce equal numbers of males and females. Adults live an average of 89 days. Adults and larvae feed on the meristematic tissue and in leaf curls.

## Effect

**Destructive stages:** Larval and adult.

**Additional plant species attacked:** None.

**Site of attack:** Both adults and larvae feed in terminal and axillary buds and along margins of leaves.

**Impact on the host:** Feeding causes leaf and tip deformation, which stunts the plant. Edges of the leaves curl inward, providing excellent shelter for adults and larvae and reducing photosynthetic surface. Damage is usually light, but dead leaves and abscising stems have been reported. Populations increase most quickly during spring and autumn.

Damage to alligatorweed caused by *A. andersoni.* (Photo credit: G. Buckingham, USDA-ARS)

**Nontarget effects:** No nontarget effects have been reported; however intensive field studies have not been conducted. An unidentified *Polygonum* sp. growing with heavily damaged alligatorweed in Gainesville, Florida, was recently observed by the author with curled leaf edges that contained larvae and adults. However, it was not determined whether they had caused the curling or had just utilized the shelter provided by leaves curled by aphids or other insects.

## Releases

**First introduced into the United States:** 1967, California, Florida, Georgia, and South Carolina.

**Established in:** Alabama, Florida, Georgia, Louisiana, Mississippi, South Carolina, and Texas.

**Habitat:** This thrips attacks both aquatic and terrestrial alligatorweed, especially along margins of waterways. Thrips survive in USDA Plant Hardiness Zone 8a and warmer, and possibly along the margin of Zone 7b. Hot summer temperatures appear to be detrimental.

**Availability:** Populations are scattered and usually low; Florida and Louisiana probably provide the best chance for successful collections.

**Stages to transfer:** Larval and adult.

**Redistribution:** Infested stems are collected and kept moist and cool until they can be placed on alligatorweed mats. Spring and autumn are the best times to collect.

## Comments

Predation, especially by true bugs (Hemiptera), appears to be an important regulating factor, although no detailed evaluation studies have been conducted. Dispersal appears to be limited by the adult winged form, which is usually short-winged and flightless. There has been little redistribution of this species. It is the most cold-tolerant of the three alligatorweed biocontrol agents.

# Arcola malloi (=Vogtia malloi)

G. R. Buckingham

**Common name:** Alligatorweed stem borer.

**Type of agent:** Insect: Moth (Lepidoptera: Pyralidae).

**Native distribution:** South America, from Argentina to Guyana.

**Original source:** Argentina.

## Biology

**Generations per year:** Three to four.

**Overwintering stage:** Larval.

**Egg stage:** The white eggs are oval and 0.69 mm long. Eggs are laid singly on the undersides of apical leaves or at the juncture of the petiole and stem. They hatch in three to four days.

**Larval stage:** Young larvae are white to amber, while older larvae appear brown because of numerous light tan, longitudinal, wavy stripes. Neonates bore into the stem tip and then downward. Larvae often bore into four to eight stems during their lifetime. The

*A. malloi* larva. (Photo credit: USDA-ARS)

*A. malloi* adult. (Photo credit: W. Durden, USDA-ARS)

internal stem tissue is consumed almost to the epidermis by older larvae. The five instars develop in 24 days. The mature larva, about 14 mm long, seals the stem node beneath it with masticated and regurgitated tissue to prevent water intrusion and then spins two silken septa in the stem, one above it and one below. It then prepares an adult escape window in the stem by chewing stem tissue but leaving the epidermis intact.

**Pupal stage:** The pupa is formed in a silken cocoon in the stem. It darkens from amber to dark brown as it develops. The pupal stage lasts about 10 days.

**Adult stage:** Adults emerge through the windows in the stems. They range from 13 to 14 mm long with tan to brown forewings that blend in well with dead stems. The wings are held tightly around the body and the moth sits with the forward portion of the body held above the substrate and the hind portion on the substrate. Females lay approximately 267 eggs. The life cycle takes about 39 days. Females live six to 10 days and males five to nine days.

## Effect

**Destructive stage:** Larval.

**Additional plant species attacked:** Occasionally silverhead (*Blutaparon vermiculare*) and *Alternanthera flavescens*.

**Site of attack:** Inside the stem.

**Impact on the host:** The stem tip wilts initially, and then long portions of the stem die, bend over, and turn yellow. Entire mats can turn yellow and submerge. Dead leaves are present on the damaged stems. (In contrast, stems damaged by the flea beetle *Agasicles hygrophila* are defoliated.) Often, heavy moth damage is not observed until autumn. Terrestrial plants are attacked but heavy damage has not been reported.

**Nontarget effects:** No nontarget effects have been reported; however effects of the feeding reported on *B. vermiculare* and on *A. flavescens* have not been studied.

## Releases

**First introduced into the United States:** 1971, Florida, Georgia, North Carolina, and South Carolina; 1972, Alabama.

**Established in:** Alabama, Florida, Georgia, Louisiana, Mississippi, North Carolina, South Carolina, and Texas.

**Habitat:** This moth is most common on aquatic and shoreline plants, although terrestrial plants can be attacked. It survives winter in USDA Plant Hardiness Zone 8b and warmer and perhaps parts of 8a. Migrants can fly into Zone 7b in the Mississippi River Valley and produce summer populations.

**Availability:** Most available in midsummer and early autumn in Louisiana, Mississippi, and South Carolina.

**Stage to transfer:** Larval.

**Redistribution:** Damaged stems are collected and kept reasonably moist and cool. Both the wilted upright stems with young larvae and the horizontal older stems with older larvae and pupae should be collected. The stems are then placed onto a new mat where the larvae will move out of the old and into new stems.

Damage to alligatorweed caused by *A. malloi.* (Photo credit: USDA-ARS)

## Comments

Dispersal distances from 900 to 1,000 km have been suggested for this moth. It is credited with playing the major role in biocontrol of alligatorweed in the upper portions of the Lower Mississippi River Valley. However, it cannot compete with the flea beetle *A. hygrophila* if both populations attack a mat at about the same time. Native parasitoids were reared from eggs and larvae/pupae early in the program, but no recent studies of parasitism levels and their effects have been reported. This moth appears to tolerate cold better than the flea beetle, but most inland populations are probably immigrants from coastal winter refuges.

## *References*

Anonymous. 1981. The use of insects to manage alligatorweed. U.S. Army Engineer Waterways Experiment Station, Vicksburg, MS. Instruction Report A-81-1.

Aulbach-Smith, C. A., S. J. de Kozlowski, and L. A. Dyck. 1990. Aquatic and Wetland Plants of South Carolina. South Carolina Aquatic Plant Management Council and South Carolina Water Resources Commission, Columbia, SC.

Brown, J. L., and N. R. Spencer. 1973. *Vogtia malloi,* a newly introduced phycitine moth (Lepidoptera: Pyralidae) to control alligatorweed. Environ. Entomol. 2: 519-23.

Buckingham, G. R. 1994. Biological control of aquatic weeds. Pages 413-79 *in* D. Rosen, F. D. Bennett, and J. L. Capinera, eds. Pest Management in the Subtropics: Biological Control—the Florida Experience. Intercept Ltd., Hampshire, UK.

Buckingham, G. R. 2002. Alligatorweed. Pages 5-15 *in* R. Van Driesche, S. Lyon, M. Hoddle, B. Blossey, and R. Reardon, eds. Biological Control of Invasive Plants in the Eastern United States. USDA Forest Service Publ. FHTET-2002-04. Morgantown, WV.

Cofrancesco, A. F., Jr. 1988. Alligatorweed survey of ten southern states. U.S. Army Engineer Waterways Experiment Station, Vicksburg, MS. Misc. Paper A-88-3.

Coulson, J. R. 1977. Biological control of alligatorweed, 1959-1972. A review and evaluation. USDA-ARS Tech. Bull. 1547.

Godfrey, R. K., and J. W. Wooten. 1981. Aquatic and Wetland Plants of the Southeastern United States: Dicotyledons. Univ. Georgia Press, Athens, GA.

Holm, L., J. Doll, E. Holm, J. Pancho, and J. Herberger. 1997. World Weeds: Natural Histories and Distribution. John Wiley and Sons, NY.

Julien, M. H., and J. E. Broadbent. 1980. The biology of Australian weeds. 3. *Alternanthera philoxeroides* (Mart.) Griseb. J. Aust. Inst. Agric. Sci. 46: 150-55.

Maddox, D. M. 1968. Bionomics of an alligatorweed flea beetle *Agasicles* sp., in Argentina. Ann. Entomol. Soc. Am. 61: 1299-1305.

Maddox, D. M. 1970. The bionomics of a stem borer, *Vogtia malloi* (Lepidoptera: Phycitidae), on alligatorweed in Argentina. Ann. Entomol. Soc. Am. 63: 1267-73.

Maddox, D. M., and A. Mayfield. 1979. Biology and life history of *Amynothrips andersoni*, a thrips for the biological control of alligatorweed. Ann. Entomol. Soc. Am. 72: 136-40.

Maddox, D. M., L. A. Andres, R. D. Hennessey, R. D. Blackburn, and N. R. Spencer. 1971. Insects to control alligatorweed: an invader of aquatic ecosystems in the United States. BioScience 21: 985-91.

Pemberton, R. W. 2000. Predictable risk to native plants in weed biological control. Oecologia 125: 489-94.

Penfound, W. T. 1940. The biology of *Achyranthes philoxeroides* (Mart.) Standley. Am. Midl. Nat. 24: 248-52.

Spencer, N. R., and J. R. Coulson. 1976. The biological control of alligatorweed, *Alternanthera philoxeroides*, in the United States of America. Aquat. Bot. 2: 177-90.

Vogt, G. B., J. U. McGuire, Jr., and A. D. Cushman. 1979. Probable evolution and morphological variation in South American disonychine flea beetles (Coleoptera: Chrysomelidae) and their amaranthaceous hosts. USDA-ARS Tech. Bull. 1593.

Vogt, G. B., P. C. Quimby, Jr., and S. H. Kay. 1992. Effects of weather on the biological control of alligatorweed in the lower Mississippi Valley region, 1973-83. USDA-ARS Tech. Bull. 1766.

# Bindweeds

## Field bindweed
*Convolvulus arvensis*

Morningglory family—Convolvulaceae

*J. L. Littlefield*

**Additional common names:** Small-flowered morningglory, wild morningglory, creeping Jenny, European bindweed.

**Native range:** Europe.

**Entry into the United States:** Field bindweed apparently contaminated crop seed and was identified in Virginia as early as 1739.

## Biology

**Life duration/habit:** Field bindweed is a creeping or twining perennial herbaceous vine that can grow 0.3 to 2 m long. Bindweed often forms dense infestations consisting of one or more clones.

**Reproduction:** This species reproduces by seeds and rhizomes.

**Roots:** Taproots may extend more than 3 m into the soil. It also has an extensive system of lateral roots.

**Stems and leaves:** The prostrate stems produce leaves that are arrowhead-shaped and normally slender with sharp, pointed lobes; they measure 1.2 to 4.3 cm long. Leaves are alternately arranged along the stem.

**Flowers:** The funnel- or trumpet-shaped flowers are about 2.5 cm in diameter and vary from white to pink. Flowers are produced from late June until early fall.

**Fruits and seeds:** Seedpods are pointed and about 5 mm long. Four rough, dark brown, pear-shaped seeds usually grow in each pod. Seeds can remain viable for up to 10 years and sometimes as long as 50 years.

Field bindweed in bloom. (Photo credit: N. Poritz, bio-control.com)

## Infestations

**Worst infested states:** Western United States, especially the Great Plains. Field bindweed is present in all 48 contiguous states in the United States, plus Hawaii.

**Habitat:** Field bindweed tolerates a great range of environmental conditions and elevations and is found in all types of ground, including cultivated fields and waste places. It grows best on fertile, dry, or moderately moist soils and is extremely difficult to control.

**Impacts:** Field bindweed is a serious weed in most of the United States, especially on farmlands. It generally grows in dense, tangled mats that may reduce crop production by as much as 60%. Crop losses in 1998 in the United States were estimated to exceed $377.8 million per year. Serious economic losses were reported in 22 states, primarily in the Great Plains and western United States.

## Comments

Field bindweed is similar in appearance to other species in the family Convolvulaceae. There are approximately 20 native and introduced bindweeds (and numerous subspecies) of the closely related genus *Calystegia*. Many of these are located in coastal regions of the western United States, although some species are more eastern in distribution. Several species are considered rare and endangered at the state level and one species, *Calystegia stebbinsii*, is classified as federally endangered. Field bindweed may also be confused with morningglories (*Ipomoea*) and wild buckwheat (*Polygonum convolvulus*). Morningglories may be differentiated by their annual growth habits, longer sepals on the flowers, and flower color generally blue or purple varying to white. Wild buckwheat plants have small, greenish-colored flowers, an annual growth habit, and an inconspicuous sheath that encircles the stem at the base of the leaf.

# Hedge bindweed *Calystegia sepium (= Convolvulus sepium)*

## Morningglory family—Convolvulaceae

*P. W. Tipping*

**Additional common names:** Bellbine, bracted bindweed, devils-vine, great bindweed, lady's nightcap, larger bindweed, Rutland beauty, wind morningglory.

**Native range:** Eastern United States, but is now widely distributed.

**Entry into the United States:** Hedge bindweed was recognized as a weed in 1889.

## Biology

**Life duration/habit:** Hedge bindweed is a perennial plant that grows in hedges, lowland forests, and arable lands.

**Reproduction:** Reproduction occurs both by seeds and growth of aerial shoots, rhizomes, and underground lateral roots.

151

Hedge bindweed. (Photo credit: E. Coombs, Oregon Department of Agriculture)

**Roots:** The plant has extensive, fleshy rhizomes and lateral roots.

**Stems and leaves:** Stems climb or trail, grow up to 2.8 m long, and often wrap around other stems of the same plant. Leaves are arrow-shaped with angular basal lobes and are 5 to 10 cm long.

**Flowers:** Flowers are present from July through August; five white or pink sepals are fused into a funnel-shaped tube 6-8 cm in diameter. Two leafy bracts lie at the base of the flowers, a characteristic that helps distinguish this species from field bindweed (*Convolvulus arvensis*).

**Fruits and seeds:** The fruit is a round capsule, 5 to 8 mm in diameter, and contains two to four seeds. These seeds are 4.5 to 5 mm long, dark brown to black with alternately rounded and flattened sides. They are long-lived in the soil.

## Infestations

**Worst infested states:** Midwestern and eastern states, although *C. sepium* is reported in all 48 contiguous states.

**Habitat:** The plant prefers moist, nutrient-rich soils. It is commonly a background plant in natural areas and a weed in row crops, nurseries, and ornamental landscapes.

**Impacts:** This plant can smother crop plants and compete for nutrients, water, and space. Cultivation is problematic in nonrow crops and new infestations can start from rhizome fragments. Long-lived seeds, along with the significant regenerative capacity of the rhizomes, enable populations of this plant to be very persistent.

## Comments

This plant is considered both a native and introduced species and contains several subspecies, all of which may or may not be weedy, including *C. sepium* ssp. *americana,* ssp. *angulata,* ssp. *binghamiae,* ssp. *erratica,* ssp. *limnophila,* and ssp. *sepium.* In addition, several closely related species include *C. macounii, C. silvatica* ssp. *fraterniflora,* and *C. spithamaea.*

# Aceria malherbae

J. L. Littlefield

**Common name:** Bindweed gall mite.
**Type of agent:** Mite (Acari: Eriophyidae).
**Native distribution:** Central and southern Europe, and northern Africa.
**Original source:** Greece.

## Biology

**Generations per year:** Multiple.
**Overwintering stages:** Adult and nymph (on the root buds).
**Egg stage:** The round, translucent eggs are deposited within the galls.
**Nymphal stage:** There are two nymphal stages. The nymphs resemble the adults but lack external genitalia.
**Adult stage:** Adults are present year-round. They are minute and worm-like and have an annulate body with two pairs of legs on the combined head and thorax. Adults have yellow-white, translucent to opaque, soft bodies. Mites are very small and are best viewed with a microscope.

## Effect

**Destructive stages:** Nymphal and adult (within galls).
**Additional plant species attacked:** It may attack some *Calystegia* spp.
**Site of attack:** Galls are formed on the leaves, petioles, and stem tips. Mites form galls on actively growing leaves and stem buds. Leaf galls can be identified by the leaves, which fold or twist upward along the midrib where the mites feed. The lower surface of the gall is roughened with small papillae. When stem buds are attacked, they fail to elongate and thus form compact clusters of stunted leaves.

**Impact on the host:** Mites stunt the plant and reduce flowering. Some reduction in plant density has been reported in Texas.

**Nontarget effects:** No nontarget impacts have been reported.

Leaf galling on field bindweed (right) caused by *A. malherbae*. (Photo credit: E. Coombs, Oregon Department of Agriculture)

*A. malherbae* adults. (Photo credit: E. Coombs, Oregon Department of Agriculture)

## Releases

**First introduced into the United States:** 1989, Texas.

**Established in:** Kansas, Montana, Oklahoma, and Texas, but also recovered in Colorado, Washington, and Wyoming.

**Habitat:** Undetermined.

**Availability:** This mite is increasingly available from established field sites, especially in Texas.

**Stages to transfer:** Nymphal and adult in the galls.

**Redistribution:** Galls may be handpicked and stored for several weeks if kept cool and moist in the refrigerator. Galls may be placed on actively growing plants by transferring individual galls to the tip of the stem or by wrapping infested stems around bindweed plants. Transplanting infested plants may be of limited use because plants tend to die rapidly without adequate care. Mowing infested sites may also spread the mite. Although mites may be transferred throughout the growing season, spring or early summer releases may be preferable because they give the mite extra time to increase populations. New techniques and strategies are being developed to effectively transfer the mite.

## Comments

The mite may be difficult to establish in a field under cultivation or herbicide treatment, thus a site less aggressively managed may be a better location for release. Populations of the mite are generally slow to develop (up to three years), but once established, the mite may disperse via the wind and spread rapidly. The mite could also potentially infest native *Calystegia* spp.; thus release is not recommended for locations in which nontarget impacts may be of concern, e.g., areas having rare or endangered species. The mite has been released on hedge bindweed, but is probably less effective than on field bindweed. This mite was initially reported in the literature as *A. convolvuli*, but is now considered a separate species.

# Tyta luctuosa

J. L. Littlefield

**Common name:** Bindweed moth.

**Type of agent:** Insect: Moth (Lepidoptera: Noctuidae).

**Native distribution:** Europe from southern Scandinavia southward; Asia east to Turkestan and south into northern India; North Africa.

**Original source:** Italy.

## Biology

**Generations per year:** Two and sometimes a partial third generation in southern Europe.

**Overwintering stages:** Adult and larval (on the root buds).

**Egg stage:** Eggs are laid on stems, leaves, and flower buds.

**Larval stage:** There are five larval instars. Larvae are found in the field from May until September—on the plant during the night and both on the plant and in the debris under the plant during the day. Because of their drab color and secretive habits, they are difficult to locate on the plant during the daytime. Some may be found at the base of the field bindweed plants or stretched out along the stems.

**Pupal stage:** Pupation occurs in the soil or litter near bindweed plants.

**Adult stage:** Adult emergence begins in May and moths of the first generation are active until June. The second-generation adults fly in July until about September. Adults are present throughout the growing season of *C. arvensis*. Females deposit an average of 435 eggs with a maximum of more than 660. Adults measure about 11 mm long.

## Effect

**Destructive stage:** Larval.

**Additional plant species attacked:** It may attack some *Calystegia* species.

**Site of attack:** The larvae generally feed at night on flowers and leaves.

*T. luctuosa* larva. (Photo credit: E. Coombs, Oregon Department of Agriculture)

*T. luctuosa* adult. (Photo credit: N. Poritz, bio-control.com)

**Impact on the host:** The amount of damage caused by larval feeding on field bindweed has not yet been determined. Larval feeding does not significantly impact hedge bindweed.

**Nontarget effects:** No nontarget effects have been reported.

## Releases

**First introduced into the United States:** 1987, Arizona, Iowa, Missouri, Oklahoma, and Texas. Also released in Kansas, Maryland, Montana, and Washington.

**Established in:** Not established.

**Habitat:** This moth apparently is not restricted to any particular habitat in its native region.

**Availability:** Unavailable.

**Stages to transfer:** Larval and adult.

**Redistribution:** Hand-collect larvae and ship with leaf and stem material in cool containers. Release as soon as possible, preferably in the calm, cool part of the evening. Adults can be collected with black lights. Ship or hand-carry adults in paperboard containers provided with shredded paper on which they can crawl, and with a source of moisture.

## Comments

Although moths have been recovered, established populations of *T. luctuosa* were not known as of 2002 despite numerous releases. Mass-rearing of the moth has been hampered because of larval mortality associated with nutritional problems.

# *References*

Austin, D. F. 1986. Convolvulaceae, The Morning Glory Family. Pages 652-61 *in* T. Barkley, ed. Flora of the Great Plains. Univ. Press Kansas, Lawrence.

Boldt, P. E., and R. Sobhian. 1993. Release and establishment of *Aceria malherbae* (Acari: Eriophyidae) for control of field bindweed in Texas. Environ. Entomol. 22: 234-37.

Boldt, P. E., S. R. Rosenthal, and R. Srinivasan. 1998. Distribution of field bindweed and hedge bindweed in the U.S.A. J. Prod. Agric. 11: 377-81.

Brummitt, R. K. 1981. Further new names in the genus *Calystegia (Convolvulaceae)*. Kew Bull. 35: 327-34.

Callihan, R. H., C. V. Eberlein, J. P. McCaffrey, and D. C. Thill. 1990. Field bindweed - biology and management. Univ. Idaho Bull. 719.

Friend, E. J., J. P. Kelly, C. W. Friend, and T. F. Peeper. 2002. Biological control of field bindweed in Oklahoma. Oklahoma Coop. Ext. Serv. PT 2002-22. Vol. 14, No. 22.

McClay, A. S., J. L. Littlefield, and J. Kashefi. 1999. Establishment of *Aceria malherbae* (Acari: Eriophyidae) as a biological control agent for field bindweed (Convolvulaceae) in the Northern Great Plains. Can. Entomol. 131: 541-47.

Rosenthal, S. S., S. L. Clement, N. Hostettler, and T. Mimmocchi. 1988. Biology of *Tyta luctuosa* (Lep.: Noctuidae) and its potential value as a biological control agent for the weed *Convolvulus arvensis*. Entomophaga 33: 185-92.

Tipping, P. W., and G. Campobasso. 1997. Impact of *Tyta luctuosa* (Lepidoptera: Noctuidae) on hedge bindweed (*Calystegia sepium*) in corn (*Zea mays*). Weed Technol. 11: 731-33.

Wang, R., and L.T. Kok. 1985. Bindweeds and their biological control. Biocontrol News Info. 6: 303-10.

Weaver, S. E., and W. R. Riley. 1982. The biology of Canadian weeds. 53. *Convolvulus arvensis* L. Can. J. Plant Sci. 62: 461-72.

# Brooms

## E. M. Coombs and M. J. Pitcairn

The broom complex consists of a number of exotic leguminous shrubby species that were originally used as ornamentals, but have since escaped cultivation and become serious invasive pests of rangelands, forests, and rights-of-way in the western United States. French broom (*Genista monspessulana*), Scotch broom (*Cytisus scoparius*), Spanish broom (*Spartium junceum*), and Portuguese broom (*Cytisus striatus*) were the primary targets for biological control in the United States as of 2004.

## French broom                                          *Genista monspessulana*

### Pea family—Fabaceae (Leguminosae)

### *M. J. Pitcairn*

**Additional common names:** Canary broom, Montpellier broom, soft broom.

**Native range:** Mediterranean region of Europe, northern Africa, and Azores Islands.

**Entry into the United States:** French broom was originally introduced as a landscape ornamental.

### Biology

**Life duration/habit:** French broom is a 1- to 3-m-tall shrub adapted to humid and sub-humid habitats. It is a nitrogen fixer, allowing it to invade low-fertility soils.

**Reproduction:** Reproduction occurs primarily by seed. French broom can resprout from the crown when aboveground growth is removed by cutting, freezing, or sometimes fire.

French broom. (Photo credit: J. DiTomaso, University of California)

**Roots:** Plants have a deep, branching taproot with nitrogen-fixing bacteria.

**Stems and leaves:** Stems are erect, dense, and green. Stems are typically leafy (as compared to Scotch broom which has few leaves), eight- to 10-ridged and round in cross section, and densely covered with silky, silvery hairs. Leaves are compound with three oblong leaflets 10 to 20 mm long. Leaves are alternate on the stem and deciduous.

**Flowers:** Bright yellow flowers are produced from March to May in California. The fragrant flowers grow in clusters (four to 10 per cluster) at the ends of short axillary pedicels. The pea-like flowers (approximately 1 cm long) are two-lipped, the top lip being lobed near the middle; the lower lip is shallow. French broom is insect-pollinated.

**Fruits and seeds:** Seeds are borne in pods 15 to 25 mm long by 5 mm wide, and densely covered with flattened, long, silky hairs. Pods contain five to eight seeds. The smooth seeds are ovoid, shiny, black, and 2 to 3 mm long. They are hard-coated and long-lived under field conditions. Fire appears to stimulate seed germination. Where seeds are present in the soil, a large flush of seedlings may emerge on newly burned areas. Pods are often produced abundantly.

## Infestations

**Worst infested states:** California, southwestern Oregon, and Washington.

**Habitat:** French broom colonizes open disturbed sites such as logged or burned areas, roadsides, pastures, and road and utility rights-of-way. It can invade undisturbed grasslands, coastal scrub, oak woodlands, and open forests. It does not tolerate heavy shade but can invade the understory of some forest communities. In California, it appears drought resistant and grows reasonably well in alkaline soils with pH 8. It does not tolerate freezing and is usually found below 800 m elevation.

**Impacts:** French broom is a strong competitor and forms dense, monospecific stands. It displaces native plant and forage species and impedes reforestation. French broom foliage and seeds are toxic to livestock. It burns readily and increases both the intensity and frequency of fires. French broom growing along roadsides obstructs views and requires expensive ongoing roadside maintenance. Its long-lived seed bank makes it difficult to eradicate.

## Comments

French broom is one of several exotic leguminous shrubs (such as Scotch broom, Spanish broom, Portuguese broom, and gorse [*Ulex europaeus*]) that have become serious pests in the western United States. Foreign exploration for biological control agents against French broom began in 2000. Several promising agents have been found, including a psyllid that attacks the stem and a weevil that destroys the seeds. As of 2003, these insects were undergoing host specificity tests to determine their safety and potential nontarget impacts.

French broom is sometimes confused with Scotch broom, which has pods with hairs only at the seam and green stems that are five-angled and ridged, flowers that are larger than 1 cm, and only about 55% of the total green tissue occurs as leaves.

159

The native moth *Uresiphita reversalis* (Lepidoptera: Pyralidae), which is known as "the genista caterpillar," has been observed to build up high numbers on local populations of French broom in central California. The larvae are orange and black spotted and feed externally on the foliage. This moth is gregarious and high populations can be very damaging to local populations of French broom. Most populations of the moth are transitory, lasting only two to three years in the local area, and do not provide long-term control of French broom.

# Scotch broom *Cytisus scoparius*

Pea family—Fabaceae (Leguminosae)

## E. M. Coombs, G. P. Markin, and T. G. Forrest

**Additional common names:** English broom, Scot's broom.

**Native range:** Central and southern Europe.

**Entry into the United States:** Southern broom was originally introduced as an ornamental.

### Biology

**Life duration/habit:** Scotch broom is a woody, bushy perennial growing 1 to 3 m tall. It is a nitrogen fixer, but its advantage for other plant species is questionable.

**Reproduction:** Scotch broom reproduces by seed.

**Roots:** This plant possesses a forked taproot and nodules with nitrogen-fixing bacteria.

**Stems and leaves:** It has many slender, upright, green to brownish-green angled stems that are hairy when young and smooth when older. Clover-like leaves develop singly or in clusters at the nodes during the growing season.

**Flowers:** During May and June, bright yellow pea-like flowers are produced in large numbers along the stems, the flowers arising singly or in pairs from the leaf axils.

Scotch broom infestation. (Photo credit: E. Coombs, Oregon Department of Agriculture)

**Fruits and seeds:** Pea-like, brownish-black (when mature), flattened, hairy-margined pods contain up to eight seeds that may remain viable for more than 80 years.

## Infestations

**Worst infested states:** This plant is distributed from British Columbia, Canada, southward to central California, from the coast to the inland valleys, primarily west of the Cascade Mountains in Oregon and Washington, in some areas on the western slopes of the Sierra Nevada Mountains, and sparingly in the northern Rocky Mountains. It also occurs along the northeastern seaboard of the United States with larger infestations in North Carolina.

**Habitat:** Scotch broom invades pastures, hillsides, and road and utility rights-of-way, and has become a serious pest in logged areas replanted with conifer seedlings. The plant is well adapted to dry hillsides, pastures, and forest clearings. This plant does not do well in areas with very cold winters.

Scotch broom plant. (Photo credit: Oregon Department of Agriculture)

**Impacts:** Scotch broom displaces forage and native plants and is unpalatable to most livestock. An economic analysis in Oregon estimated an annual loss of $47 million due to Scotch broom, mostly from impacts on reforestation.

## Comments

Because of its profuse yellow flowers in the spring, it is difficult for many people to think of Scotch broom as a serious pest. Several other closely related broom species (e.g., French, Portuguese, and Spanish broom) are also becoming serious pests in the western United States.

Biological control of Scotch broom began in 1960. Three insects have been introduced as biological control agents, and numerous accidental introductions of natural enemies have also been reported, including *Agonopterix nervosa*, *Arytainilla spartiophila*, *Bruchidius villosus*, *Dictyonota fuligosa*, *Gargara genistae*, *Leucoptera spartifoliella*, *Melanotrichus concolor*, and *M. virescens*. *Bruchidius villosus* was later approved for use as a biological control agent for Scotch broom in Oregon and Washington. *Leucoptera spartifoliella* was found to be adventive on the U.S. West Coast after it was officially introduced. Use of unapproved natural enemies for biological control of Scotch broom is not recommended. Scotch broom is unique in that it is heavily attacked by a number of native arthropods, particularly aphids and other sucking insects.

Many Scotch broom plants in older infestations are dying due to a plant pathogen suspected to be *Selenophoma juncea*. It is widespread and seems to be the most significant mortality factor. Because it has not been approved as a biological control agent, it should not be redistributed.

161

# Bruchidius villosus

E. M. Coombs, T. G. Forrest, and G. P. Markin

**Common name:** Scotch broom bruchid.

**Type of agent:** Insect: Beetle (Coleoptera: Bruchidae).

**Native distribution:** Western and central Europe.

**Original source:** North Carolina.

## Biology

**Generations per year:** One, in late spring.

**Overwintering stage:** Adults generally overwinter away from the host plant.

**Egg stage:** Between two and 12 (generally about 10) small, white, round eggs are cemented onto the pods, where they hatch in one to two weeks.

**Larval stage:** Larvae hatch from the underside of the egg and tunnel in the pod until they reach a seed whereupon they bore into the seed and complete development after four larval instars. Several larvae may attack a single seed, but one usually prevails. Larvae are very small (1 to 2 mm), creamy white with brownish head capsules, comma-shaped, and have poorly developed legs.

**Pupal stage:** The pupal period lasts from 10 to 20 days, depending on the temperature. Larvae pupate within the seed coat. They are creamy white, becoming darker as they mature. Pupae somewhat resemble adults. Several pupae often occur together in the same pod, occasionally more than one in the same seed.

**Adult stage:** The adults are freed from the pods when the pods mature and split open. Adults are 2 mm long with a dark gray body; elytra are short and broadened. Overwintering adults become active in spring when broom plants flower. (The exotic vetch bruchid, *Bruchus brachialis*, may also occur on Scotch broom during bloom, leading to some confusion, but it is larger and the antennae and front pair of legs are yellowish.)

*B. villosus* larvae. (Photo credit: E. Coombs, Oregon Department of Agriculture)

*B. villosus* adult. (Photo credit: E. Coombs, Oregon Department of Agriculture)

## Effect

**Destructive stage:** Larval.

**Additional plant species attacked:** Portuguese broom, Spanish broom, and French broom.

**Site of attack:** Larvae feed in seeds in pods and adults feed on pollen.

**Impact on the host:** *Bruchidius villosus* effectively reduces Scotch broom seed production and perhaps decreases the rate of spread. In North Carolina, seed destruction is often more than 80%. In several sites in Oregon, seedpod attack and seed reduction have been 10 to 25%. This rate is expected to increase as the beetles build up populations and later spread into infested areas throughout the U.S. West Coast.

**Nontarget effects:** No nontarget effects are known. The beetle was tested on native lupines from the West Coast.

## Releases

**First introduced into the United States:** Accidentally introduced, originally collected in Massachusetts in 1918. First released on the West Coast in Oregon in 1998 and later in Idaho and Washington.

**Established in:** Eastern United States, Oregon, and Washington.

**Habitat:** *Bruchidius villosus* prefers meadows and hillsides with southern exposures, while cold, damp, and heavily shaded areas, north-facing hillsides, and areas close to the ocean and elevations above 305 m are undesirable. The beetle establishes well on the U.S. West Coast. Populations take about four to five years to build up.

**Availability:** The beetle is well established in North Carolina and can be collected in the spring during early flowering. Several populations became collectible in Oregon in 2003.

**Stage to transfer:** Adult (after mating).

**Redistribution:** Collect and redistribute adults after they have mated. Moving adults in seed pods by using bouquets of Scotch broom has proven to be only marginally successful. Heavy-duty sweep nets or a beating sheet with a beat stick can be used to dislodge adults from plants. Collecting adults late in the flowering season on the West Coast is difficult because the presence of numerous aphids, psocids, and other insects makes it difficult to sort out the beetles. Also, the honeydew excreted by these insects makes nets very sticky.

Releases of 100 to 250 adults are normally sufficient to colonize a site. Populations usually become collectible by the fourth or fifth year following the initial release. Adults can be stored for a week if kept cool and dry. Adults survive well in containers provided with shredded tissue paper for them to crawl on.

## Comments

Even though *B. villosus* is an accidentally introduced natural enemy of Scotch broom, it was tested for host specificity and environmental safety using the USDA-APHIS Technical Advisory Group for Biological Control Agents of Weeds (TAG) protocol before being released on the West Coast.

In 1998, the natural enemy of *B. villosus* and *E. fuscirostre*, a small parasitic wasp (*Pteromalus sequester*), was discovered in Oregon. It has the potential to undo the biocontrol impacts of both beetles.

## *Exapion fuscirostre (= Apion fuscirostre)*

E. M. Coombs and G. P. Markin

**Common name:** Scotch broom seed weevil.

**Type of agent:** Insect: Beetle, weevil (Coleoptera: Brentidae [= Apionidae]).

**Native distribution:** Central and southern Europe.

**Original source:** Italy.

## Biology

**Generations per year:** One.

**Overwintering stage:** Adult (in duff near host).

**Egg stage:** Females must feed on flowers in the spring to stimulate egg production. Between five and 10 small, white to yellowish, round eggs are inserted into the pods where they hatch in five to 15 days.

**Larval stage:** Larvae feed on developing seeds inside pods where they complete development in 20 to 40 days. Larvae are creamy white with brownish head capsules, comma-shaped, and have poorly developed legs.

*E. fuscirostre* larva within a Scotch broom seed (third from left). (Photo credit: E. Coombs, Oregon Department of Agriculture)

**Pupal stage:** The pupal period lasts from 10 to 20 days, depending on the temperature. Larvae pupate within the seedpod without forming a pupal case. They are creamy white, becoming darker as they mature. They resemble the adult weevil with the legs held close to the body. Several pupae often occur together in the same pod, but rarely more than one per seed.

**Adult stage:** Adults can be found throughout the year, including on warm winter days, feeding on terminal twigs. Overwintering adults may still be active when the first-generation adults begin to emerge. The adult weevils are freed from the pods when they dry and split open. In some areas, high temperatures may kill adults that cannot escape from the mature black pods. Adults feed on terminal stem growth, giving the shoots a grayish, mottled appearance. Adults fly to the yellow flowers of Scotch broom in late winter and spring. They are active walkers, searching rapidly up and down the stems. Adults are 2 to 3 mm long and are laterally compressed. Each has a long curved beak, a dark gray body, a wide, dark gray band that extends down both sides of the back, and light brown legs.

## Effect

**Destructive stage:** Larval.

**Additional plant species attacked:** None.

**Site of attack:** Larvae feed on seeds in pods and adults feed on flowers. In large numbers, adults damage tips of twigs.

**Impact on the host:** *Exapion fuscirostre* effectively reduces Scotch broom seed production and perhaps decreases the rate of spread. In California, seed has been reduced by 60%. In Oregon, 40 to 60% of the seedpods are attacked. Seed reduction in attacked pods averages 85%. Extensive feeding by adults on small twigs has caused terminal shoot dieback. The insect's overall effectiveness in reducing Scotch broom stand densities is questionable.

**Nontarget effects:** There are no known or reported nontarget effects caused by *A. fuscirostre*.

## Releases

**First introduced into the United States:** 1964, California.

**Established in:** California, Oregon, and Washington.

**Habitat:** *Exapion fuscirostre* prefers meadows and hillsides with southern exposures, while cold, damp, and heavily shaded areas, north-facing hillsides, and areas close to the ocean are undesirable. Weevils are easily established in mild climatic areas west of the Cascades; the adults disperse rapidly, up to 2 km a year from the release point.

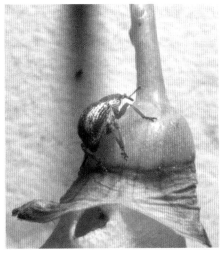

*E. fuscirostre* adult. (Photo credit: E. Coombs, Oregon Department of Agriculture)

165

**Availability:** The weevil is well established at many locations and can be collected in large numbers in the spring during early flowering.

**Stage to transfer:** Collect and redistribute adults after they have mated (mid to late flower). Moving adults in seedpods by using bouquets of Scotch broom has proven to be only marginally successful.

**Redistribution:** Heavy-duty sweep nets or a beating sheet with a beat stick can be used to dislodge adults from plants. Most collections are made in April and May. Collecting adults late in the flowering season is difficult because the presence of numerous aphids, psyllids, and other insects makes it difficult to sort out the weevils. Also, the honeydew excreted by these insects makes nets very sticky.

Releases of 100 to 250 adults are normally sufficient to colonize a site. Populations usually become collectible by the third or fourth year following release. Adults can be easily stored for a week or two if kept cool and dry. Adults do well in containers provided with shredded tissue paper for the insects to crawl on and a small amount of plant material for food.

## Comments

This insect was not distributed in all Scotch broom-infested areas initially because Scotch broom was used as a bank stabilizer. Now the plant is considered a weed because of its invasiveness and competitive impact on reforestation.

The weevil is now widespread in California, Oregon, and Washington. In 1998, a natural enemy of *E. fuscirostre* and *Bruchidius villosus*, a small parasitic wasp (*Pteromalus sequester*), was discovered in Oregon. It has great potential to undo the biocontrol impacts of both beetles.

This weevil has been recently placed in the genus *Exapion* and the family Brentidae.

## *Leucoptera spartifoliella*

E. M. Coombs and G. P. Markin

**Common name:** Scotch broom twig miner.
**Type of agent:** Insect: Moth (Lepidoptera: Lyonetiidae).
**Native distribution:** Central and southern Europe.
**Original source:** France.

### Biology
**Generations per year:** One.
**Overwintering stage:** Larval.

**Egg stage:** Eggs are laid singly on the plant's post-bloom stem growth during the spring. The incubation period lasts 15 to 18 days.

**Larval stage:** Larvae burrow beneath the epidermis of young shoots and tunnel up and down their length to feed. Brownish discolored larval mines are readily visible; the mines become larger as the larvae grow. The larval stage lasts 10 to 11 months. The larvae are flattened, about 3 to 4 mm long, and light greenish-brown. They are generally not observed unless they are carefully dissected from the tunnels. Mature larvae exit the tunnels before spinning pupal cocoons.

**Pupal stage:** Cocoons can be found on the underside of leaf petioles or small stems. Cocoons are white, narrow, and 4 to 5 mm long. The anterior portion of the cocoon is curved backward where the silk is attached to the leaf petiole or stem.

**Adult stage:** Adults are present in late spring and summer. They are so small that they are not often seen or recognized. They are mostly white, 3 to 5 mm long, and have golden blotches near the feathery wing tips.

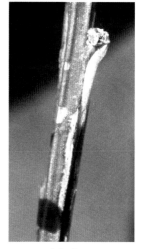

*L. spartifoliella* pupa and damage. (Photo credit: E. Coombs, Oregon Department of Agriculture)

## Effect

**Destructive stage:** Larval.

**Additional plant species attacked:** None.

**Site of attack:** Stems.

**Impact on the host:** The effectiveness of this natural enemy is questionable. Large numbers of larvae may deform plants and cause stem dieback. However, flowering and regrowth often occur below the attacked shoots. Density changes in Scotch broom infestations attributable to this agent have not been documented. The moth also is heavily parasitized in Oregon and Washington.

**Nontarget effects:** There are no known reported cases of effects on nontarget plants.

*L. spartifoliella* adult. (Photo credit: E. Coombs, Oregon Department of Agriculture)

## Releases

**First introduced into the United States:** 1960, California. In addition, it is believed to have been accidentally imported into Washington prior to 1940.

**Established in:** California, Oregon, and Washington.

**Habitat:** This species does well in Scotch broom infestations located below 800 m west of the Cascade Mountains of Oregon and Washington and on the western slopes of the Sierra Nevada of California, but is not abundant near the ocean at elevations over 1,000 m or in very hot and dry sites.

**Availability:** Although this species is available for mass collections, it is already established at most Scotch broom infestations in the Pacific Northwest.

**Stages to transfer:** Because of high parasitism, it is best to release adults reared from caged cocoons to avoid moving parasitoids with the agent. For local sites, bouquets of twigs containing pupae can be collected from existing infestations and placed at new release sites, or rear adults in cages to prevent moving parasitoids.

**Redistribution:** Redistribution of this agent can be made by light-trapping adults, rearing adults from caged cocoons, or redistributing pupae-infested twigs or whole plants in early spring. Adults can be stored for several days in containers with a resting substrate material such as excelsior or tissue paper. They must be provided with a moistened piece of sponge. Larvae and cocoons on twigs collected in late winter and early spring may be kept at ambient temperature out of direct sun until the adult moths are ready to emerge from the pupae.

## Comments

This insect was apparently present in the Pacific Northwest before it was deliberately released into California. It was probably accidentally imported in infested plants brought from Europe as ornamentals to the United States and Canada. Heavy parasitism of the moth has greatly reduced its effectiveness as an agent. It is not known whether this parasitoid entered with the accidentally introduced moths or whether it is a native species that has adapted to a new host. Because of the moth's wide distribution, little work is being done on this insect as of 2003. There is some speculation that larval mining may make the plant prone to attack by opportunistic pathogens. Many Scotch broom plants frequently are found dead or dying in late spring. The primary suspect pathogen is *Selenophoma juncea*.

## References

Andres, L. A., R. B. Hawkes, and A. Rizza. 1967. *Apion* seed weevil introduced for biological control of Scotch broom. Calif. Agric. 21: 13.

Andres, L. A. and E. M. Coombs. 1995. Chapter 79: Scotch broom, *Cytisus scoparius* (L.) Link (Leguminosae). Pages 303-5 *in* J.R. Nechols, L.A. Andres, J.W. Beardsley, R. Goeden, and C.G. Jackson, eds. Biological Control in the U.S. Western Region: Accomplishments and Benefits of Regional Research Project W-84, 1964-1989. Univ. Calif. Div. Agric. Nat. Res. Pub. 3361. Oakland, CA.

Bossard, C. C., J. M. Randall, and M. C. Hoshovsky. 2000. Invasive Plants of California's Wildlands. Univ. California Press, Berkeley, CA.

Frick, K. E. 1964. *Leucoptera spartifoliella*, an introduced enemy of Scotch broom in the western United States. J. Econ. Entomol. 57: 589-91.

Isaacson, D. L., G. A. Miller, and E. M. Coombs. 1995. Use of geographic systems (GIS) distance measures in managed dispersal of *Apion fuscirostre* for control of Scotch broom (*Cytisus scoparius*). Pages 695-99 *in* E. S. Delfosse and R. R. Scott, eds. Proc. VIII Int. Symp. Biol. Contr. Weeds, 2-7 February 1992. Lincoln Univ., Canterbury, New Zealand. DSIR/CSIRO, Melbourne, Australia.

Redmon, S. G., T. G. Forrest, and G. P. Markin. 2000. Biology of *Bruchidius villosus* (Coleoptera: Bruchidae) on Scotch broom in North Carolina. Florida Entomol. 83: 242-53.

Syrett, P., S. V. Fowler, E. M. Coombs, J. R. Hosking, G. P. Markin, Q. E. Paynter, and A. W. Sheppard. 1999. The potential for biological control of Scotch broom (*Cytisus scoparius* (L.) Link) (Fabaceae) and related weedy species. Biocontrol News Info. 20: 17-34.

Waloff, N. 1968. Studies on the insect fauna on Scotch broom *Sarothamnus scoparius* (L.) Wimmer. Advan. Econ. Res. 5: 87-208.

# Eurasian watermilfoil

*Myriophyllum spicatum*

Watermilfoil family—Haloragaceae

## G. R. Buckingham

**Additional common name:** Spike watermilfoil.

**Native range:** Africa and Eurasia.

**Entry into the United States:** Eurasian watermilfoil possibly entered U.S. waters in ships' ballast.

## Biology

**Life duration/habit:** Eurasian watermilfoil is a submersed aquatic perennial rooted in the bottom of bodies of water. Multiple long, cylindrical shoots grow to the water surface from the crown.

**Reproduction:** Reproduction is primarily vegetative from broken stems and rhizomes. The plant undergoes one or two autofragmentation periods after flowering when pieces of stems are released into the water. Seeds are produced in abundance, but their role has not been determined.

**Roots:** The roots are slender and fragile.

**Stems and leaves:** Stems are slender and can grow up to 10 m to the surface. Branching increases near the surface and is profuse at the surface. Stems have large air-filled cavities and often bleed air when broken underwater. Leaves, 2 to 4 cm long, are whorled in groups of four. Each leaf has 5 to 24 (usually at least 12) pairs of opposite, thread-like segments arranged similar to a feather with a terminal segment.

**Flowers:** Flowers are on emersed terminal spikes, 2.5 to 10 cm long, that curl back into the water after pollination. The stem widens below the flower spike, which increases buoyancy to hold the spike aloft. This characteristic helps distinguish *M. spicatum* from the native northern watermilfoil, *M. sibiricum*, of Canada and the

Eurasian watermilfoil. (Photo credit: G. Buckingham, USDA-ARS)

169

Eurasian watermilfoil infestation. (Photo credit: U.S. Army Corps of Engineers)

northern United States. Eurasian watermilfoil flowers are whorled in groups of four to five small, reddish petals. Female flowers are found on the lower portion of the spike and male flowers on the terminal portion. The female flowers mature before most of the male flowers. There is usually one flowering period in the northern United States, late July to September, and two in the southern United States, June to July and September to October.

**Fruits and seeds:** The fruits have four seeds, each 2 to 3 mm in diameter.

## Infestations

**Worst infested states:** Eurasian watermilfoil is distributed throughout eastern and midwestern North America with infestations scattered throughout western North America. The range includes at least 37 states (including Alaska and Florida) and three Canadian provinces and is still expanding.

**Habitat:** Eurasian watermilfoil is found in many types of waterways, especially impoundments and natural lakes, but also is present in spring-fed rivers and in brackish coastal waters. There appears to be no major climatic limit to its spread except perhaps the extremes of USDA Plant Hardiness Zones 1 and 2 in the north and 10 and 11 in the south. Many southern waterbodies, especially shallow ones, are probably too warm, but spring waters are not.

**Impacts:** Eurasian watermilfoil can grow almost as a monoculture and dominate waterways. Heavy mats disrupt water flow, interfere with human activities, shade and compete with native plants and animals, contribute to flooding, and increase siltation.

## Comments

Eurasian watermilfoil has been variously thought to have invaded at the turn of the 19th century or as late as 1942. The latter date was based on an examination of herbarium specimens. A native weevil, *Euhrychiopsis lecontei*, has been commercialized for biological control of this important weed.

# Acentria ephemerella (= Acentria nivea)

G. R. Buckingham

**Common name:** Watermilfoil moth.

**Type of agent:** Insect: Moth (Lepidoptera: Pyralidae).

**Native distribution:** Europe.

**Original source:** Europe.

## Biology

**Generations per year:** Two.

**Overwintering stages:** Larval and pupal.

**Egg stage:** The yellowish, ovate eggs are laid in masses on leaves and stems.

**Larval stage:** Larvae lack the gills found on many other aquatic caterpillars. Larvae bore into the stems and also build loose shelters by tying leaves or stem sections together. The shelters are water-filled and the larvae abandon them when they move to new stems. Larvae feed on leaves and stems, which they girdle, causing them to break. During winter, larvae remain on stems and leaves in cocoon-like, thick-walled, water-filled structures called hibernacula.

**Pupal stage:** Pupae are formed on stems in air-filled cocoons. Pupae die if water replaces air in the stems, as often happens with cocoons collected on short, broken stems. Cocoons are present both during the growing season and in winter.

**Adult stage:** Males are small, grayish-white moths, with a 12 mm wingspan, that somewhat resemble caddisflies. Flightless females with reduced wings are most common, but long-winged females are occasionally produced and join males in dispersal flights. Males fly low across the water and mate with submersed females that extend their abdomens above the surface or with females resting on the surface. Females in captivity die within two to three days. Field longevity is unknown.

171

*A. ephemerella* larva.
(Photo credit: C. Bennett, USDA-ARS)

## Effect

**Destructive stage:** Larval.

**Additional plant species attacked:** Northern watermilfoil (*M. sibiricum*), coontail (*Ceratophyllum demersum*), pondweeds (*Potamogeton* spp.), waterweed (*Elodea canadensis*), and water-star grass (*Zosterella dubia*).

**Site of attack:** Larvae feed both in and on the stems and on leaves.

**Impact on the host:** The concurrent girdling of leaves and stems during feeding and case-making is probably more damaging than feeding. The leaves and stems drop

Adult *A. ephemerella*. (Photo credit: G. Buckingham, USDA-ARS)

off the plant. Heavy larval damage has been associated with decline of Eurasian watermilfoil at sites in Canada and New York.

**Nontarget effects:** Larvae feed on a variety of native plant species, but population damage similar to that on watermilfoil was not observed on other species during field studies in New York. Heavy localized feeding on pondweeds has been reported in Europe.

## Releases

**First introduced into the United States:** Accidental introduction prior to 1927 (first found in Quebec, Canada).

**Established in:** Iowa, Massachusetts, Michigan, Minnesota, New Hampshire, New York, Vermont, and Wisconsin.

**Habitat:** Impoundments, lakes, and rivers. This moth is primarily a species of northern and middle Europe, which suggests a continuing northern distribution in North America. However, caterpillars might be able to live in spring waters of the southeastern United States.

**Availability:** Availability is limited. New York and Vermont have the highest reported populations of this moth.

**Stages to transfer:** Larval or pupal.

**Redistribution:** Infested stems should be collected. Containers with larvae in water must be aerated before or during periods without light if they are densely packed with plants or the larvae will become immobile and soon die.

## Comments

This species, along with the native weevil *Euhrychiopsis lecontei*, is associated with plant declines. It appears to be difficult to separate the effects of the two species; researchers disagree on the relative importance of each. If flying adults are collected for redistribution, the sex should be checked to confirm that females are present because most females are flightless.

## References

Aiken, S. G., P. R. Newroth, and I. Wile. 1979. The biology of Canadian weeds. 34. *Myriophyllum spicatum* L. Can. J. Plant Sci. 59: 201-15.

Batra, S. W. T. 1977. Bionomics of the aquatic moth, *Acentropus niveus* (Oliver), a potential biological control agent for Eurasian watermilfoil and Hydrilla. J. NY Entomol. Soc. 85: 143-52.

Buckingham, G. R. 1994. Biological control of aquatic weeds, Chapter 22. Pages 413-79 *in* D. Rosen, F. D. Bennett, and J. L. Capinera, eds. Pest Management in the Subtropics: Biological Control—the Florida Experience. Intercept Ltd., Hampshire, UK.

Buckingham, G. R., and B. M. Ross. 1981. Notes on the biology and host specificity of *Acentria nivea* (=*Acentropus niveus*). J. Aquat. Plant Manage. 19: 32-36.

Couch, R., and E. Nelson. 1986. *Myriophyllum spicatum* in North America. Pages 8-18 *in* L. W. J. Anderson, ed. Proc. First Int. Symp. on Watermilfoil *(Myriophyllum spicatum)* and Related Haloragaceae Species, July 23 and 24, 1985, Vancouver, BC, Canada. Aquatic Plant Manage. Soc., Vicksburg, MS.

Creed, R. P., Jr., and S. P. Sheldon. 1994. The effect of two herbivorous insect larvae on Eurasian watermilfoil. J. Aquat. Plant Manage. 32: 21-26.

Godfrey, R. K., and J. W. Wooten. 1981. Aquatic and Wetland Plants of the Southeastern United States: Dicotyledons. Univ. Georgia Press, Athens, GA.

Gross, E. M., R. L. Johnson, and N. G. Hariston, Jr. 2001. Experimental evidence for changes in submersed macrophyte species composition caused by the herbivore *Acentria ephemerella* (Lepidoptera). Oecologia 127: 105-14.

Johnson, R. L., and B. Blossey. 2002. Eurasian watermilfoil. Pages 79-90 *in* R. Van Driesche, S. Lyon, B. Blossey, M. Hoddle, and R. Reardon, eds. Biological Control of Invasive Plants in the Eastern United States. USDA Forest Service. Publ. FHTET-2002-04. Morgantown, WV.

Johnson, R. L., E. M. Gross, and N. G. Hairston, Jr. 1998. Decline of the invasive submersed macrophyte *Myriophyllum spicatum* (Haloragaceae) associated with herbivory by larvae of *Acentria ephemerella* (Lepidoptera). Aquat. Ecol. 31: 273-82.

Johnson, R. L., P. J. Van Dusen, J. A. Toner, and N. G. Hairston, Jr. 2000. Eurasian watermilfoil biomass associated with insect herbivores in New York. J. Aquat. Plant Manage. 38: 82-88.

Newman, R. M., and D. D Biesboer. 2000. A decline of Eurasian watermilfoil in Minnesota associated with the milfoil weevil, *Euhrychiopsis lecontei*. J. Aquat. Plant Manage. 38: 105-11.

Newman, R. M., D. W. Ragsdale, A. Milles, and C. O'brien. 2001. Overwinter habitat and the relationship of overwinter to in-lake densities of the milfoil weevil, *Euhrychiopsis lecontei*, a Eurasian watermilfoil biological control agent. J. Aquat. Plant Manage. 39: 63-67.

Painter, D. S., and K. J. McCabe. 1988. Investigation into the disappearance of Eurasian watermilfoil from the Kawartha Lakes. J. Aquat. Plant Manage. 26: 3-12.

Scholtens, B. G., and G. J. Balogh. 1996. Spread of *Acentria nivea* (Lepidoptera: Pyralidae) in central North America. Great Lakes Entomol. 29: 21-24.

# Giant salvinia

*Salvinia molesta*

Salvinia family—Salviniaceae

## P. W. Tipping

**Additional common names:** African pyle, aquarium watermoss, Kariba weed, koi kandy, Australian azolla, water spangles, water fern, floating fern.

**Native range:** Southeastern Brazil.

**Entry into the United States:** First reported outside of cultivation in 1995 at a pond in southeastern South Carolina. It was eradicated before it spread. In 1998, a new infestation was found in Texas.

### Biology

**Life duration/habit:** Giant salvinia is a free-floating, perennial aquatic fern adapted to stagnant or slightly flowing freshwater systems in tropical to temperate climates.

**Reproduction:** Incapable of sexual reproduction. Plants spread through clonal growth and fragmentation.

**Roots:** The plant does not have true roots, but a highly modified leaf serves many of the same functions.

**Stems and leaves:** Different growth forms of this plant exhibit oval leaves ranging from 15 to 60 mm wide depending on the growth stage. Leaves are connected with a horizontal rhizome and at each internode there is a pair of above-water leaves, a submerged "root," and associated buds, forming a ramet. Leaf hairs on the lower leaf surface are joined together at the tip, forming a diagnostic "eggbeater" shape.

**Flowers:** This plant has no flowers but does produce sporocarps within the "root" area.

**Fruits and seeds:** Giant salvinia is a pentaploid, and has a chromosome number of 45. Spores are formed, but they are deformed and nonviable.

### Infestations

**Worst infested states:** Louisiana and Texas, with lesser infestations in Alabama, Arizona, California, Florida, Georgia, Hawaii, Mississippi, and North Carolina.

Giant salvinia. (Photo credit: R. Helton, USDA-ARS)

**Habitat:** The plant favors stagnant to slow-moving freshwater sites including ponds, lakes, marshes, and canals. Plants are limited to areas where ice does not form regularly. Temperatures below -3° C will kill the plant.

**Impacts:** Problems caused by this plant and the thick mats it forms include disrupting or preventing recreational activities such as boating and fishing; blocking drains, spillways, and intakes for irrigation and electrical power generation; providing harborages for pathogen-carrying insects such as mosquitoes; crowding out native aquatic plants; and reducing the oxygen content of the water, resulting in degraded fisheries.

## Comments

This is a relatively new weed to the United States despite its use in the aquarium and aquatic plant trade for many years, practices that continue today. Although it can be spread by animals like turtles and alligators over short distances, human activities are responsible for its rapid spread throughout the southeastern and southwestern United States, particularly by boat trailers that transport viable plant fragments to distant bodies of water.

## *Cyrtobagous salviniae*

P. W. Tipping

**Common name:** Salvinia weevil.

**Type of agent:** Insect: Weevil (Coleoptera: Curculionidae).

**Native distribution:** South America.

**Original source:** Brazil.

## Biology

**Generations per year:** The Florida population (see below) appears to have one generation on common salvinia (*Salvinia minima*).

175

**Overwintering stage:** Unknown, but probably the adult in more temperate climates. In tropical areas, dry and rainy conditions may regulate population growth.

**Egg stage:** Eggs measure 0.5 by 0.24 mm and are laid singly in cavities excavated by adults in lower leaves, developing leaves, rhizomes, and "roots." Hatching occurs in about 10 days at 25.5° C.

**Larval stage:** Larvae are white, 4.0 mm long, gently curved and tapered with setae modified as hooks or trailing hairs. Larvae complete three instars in about 23 days.

**Pupal stage:** Pupation occurs in a 2.0-by-2.6-mm-long cocoon that is woven from "root" hairs and attached underwater to the "roots," rhizomes, or leaf bases.

**Adult stage:** Newly emerged adults are light brown and turn darker after five to seven days. Adults range in length from 1.8 to 2.2 mm, the males being slightly smaller than the females. Adults can be long-lived, with female longevity exceeding 38 weeks, laying more than 300 eggs. Adults prefer to feed on new buds but will also attack leaves and rhizomes.

### Effects

**Destructive stage:** Larval and to lesser extent adult.

**Additional plant species attacked:** Common salvinia (*S. minima*) is readily attacked by both populations (see below).

**Site of attack:** Larvae feed externally on buds before burrowing into rhizomes.

**Impact on the host:** Attacked plants turn brown and become more brittle, leading to a waterlogged appearance, and eventually sink.

**Nontarget effects:** This insect feeds and reproduces only within the Salviniaceae. Host range tests found survival only on New World species of salvinia. Both populations of *C. salviniae* have never been found in the field on species other than giant and common salvinia.

### Releases

**First introduced into the United States:** The first release of the Brazilian population was in 2001 in Louisiana and Texas. The Florida population was first detected in 1960; it was probably accidentally introduced.

**Established in:** The Brazilian population is considered established at selected locations in Texas and Louisiana. The Florida population is well established throughout Florida where it feeds on common salvinia. It is currently being redistributed to Louisiana.

**Habitat:** The weevil tends to perform better on plants with higher nitrogen levels. The climatic limits of this insect are unknown, but are probably similar to the plant.

*C. salviniae* on a giant salvinia leaf. (Photo credit: S. Bauer, USDA-ARS)

**Availability:** Release of the Brazilian population is limited to certain counties in east Texas and western Louisiana, while the Florida population is permitted in all areas of Texas and Louisiana. Individuals may apply for permits for other states.

**Stages to transfer:** Adult individually; adult and larval in infested plant material.

**Redistribution:** Adults are collected from salvinia using Berlese funnels or by submerging infested plant material and collecting them as they rise to the surface. Adults should be held on ramets of salvinia in ventilated containers and then released directly or into infested plant material onto uninfested mats of salvinia.

## Comments

Morphologically, there are no differences between the Brazil and Florida populations of the salvinia weevil except for the slightly smaller size of the Florida population adults. Recent molecular work has found the two populations group close together when compared to *C. singularis*, a separate species that also feeds on the Salviniaceae in South America. Studies are underway to determine whether the two populations can interbreed. The Brazil population has survived two winters in Texas and Louisiana and is considered established. In July 2003, a site in Louisiana was completely cleared of giant salvinia by the weevil. Giant salvinia continues to decline at another site in Texas as weevil numbers increase. In tank studies, adults from the Florida population were more effective than the Brazil population against both common and giant salvinia. Currently the Brazil population is being released on giant salvinia and the Florida population is used against common salvinia.

## *References*

Forno, I. W., D. P. A. Sands, and W. Sexton. 1983. Distribution, biology and host specificity of *Cyrtobagous singularis* Hustache (Coleoptera: Curculionidae) for the biological control of *Salvinia molesta*. Bull. Entomol. Res. 73: 85-95.

Harley, K. L. S., and D. S. Mitchell. 1981. The biology of Australian weeds. 6. *Salvinia molesta* D. S. Mitchell. J. Aust. Inst. Agric. Sci. 47: 67-76.

Julien, M. H., T. D. Center, and P. W. Tipping. 2002. Floating fern (Salvinia). Pages 17-32 *in* R. Van Driesche, S. Lyon, B. Blossey, M. Hoddle, and R. Reardon, eds. Biological Control of Invasive Plants in the Eastern United States. USDA Forest Serv. Pub. FHTET-2002-04. Morgantown, WV.

May, B. M., and D. P. A. Sands. 1986. Descriptions of larvae and biology of *Cyrtobagous* (Coleoptera: Curculionidae): agents for biological control of *Salvinia*. Proc. Entomol. Soc. Wash. 88: 303-12.

Mitchell, D. S., P. Tomislav, and A. B. Viner. 1980. The water-fern *Salvinia molesta* in the Sepik River, Papua New Guinea. Environ. Conserv. 7: 115-22.

Room, P. M. 1990. Ecology of a simple plant-herbivore system: biological control of *Salvinia*. TREE 5: 74-79.

Tipping, P. W., and T. D. Center. 2003. *Cyrtobagous salviniae* (Coleoptera: Curculionidae) successfully overwinters in Texas and Louisiana. Florida Entomol. 86: 92-103.

**177**

# Gorse

*Ulex europaeus*

## Pea family—Fabaceae (Leguminosae)

*E. M. Coombs, G. P. Markin, P. D. Pratt, and B. Rice*

**Additional common names:** Prickly or thorn broom, furze, common gorse.

**Native range:** Western Europe.

**Entry into the United States:** Gorse was introduced from Scotland in the late 1800s as an ornamental or hedge plant to contain livestock, but soon escaped from cultivation.

## Biology

**Life duration/habit:** Gorse is a branched, spiny, evergreen perennial shrub that grows 1 to 5 m tall.

**Reproduction:** Seeds.

**Roots:** Extensive woody, multi-branched root system contains nitrogen-fixing root nodules.

**Stems and leaves:** The plants are leafless except as seedlings when they have the typical trifoliate leaves of many legumes. Older plants are a solid mass of dense, sharp spines 4.5 to 6.5 cm long. Photosynthesis occurs mainly in the epidermis of the stems and spines.

**Flowers:** Gorse has pea-like bright yellow flowers that are borne on second-year twigs. Flowering occurs in March and April, or May to July (depending upon location), with a minor secondary bloom in late fall.

**Fruits and seeds:** Gorse produces three to eight seeds in small, hairy pods that are 15 to 18 mm long. Mature pods become black and, upon drying, eject the seeds several meters from the plant. Seeds remain viable in the soil for up to 20 years.

Gorse infestation. (Photo credit: California Department of Food and Agriculture)

## Infestations

**Worst infested states:** Western California, Oregon, and Washington. Infestations have also been reported along the east coast of the United States from Virginia to Massachusetts, and on two islands of Hawaii.

**Habitat:** Gorse has become a serious pest of coastal habitats, pastures, logged areas, utility rights-of-way, and around municipalities. Gorse does not grow well in areas with severe winters. On the U.S. West Coast, it is primarily restricted to counties bordering the Pacific Ocean.

**Impacts:** Gorse is an invasive pioneer species that rapidly excludes desirable vegetation. It has replaced much of the vegetation in salt spray meadows along the Pacific Coast, where many sensitive and threatened plants grow. It is impossible to walk through spiny gorse stands at several state parks along the ocean. At inland sites, gorse quickly invades meadows and pastures, preventing livestock and wildlife access. In logged areas, gorse can crowd out conifer seedlings, greatly affecting reforestation efforts. Gorse is difficult to control chemically and often infests sensitive habitats or rugged terrain where herbicide applications and other treatments are cost-prohibitive.

## Comments

Gorse is highly flammable and has an oil content of 2 to 4%. In 1936, a wildfire fueled by gorse burned down the town of Bandon, Oregon, killing 13 people.

*Agonopterix nervosa*, an accidentally introduced moth that feeds on Scotch broom (*Cytisus scoparius*) flowers, acts as a twig/leaf-tier on gorse, in some areas damaging more than 50% of growing twigs, causing short-term stunting and making plants appear more full and bushy. Several other species of biocontrol agents have been released against gorse in Hawaii and New Zealand: the shoot moth *Pempelia genistella*, the shoot tip moth *A. ulicitella*, and a thrips, *Sericothrips staphylinus*. Host specificity studies on these organisms have been partially completed (as of 2003) to determine whether they are safe to release on the U.S. mainland.

# *Exapion ulicis (= Apion ulicis)*

E. M. Coombs and G. P. Markin

**Common name:** Gorse seed weevil.

**Type of agent:** Insect: Beetle, weevil (Coleoptera: Bretidae [= Apionidae]).

**Native distribution:** Western Europe.

**Original source:** Southern France.

## Biology

**Generations per year:** One.

**Overwintering stage:** Adult.

**Egg stage:** During the spring, females insert three to five eggs into growing seedpods. The round eggs are 0.2 to 0.3 mm in diameter and a light, clear-yellowish color.

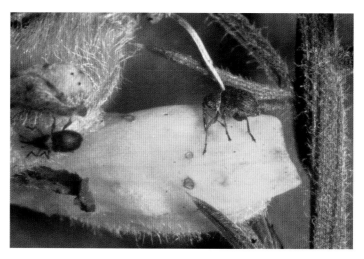

*E. ulicis* adult. (Photo credit: E. Coombs, Oregon Department of Agriculture)

**Larval stage:** The larvae are found in the field from April to June. They feed on the growing seeds (usually one per seed) inside the pods and complete their development in six to eight weeks. The larvae are 2 to 3 mm long, comma shaped and cream colored.

**Pupal stage:** Pupation occurs within the seedpods from May to July. Pupae are 3 mm long and at first are a cream to light gray color but slowly darken.

**Adult stage:** Adults can be found all year long. Overwintering adults generally die after the spring flowering period. New adults emerge from June to August. Adults are 2 to 3 mm long, light gray with long, curved snouts that are about half the body length. Adults use their snouts to probe into mature green gorse spines and stems to feed on tissue under the cuticle. Their feeding causes round, brownish scars 1 to 2 mm in diameter.

## Effect

**Destructive stages:** Larval (on seeds), and to a lesser extent adult (on spines).

**Additional plant species attacked:** None.

**Site of attack:** Seeds are destroyed by larvae, while the green photosynthetic tissue in the spines and stem is destroyed by the adults.

**Impact on the host:** From 30 to 95% of the seedpods are attacked in California, Oregon, and Washington. The weevil may retard the spread of the plant, but does not reduce established stand density.

**Nontarget effects:** There are no known nontarget effects caused by this weevil.

## Releases

**First introduced into the United States:** 1953, California.

**Established in:** California, Hawaii, Oregon, and Washington, wherever gorse is found. It has not been reported on the U.S. East Coast.

**Habitat:** Open, sunny pastures and hillsides are favored while scattered plants, shaded plants, and plants within the salt spray zone along coastal areas are not. Areas with hard winter freezes limit the weevil and the host plant.

**Availability:** The weevil is easily collected in large numbers in the spring.

**Stage to transfer:** Adult.

**Redistribution:** Adults can be collected in early spring by holding a beating sheet under a bush and hitting the branches with a stick. Adults can easily be stored for several weeks if kept cool and dry.

## Comments

The seed weevil occurs at nearly all gorse infestations, so redistribution is unnecessary except where large areas of the gorse (and the weevil) have been destroyed by fire. Gorse will readily regrow from root crowns and seeds, but the weevil may take several years to recover.

The weevils may reduce gorse invasiveness. Gorse seeds missed by the weevils can survive up to 30 years in the soil. Adults generally do not cause much damage to plants.

The weevil has also been known as *Apion ulicis*. It has been recently placed in the family Brentidae.

## *Tetranychus lintearius*

E. M. Coombs, P. D. Pratt, G. P. Markin, and B. Rice

**Common name:** Gorse spider mite.

**Type of agent:** Mite: Spider mite (Acari: Tetranychidae).

**Native distribution:** Western Europe.

**Original sources:** England, Spain, and Portugal (via New Zealand).

### Biology

**Generations per year:** Four to six.

**Overwintering stages:** Adult, nymphal, or egg.

**Egg stage:** Females can produce one to four eggs per day all year long. Eggs are laid on shoots infested by the mite colony. The colony moves on, leaving eggs and mature males behind. Eggs are very small, round, and clear white.

**Nymphal stage:** The immatures complete four growth stages. First-stage mites (larvae) have four legs, often congregate at shoot tips, and are dispersed by the wind. The mites in the second to fourth growth stage (nymphs) resemble the adult mites but are smaller, brownish, and have eight legs.

**Adult stage:** Adult mites live in large colonies on terminal branches and produce large quantities of webbing. Their life span is between 17 and 28 days. Adults resemble other spider mite species and are 0.5 mm long and brick red.

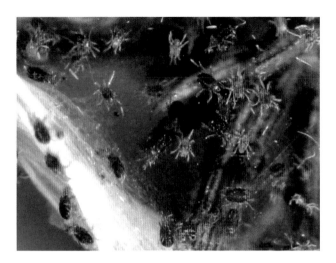

*T. lintearius* colony. (Photo credit: E. Coombs, Oregon Department of Agriculture)

## Effect

**Destructive stages:** Adult and nymphal.

**Additional plant species attacked:** None.

**Site of attack:** Stems and spines.

**Impact on the host:** Mites pierce and extract cell contents from spines and stem tissue. Heavy damage on stems reduces flowering the following year. Some stunting of branches may also occur.

**Nontarget effects:** No nontarget effects have been reported.

## Releases

**First introduced into the United States:** 1994, California and Oregon; 1995, Hawaii and Washington.

**Established in:** California, Hawaii, Oregon, and Washington.

**Habitat:** Favored areas are open gorse patches, away from the ocean; damp areas close to the ocean or heavily shaded gorse plants in forested areas are not attacked. The mites are more productive at warmer temperatures. The mite is somewhat cold-hardy but is susceptible to severe winter weather.

**Availability:** Available in California, Oregon, and Washington.

**Stages to transfer:** Nymphal and adult.

**Redistribution:** Inoculate uninfested plants with adult colonies on branches clipped from infested plants. The mites can be stored for several months if kept cool and dry on living host material.

## Comments

*Tetranychus lintearius* is the first tetranychid mite to be used for biological control of a weed in the United States. Mites from several different climatic zones in Europe were introduced into similar zones in the western United States. The mass movement of large colonies of adults is unique. Extreme webbing can result in spectacular scenes.

*T. linearius* webbing and colonies. (Photo credit: E. Coombs, Oregon Department of Agriculture)

The host specificity and reproductive isolation of the gorse spider mite has been thoroughly tested and found to be safe. Several members of the genus *Tetranychus* are serious crop pests. In New Zealand tests, none of the pest species were able to hybridize and produce fertile eggs with the gorse spider mite. All the offspring that were produced were males, which are the result of unfertilized eggs.

Severe predation of the gorse spider mite has occurred in the Bandon, Oregon, area. The predatory mite *Phytoseiulus persimilis* reduced extensive gorse mite colonies in one season by more than 95%. Do not transport gorse spider mites from areas where *P. persimilis* occurs (south of Bandon).

At many sites without the predatory mite, a predatory coccinellid (*Stethorus punctillum*)—a small, round, shiny ladybird beetle—has also caused severe declines in gorse spider mite populations. As a result of so much predation, the mite has been rendered a more or less ineffective biological control agent in most areas.

## References

Davies, W. M. 1928. The bionomics of *Apion ulicis* Forst. (gorse weevil), with special reference to its role in the control of *Ulex europaeus* in New Zealand. Ann. Appl. Biol. 15: 263-86.

Hill, R .L., and D. J. O'Donnell. 1991. Reproductive isolation between *Tetranychus lintearius* and two related mites, *T. urticae* and *T. turkenstani* (Acarina: Tetranychidae). Exp. Appl. Acarol. 11: 241-51.

Hill, R. L., and D. J. O'Donnell. 1991. The host range of *Tetranychus lintearius* (Acarina: Tetranychidae). Exp. Appl. Acarol. 11: 253-69.

Markin, G. P., E. R. Yoshioka, and R. E. Brown. 1995. 78. Gorse, *Ulex europaeus* L. (Fabaceae). Pages 299-302 *in* J. R. Nechols, L. A. Andres, J. W. Beardsley, R. D. Goeden, and C. G. Jackson, eds. Biological Control in the Western United States: Accomplishments and Benefits of Regional Research Project W-84, 1964-1989. Univ. Calif. Div. Agric. Nat. Res. Pub. 3361. Oakland, CA.

Pratt, P. D., E. M. Coombs, and B. A. Croft. 2003. Predation by phytoseiid mites on *Tetranychus lintearius* (Acari: Tetranychidae), an established weed biological control agent of gorse (*Ulex europaeus*). Biol. Control 26: 40-47.

Stone, C. 1986. An investigation into the morphology and biology of *Tetranychus lintearius* Dufour (Acari: Tetranychidae). Exp. Appl. Acarol. 2: 173-86.

# Hydrilla

*Hydrilla verticillata*

Frog's-bit family—Hydrocharitaceae

## G. R. Buckingham and M. J. Grodowitz

**Additional common names:** Florida elodea (early name before it was identified), waterthyme.

**Native range:** Africa, Asia, Australia, and relict populations in Europe.

**Entry into the United States:** Hydrilla was probably introduced via the aquarium trade prior to 1960.

## Biology

**Life duration/habit:** Hydrilla is a submersed aquatic perennial with long cylindrical stems and a small root system. Multiple stems grow from the crown up to 8.5 m to the water surface (15 m in a clear spring). The stems branch more as they near the surface so that most of the biomass is close to the surface even in deep water.

**Reproduction:** Reproduction is mostly vegetative although seeds are present in some regions. Dioecious female plants are found in the southeastern United States and in California. No dioecious male plants have been reported. Monoecious plants (both sexes on the same plant) with male flowers on the stems below the female flowers have been reported in several northern states and California. Hydrilla can grow a new shoot from a fragment of only one whorl of leaves. Turions, or swollen leaf buds, are formed both in the leaf axils of stems and on the tips of rhizomes. The subterranean turions on the rhizomes are often referred to as tubers. Both types of turions break off and form new shoots. Stolons, or horizontal stems that grow on the surface of the substrate, can also form new plants.

**Roots:** The root system is usually very small. Thread-like adventitious roots grow from multiple white rhizomes and green stolons. Interestingly, plants growing in Lake Tanganyika, Africa, along a shore with turbulent wave action had large masses of yellow-orange, leathery roots.

Hydrilla in Florida. (Photo credit: G. Buckingham, USDA-ARS)

Hydrilla infestation in Florida. (Photo credit: C. Bennett, USDA-ARS)

**Stems and leaves:** The horizontal rhizomes and stolons have small, opposite, ovate leaves. Ascending stems are slender, usually less than 4 mm wide, with whorls of 2 to 10 leaves. The narrow, elongate leaves are mostly less than 15 mm long and 6 mm wide. Margins of the leaves are serrate and the lower surface of the midrib has small teeth. These serrations and teeth can be quite noticeable; however teeth are difficult to detect on plants growing in shade or acidic water.

**Flowers:** Flowers are unisexual with only female flowers in the United States on the dioecious (or single sex) strain. The 4- to 8-mm-wide white female flowers are formed near the tips of stems and float on the water surface. The floral tube is 4 to 5 cm long and arises from a leaf axil. The male flower, enclosed in a spherical spathe, is formed in the leaf axil near the tip of the stem of the monoecious strain. The flower is stalkless, but it breaks loose and floats to the surface where it explosively opens, shooting pollen into the air. The pollen rains on the nearby female flowers, thereby fertilizing them.

**Fruits and seeds:** Fruits, which are attached in the leaf axils near the stem tips, are elongate, 5 to 15 mm long, widest at the base and narrow at the top. They can be smooth or have long lateral processes. There are two to six oblong-ellipsoidal seeds, each 2.5 mm long.

## Infestations

**Worst infested states:** Alabama, Florida, Georgia, Louisiana, North Carolina, South Carolina, and Texas.

**Habitat:** Hydrilla invades all types of water bodies, from the smallest shallow ditch to deep springs. It may also be found in acidic, highly calcareous, or brackish waters. The extensive native distribution of hydrilla, from the Asian tropics to Siberia, suggests that climate should not prevent it from invading the entire United States and at least southern Canada.

**Impacts:** Hydrilla is currently the major aquatic weed in the southeastern United States. It disrupts drainage in canals causing flooding, impedes the flow of irrigation water, interferes with navigation and recreation, and impacts fish and native plant

185

populations. Control costs on public waters in Florida alone in FY99 were about $12.7 million.

## Comments

Hydrilla, mostly the monoecious strain, is beginning to invade the northern United States. Connecticut, Delaware, Indiana, Maryland, Massachusetts, Virginia, and Washington have infestations.

# Bagous affinis

G. R. Buckingham and M. J. Grodowitz

**Common name:** Indian hydrilla tuber weevil.

**Type of agent:** Insect: Beetle, weevil (Coleoptera: Curculionidae).

**Native distribution:** Pakistan to Thailand.

**Original source:** Karnataka, India.

## Biology

**Generations per year:** Multiple, probably two to three.

**Overwintering stage:** Adult.

**Egg stage:** In the laboratory, the eggs were laid out of water and singly into moist, rotten wood, drying hydrilla stems, rhizomes, and sand. No eggs were placed into tubers. Rotten wood received the overwhelming majority of eggs, which were often packed together in the smaller pieces. The whitish eggs were ovate, 0.5 mm long and 0.35 mm in diameter. Development took three to five days.

**Larval stage:** All three instars are whitish with tan heads. Mature larvae are about 5 mm long. Hatching larvae crawl through the exposed soil to the tubers, which they bore into to feed. One or two larvae can complete development inside a tuber, but larger numbers destroy the tuber and then enter new ones. Larvae cannot survive submersion.

**Pupal stage:** White pupae are formed inside the tuber unless it is destroyed and they are forced to pupate in the soil. They turn yellow as they age. Pupae average 3.75 mm long and develop in four to six days. Pupae in tubers can survive up to five days of submersion.

**Adult stage:** Adults feed on hydrilla stems and leaves exposed by receding waters. Adults are small,

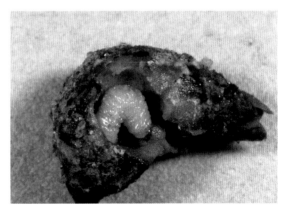

*B. affinis* larva in a hydrilla tuber. (Photo credit: G. Buckingham, USDA-ARS)

3- to 4-mm-long, mottled brown weevils. The wing muscles of ovipositing females are undeveloped, but migrating beetles are collected at blacklights. Females lay an average of 232 eggs and live an average of 117 days. Adults are unable to swim, but can survive three days of submersion. Total development takes only 22 days, but females have about a one-month average preoviposition period.

*B. affinis* adults on a hydrilla tuber. (Photo credit: G. Buckingham, USDA-ARS)

## Effect

**Destructive stages:** Larval and, to a lesser extent, adult.

**Additional plant species attacked:** None.

**Site of attack:** Larvae feed inside tubers (the subterranean turions) out of water. Adults are destructive to stems of stranded plants, which eventually dry.

**Impact on the host:** Heavily eaten tubers do not produce shoots. In Pakistan, hydrilla was reduced up to 99% after heavy weevil infestations of tubers.

**Nontarget effects:** No nontarget effects have been reported, but the weevil is not known to be established in the United States.

## Releases

**First introduced into the United States:** 1987, Florida.

**Established in:** The weevil was recovered in California and Florida but permanent establishment has not been reported.

**Habitat:** This weevil needs dry or drying water bodies. Ideal sites would be areas with gently sloping banks that are slowly exposed as the water recedes during dry periods. Reservoir drawdowns would be ideal if conducted during the warm season and for two or more months to allow preoviposition and immature development. The weevil was found in India at sites with both light and heavy clay soils. The climatic limits in the United States are unknown, but USDA Plant Hardiness Zones 9 and 10 are closest to the temperatures of the native range. The native range, however, has no summer rainfall.

**Availability:** This weevil is not available in the United States.

**Stage to transfer:** Adult.

**Redistribution:** Adults can be collected from hydrilla stranded by receding water or at a blacklight set up along the drying water body. Adults are easily shipped in cardboard containers (preferably unwaxed mailing tubes) containing moist wood excelsior and small amounts of hydrilla stems or tubers.

## Comments

This species is adapted to the monsoon climate of the Indian subcontinent with predictable annual dry and wet seasons. It must have periods of drought or artificial drawdowns to develop. No permanent establishment has been reported, but Florida release sites have not been surveyed during dry periods for several years after the initial releases.

# Bagous hydrillae

G. R. Buckingham and M. J. Grodowitz

**Common name:** Australian hydrilla stem-boring weevil.

**Type of agent:** Insect: Beetle, weevil (Coleoptera: Curculionidae).

**Native distribution:** Australia.

**Original source:** Queensland, Australia.

## Biology

**Generations per year:** Multiple, possibly four to five.

**Overwintering stage:** Adult.

**Egg stage:** Eggs are laid singly in stems, usually near a node. They are whitish, ovate, 0.52 mm long, and hatch in two to three days.

**Larval stage:** Larvae bore into the stem, which turns black where damaged. Adult and larval feeding cause stems to break and fragments to float to shore where they become stranded. The mature larva exits and pupates among the stranded stems or in the soil. Undoubtedly, some larvae remain in unbroken stems until the water body dries and the hydrilla is stranded in place. Larvae do not pupate in submersed hydrilla. Larvae develop in about eight days and are prepupae for 1.5 to 3.5 days. There are three instars.

**Pupal stage:** The whitish pupae form in stranded hydrilla or in the soil. They develop in three to five days.

**Adult stage:** Adults are small (about 3 mm long), elongate, brown weevils. They emerge on shore, fly to submersed hydrilla that has reached the water's surface, the wing

Damage to a submersed hydrilla stem caused by *B. hydrillae.* (Photo credit: G. Buckingham, USDA-ARS)

muscles atrophy, and they begin feeding and oviposition. They cannot swim, but they are at home in the water and crawl on the stems. Females lay an average of almost 300 eggs during their lifetime. Adults can live for more than three months in the laboratory, although males average only 33 days and females 39 days. Development, from egg to adult, takes 12 to 14 days.

*B. hydrillae* adult. (Photo credit: G. Buckingham, USDA-ARS)

## Effect

**Destructive stages:** Larval and adult.

**Additional plant species attacked:** None.

**Site of attack:** Adults feed externally on leaves and stems and larvae feed within stems.

**Impact on the host:** During heavy weevil infestations in Australia, hydrilla stems near the surface appeared to have been mowed. The cut stems wash up on shore, sometimes in windrows, where they are killed by the larvae or by drying.

**Nontarget effects:** No nontarget effects have been reported, but this species is not known to be established in the United States.

## Releases

**First introduced into the United States:** 1991, Florida.

**Established in:** Tentative establishment was reported in Florida and in Texas in 1996, but permanent establishment has not been confirmed.

**Habitat:** Habitats with periods of drying are ideal. Water bodies with sloping, unvegetated shorelines allow floating hydrilla to accumulate in windrows where the larvae can pupate. If this species eventually establishes, it will be in the southern climates, probably USDA Plant Hardiness Zone 8 or warmer.

**Availability:** This weevil is generally not available in the United States. (Available in Australia.)

**Stage to transfer:** Adult.

**Redistribution:** Adults can be collected from hydrilla stranded along shore or at a blacklight set up along the water body. Adults are easily shipped in cardboard containers (preferably unwaxed mailing tubes) containing moist wood excelsior and small amounts of hydrilla stems.

*B. hydrillae* larva in a stem. (Photo credit: G. Buckingham, USDA-ARS)

## Comments

Great effort was made to establish *B. hydrillae*, which was released by the thousands at many sites. Unfortunately, most habitats in the current range of hydrilla in the United States are not suitable for this species because periods of water body drying or the formation of hydrilla windrows along shore are necessary for development.

## *Hydrellia balciunasi*

G. R. Buckingham and M. J. Grodowitz

**Common name:** Australian hydrilla leaf-mining fly.

**Type of agent:** Insect: Fly (Diptera: Ephydridae).

**Native distribution:** Australia.

**Original source:** Queensland, Australia.

### Biology

**Generations per year:** Multiple, possibly five to six.

**Overwintering stages:** Larval and possibly pupal.

**Egg stage:** Females deposit eggs singly or in small groups on leaves and stems at or above the water surface. Eggs cannot survive long periods of submersion. The eggs are shiny white, elongate, 0.45 mm long and 0.14 mm in diameter, with longitudinal ridges. The eggs hatch in three days.

**Larval stage:** Neonates wander a little before entering a leaf, usually on the upper surface. Almost all of the leaf is mined before the larva exits and enters a neighboring leaf. Each mines a total of four to nine leaves. Three instars develop in an average of 12 days. Larvae are green to yellow or white with two noticeable anal spines that apparently puncture plant tissue to obtain air. The mature larva is about 3.6 mm long.

**Pupal stage:** The mature larva anchors itself in the leaf axil by inserting the anal spines into the stem. The larval skin then hardens to form the puparium and the pupa is

*H. balciunasi* larva in a hydrilla leaf. (Photo credit: G. Buckingham, USDA-ARS)

formed inside. The puparium fills with air from the stem and remains attached to the stem even if the damaged leaf disintegrates. The new puparium is a semi-transparent green, but it turns brown as it ages. It closely resembles a leaf bud. Total time in the puparium is seven to 10 days, but the actual pupal stage is probably five to eight days.

**Adult stage:** The adult emerges from the puparium and floats to the surface in a bubble of air. The small, 1.5-mm-long, dark gray flies have shiny golden, sometimes bronze, faces. Females oviposit the same day they emerge and live about 20 days. They lay an average of 35 eggs with a maximum of 83. Adults often sit on broad-leaved plants, for example waterlilies, where they mate and interact. They often cluster on dead brethren and other insects as if feeding, but feeding has not been proven.

*H. balciunasi* adult on hydrilla. (Photo credit: G. Buckingham, USDA-ARS)

However, they ate a high-protein diet during laboratory rearing.

## Effect

**Destructive stage:** Larval.

**Additional plant species attacked:** None.

**Site of attack:** Larvae mine the leaves.

**Impact on the host:** Mined leaves decay and may remain attached to the stems for long periods. With heavy infestations, large areas of the stems appear brown due to the high numbers of decaying leaves. After a period of time, the damaged leaves may fall off the stems. In the laboratory, some heavily damaged stems produced new leaves but others died. No major impact has been observed in the field since population numbers have remained low and range expansion has been limited.

**Nontarget effects:** No nontarget effects have been reported.

## Releases

**First introduced into the United States:** 1989, Florida.

**Established in:** Texas.

**Habitat:** This fly does best in water bodies where hydrilla forms surface mats. The climatic limits have not been established, but based upon the native range in Australia, this fly will probably do best in USDA Plant Hardiness Zone 8 or warmer.

**Availability:** Availability limited to a few Texas sites.

**Stage to transfer:** Larval.

**Redistribution:** Infested stems can be collected and carried wrapped in moist paper towels to the release site or the adults can be allowed to emerge and oviposit. Egg-laden stems should be held until larvae reach the second or third instar to ensure maximum survival and then transported in enclosed containers. If stems are held in water, they should not be kept in the dark longer than overnight, especially if the containers are heavily packed with plants, since the plants might not produce enough oxygen for the larvae. Infested stems can be wrapped in moistened paper towels, placed in plastic bags, and shipped in insulated containers. Shipping of pupae should be avoided, if possible. For short-distance transfers, adults can be collected with mechanical aspirators from styrofoam floats placed on the water surface in an infested mat. Adults can be stored under cool temperatures and high humidity and carried to nearby release areas.

## Comments

This species has been hard to establish with confirmed establishment after release in only two sites. Recently, this species has been found at several nonrelease sites in northeastern Texas. During laboratory studies, larvae required softer leaves to do well than did *H. pakistanae*. However, higher protein content of the leaves also has been suggested as an explanation.

## *Hydrellia pakistanae*

G. R. Buckingham and M. J. Grodowitz

**Common name:** Indian hydrilla leaf-mining fly.

**Type of agent:** Insect: Fly (Diptera: Ephydridae).

**Native distribution:** Pakistan to China.

**Original source:** Karnataka, India.

### Biology

**Generations per year:** Multiple, possibly five to six.

**Overwintering stages:** Larval and possibly pupal.

**Egg stage:** Like *H. balciunasi*, females deposit eggs singly or in small groups on leaves and stems at or above the water surface. Eggs cannot survive long periods submersed. The eggs are similar to those of *H. balciunasi*: shiny white, elongate, 0.54 mm long and 0.16 mm in diameter, with longitudinal ridges. The eggs hatch in about three days.

**Larval stage:** Larvae are similar in appearance and behavior to *H. balciunasi*. They mine more leaves than *H. balciunasi* with an average total of 12 leaves (range 6 to 21). The three instars develop in about 13 days.

**Pupal stage:** The puparium is attached to the stem at the leaf axil. Total time in the puparium is about seven days.

**Adult stage:** The small, 1.5-mm-long, dark gray flies have shiny golden, sometimes silver, faces. They are similar in appearance and behavior to *H. balciunasi* but can be distinguished by the male and female genitalia. Males have a pair of large bristles on the genitalia that are needle-like in *H. pakistanae*, but broadly spatulate in *H.*

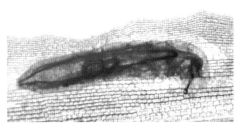

*H. pakistanae* larva in a hydrilla leaf. (Photo credit: M. Thomas, USDA-ARS)

*balciunasi.* Females have L-shaped cerci compared to the triangular cerci of *H. balciunasi.* Females oviposit the day they emerge and live an average of 10 days (range 6 to 21 days). They average 68 eggs with 107 eggs maximum.

## Effect

**Destructive stage:** Larval.

**Additional plant species attacked:** None.

**Site of attack:** Leaves.

**Impact on the host:** Larval mining destroys the leaf's contents. Recent research has indicated substantial reductions in photosynthesis even with low to moderate leaf damage. Tuber number is also substantially reduced under moderate feeding damage as observed under controlled experiments and at field sites. When leaf damage approaches 25 to 35%, holes have been observed developing in the hydrilla canopy by higher fragmentation at insect feeding areas, the apparent loss of plant buoyancy, and subsequent sinking of the hydrilla. In areas with continual feeding pressure by *H. pakistanae,* there is often a slow but perceptible change in the plant community moving from a hydrilla monoculture to a mixed bed of native plants and limited hydrilla.

**Nontarget effects:** No nontarget effects have been reported.

## Releases

**First introduced into the United States:** 1987, Florida.

**Established in:** Alabama, Arkansas, Florida, Georgia, Louisiana, and Texas.

**Habitat:** This fly inhabits all water bodies where hydrilla forms surface mats, including spring runs. It will probably be limited to USDA Plant Hardiness Zone 8 or warmer. If flies are needed for northern hydrilla infestations, a population from extreme northern China should survive in Zones 7 and 6 or colder.

**Availability:** Widely available in Florida and Texas during late spring, summer, and early fall.

**Stage to transfer:** Larval.

**Redistribution:** (Refer to the redistribution notes for *H. balciunasi.*) Infested stems can be collected and carried wrapped in moist paper towels to the release site. Adults can be allowed to emerge and oviposit and the stems with second or third instars can be transported in

*H. pakistanae* adult on hydrilla. (Photo credit: G. Buckingham, USDA-ARS)

193

Hydrilla damaged by *H. pakistanae*
larvae (right). (Photo credit:
G. Buckingham, USDA-ARS)

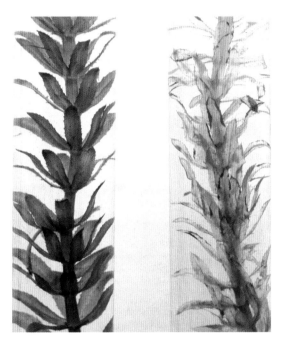

enclosed containers. For shipping, infested stems can be wrapped in moistened paper towels, placed in plastic bags, and shipped in insulated containers. For short-distance transfers, adults can be stored under cool temperature and high humidity and carried to nearby release areas.

## Comments

This species established readily and rapidly expanded its range. Hydrilla has declined at many of the release sites, but cause and effect have been difficult to establish. Hydrilla declines have been correlated with increases in fly populations at small rearing ponds in Alabama, Florida, and Texas. More controlled experimentation has shown that even moderate leaf damage by *Hydrellia* spp. feeding causes declines in biomass and tuber numbers. Photosynthesis is also severely impacted at relatively low feeding levels. High rates of parasitism by a native diapriid wasp, *Trichopria columbiana*, have been observed late in the growing season at some sites in Texas and may be an important population regulator. Additional populations from Punjab, Pakistan (1990) and Beijing and Liaoning, Peoples Republic of China (1992) were also released in Florida and elsewhere.

## *References*

Balciunas, J. K., and D. W. Burrows. 1996. Distribution, abundance and field host-range of *Hydrellia balciunasi* Bock (Diptera: Ephydridae) a biological control agent for the aquatic weed *Hydrilla verticillata* (L.f.) Royle. Aust. J. Entomol. 35: 125-30.

Balciunas, J. K., and M. F. Purcell. 1991. Distribution and biology of a new *Bagous* weevil (Coleoptera: Curculionidae) which feeds on the aquatic weed, *Hydrilla verticillata*. J. Aust. Entomol. Soc. 30: 333-38.

Balciunas, J. K., M. J. Grodowitz, A. F. Cofrancesco, and J. F. Shearer. 2002. Hydrilla. Pages 91-114 *in* R. Van Driesche, S. Lyon, B. Blossey, M. Hoddle, and R. Reardon, eds. Biological Control of Invasive Plants in the Eastern United States. USDA Forest Serv. Publ. FHTET-2002-04. Morgantown, WV.

Bennett, C. A., and G. R. Buckingham. 1991. Laboratory biologies of *Bagous affinis* and *B. laevigatus* (Coleoptera: Curculionidae) attacking tubers of *Hydrilla verticillata* (Hydrocharitaceae). Ann. Entomol. Soc. Am. 84: 420-28.

Buckingham, G. R. 1994. Biological control of aquatic weeds, Chapter 22. Pages 413-79 *in* D. Rosen, F. D. Bennett, and J. L. Capinera, eds. Pest Management in the Subtropics: Biological Control - the Florida Experience. Intercept Ltd., Hampshire, UK.

Buckingham, G. R., and E. A. Okrah. 1993. Biological and host range studies with two species of *Hydrellia* (Diptera: Ephydridae) that feed on hydrilla. U.S. Army Engineer Waterways Exp. Station, Vicksburg, MS. Tech. Rept. A-93-7.

Buckingham, G. R., E. A. Okrah, and M. Christian-Meier. 1991. Laboratory biology and host range of *Hydrellia balciunasi* (Diptera: Ephydridae). Entomophaga 36: 575-86.

Center, T. D., M. J. Grodowitz, A. F. Cofrancesco, G. Jubinsky, E. Snoddy, and J. E. Freedman. 1997. Establishment of *Hydrellia pakistanae* (Diptera: Ephydridae) for the biological control of the submersed aquatic plant *Hydrilla verticillata* (Hydrocharitaceae) in the southeastern United States. Biol. Control 8: 65-73.

Doyle, R. D., M. J. Grodowitz, R. M. Smart, and C. Owens. 2002. Impact of herbivory by *Hydrellia pakistanae* (Diptera: Ephydridae) on growth and photosynthetic potential of *Hydrilla verticillata*. Biol. Control 24: 221-229.

Godfrey, R. K., and J. W. Wooten. 1981. Aquatic and Wetland Plants of Southeastern United States. Dicotyledons. Univ. Georgia Press, Athens.

Godfrey, K. E., L. W. J. Anderson, S. D. Perry, and N. Dechoretz. 1994. Overwintering and establishment potential of *Bagous affinis* (Coleoptera: Curculionidae) on *Hydrilla verticillata* (Hydrocharitaceae) in northern California. Florida Entomol. 77: 221-30.

Grodowitz, M. J., R. Doyle, and R. M. Smart. 2000. Potential use of insect biocontrol agents for reducing the competitive ability of *Hydrilla verticillata*. U.S. Army Engineer Research and Development Center, Vicksburg, MS. ERDC/EL SR-00-1.

Grodowitz, M. J., T. D. Center, A. F. Cofrancesco, and J. E. Freedman. 1997. Release and establishment of *Hydrellia balciunasi* (Diptera: Ephydridae) for the biological control of the submersed aquatic plant *Hydrilla verticillata* (Hydrocharitaceae) in the United States. Biol. Control 9: 15-23.

Grodowitz, M. J., J. E. Freedman, H. Jones, L. Jeffers, C. Lopez, and F. Nibling. 2000. Status of waterhyacinth/hydrilla infestations and associated biological control agents in lower Rio Grande Valley cooperating irrigation districts. U.S. Army Engineer Research and Development Center, Vicksburg, MS. ERDC/EL SR-00-11.

Langeland, K. A. 1996. *Hydrilla verticillata* (L. F.) Royle (Hydrocharitaceae), "the perfect aquatic weed." Castanea 61: 293-304.

O'Brien, C. W., and I. S. Askevold. 1992. Systematics and evolution of weevils of the genus *Bagous* Germar (Coleoptera: Curculionidae), I. Species of Australia. Trans. Am. Entomol. Soc. 118: 331-452.

O'Brien, C. W., and H. R. Pajni. 1989. Two Indian *Bagous* weevils (Coleoptera, Curculionidae) tuber feeders of *Hydrilla verticillata* (Hydrocharitaceae), one a potential biocontrol agent in Florida. Florida Entomol. 72: 462-68.

Wheeler, G. S., and T. D. Center. 2001. Impact of the biological control agent *Hydrellia pakistanae* (Diptera: Ephydridae) on the submersed aquatic weed *Hydrilla verticillata* (Hydrocharitaceae). Biol. Control 21: 168-81.

# Knapweeds

## J. M. Story, G. L. Piper, and E. M. Coombs

The genus *Centaurea* (Asteraceae=Compositae) consists of nearly 500 predominantly Eurasian and Mediterranean species. Only two *Centaurea* spp. (basket flowers)—*C. americana* and *C. rothrockii*—are native to North America. *C. americana* is known from Missouri, Louisiana to Kansas, eastern Arizona, and northern Mexico, while *C. rothrockii* grows only at 2,000 to 2,600 m elevations along streams in southwestern New Mexico and southeastern Arizona to Oaxaca, Mexico.

The knapweeds targeted for biological control in North America are diffuse, spotted, squarrose, and Russian knapweed, and yellow starthistle. (Russian knapweed, once included in the genus *Centaurea*, is now accepted as *Acroptilon repens*.)

Diffuse (*Centaurea diffusa*) and spotted (*C. stoebe* ssp. *micranthos*) (= *C. maculosa*) knapweed are widespread pests in both the United States and Canada, whereas yellow starthistle (*C. solstitialis*) is primarily a weed of the western United States. Squarrose knapweed (*C. virgata* ssp. *squarrosa*), although reported in California and Oregon, is primarily a problem in Utah. Purple starthistle (*C. calcitrapa*) is considered a serious pest only in California, but it also grows in Utah and Wyoming. Meadow knapweed (*C. pratensis* or *C. jacea* x *nigra*) has become a serious pest along the U.S. West Coast.

Many of the insects introduced against spotted knapweed also attack diffuse knapweed and vice versa. Some insects released against these two plants also attack the closely related purple starthistle and meadow and squarrose knapweed. These species have the potential to become widespread range weeds in some locations.

Field surveys of arthropods and pathogens associated with the Asteraceae—particularly thistles, starthistles, and knapweeds—were begun by the USDA Agricultural Research Service and the Commonwealth Institute of Biological Control (now CABI Bioscience) during the early 1960s. The first release in the United States of a biocontrol agent against a knapweed species occurred against spotted knapweed in 1973.

# Cornflower

*Centaurea cyanus*

## Sunflower family—Asteraceae

*G. L. Piper*

**Additional common names:** Bachelor's button, bluebottle, blue caps, corn-bottle, corn-centaury, garden cornflower, hurtsickle, witches' bells.

**Native range:** Eurasia.

**Entry into the United States:** Cornflower was intentionally introduced as a garden ornamental but soon escaped from cultivation.

## Biology

**Life duration/habit:** Cornflower is a branched annual or winter annual growing 0.3 to 1 m tall.

**Reproduction:** Reproduction is by seeds.

**Roots:** Cornflower possesses an abbreviated taproot.

**Stems and leaves:** The ridged, slender, somewhat wiry stems are covered with woolly hairs that make the stems appear dull grayish-green. The alternately arranged lower leaves are very narrow, 10 cm long by 6 to 12 mm wide, and have prominent, long, narrow lobes along the margins. The undersurface of the leaves is covered with matted, white hairs, imparting a cobweb-like appearance to the leaves. The stalkless, small middle and upper leaves are narrow with sharp, pointed tips and have smooth or short-toothed margins.

**Flowers:** Flowering occurs from May to September. The flower heads are borne singly at the ends of the branches and are 4 cm in diameter. Flower color is variable, ranging from light blue to purple, maroon, pink, white, or bicolored. The larger, irregularly lobed or notched, marginal (ray) florets are sterile; the inner (disk) tubular florets are smaller and fertile.

Cornflower plant. (Photo credit: E. Coombs, Oregon Department of Agriculture)

197

**Fruits and seeds:** The oblong seeds are buff- to gray-colored with a white to cream-colored lower tip, 4 to 5 mm long, and bear an upper tuft of stiff, tawny bristles of various lengths.

## Infestations

**Worst infested states:** The plant is naturalized throughout temperate areas of the United States, with notable infestations occurring from California into the Pacific Northwest.

**Habitat:** Cornflower prefers open areas with light, well-drained soils. It readily establishes along roadsides, in pastures, Conservation Reserve Program lands, and fallowed and cultivated fields. It is still planted as an ornamental by many individuals, the seeds being readily available from many commercial sources. The plant is often a component of many wildflower seed mixes marketed for highway median strips and meadow beautification. It is also commonly used in dried flower arrangements.

**Impacts:** The plant is sometimes a problem in grain crops.

## Comments

Several European insects introduced into North America for the biological control of diffuse, meadow, spotted, squarrose, and other knapweeds and yellow starthistle destroy the seeds of cornflower. *Urophora quadrifasciata* larvae form galls within the seed heads. Cornflower also serves as an early season host for the seed head fly *Chaetorellia australis*, introduced to attack the seeds of yellow starthistle. Both biocontrol agents are well established on *C. cyanus* in the Pacific Northwest.

# Diffuse knapweed                                     *Centaurea diffusa*

## Sunflower family—Asteraceae

### G. L. Piper and J. M. Story

**Additional common names:** None widely accepted.

**Native range:** Diffuse knapweed grows in the eastern Mediterranean area and in western Asia to western Germany. North American introductions appear to have been from the Mediterranean region.

**Entry into the United States:** It was first recorded in 1907 in Washington; there may have been multiple introductions. One theory suggests that this plant entered the United States as a contaminant of alfalfa seed.

## Biology

**Life duration/habit:** Diffuse knapweed is usually a biennial but can be a short-lived perennial.

**Reproduction:** The plant reproduces exclusively by seed.

**Roots:** Diffuse knapweed is a taprooted plant.

**Stems and leaves:** Plants generally grow 15 to 60 cm tall. The leaves are deeply divided into many toothed parts. Lower leaves are 5 to 10 cm long, and the upper leaves are smaller and have a few slender lobes. Rosettes are formed the first year.

Diffuse knapweed flower heads. (Photo credit: S. Dewey, Utah State University)

**Flowers:** Flowering occurs from July to September. The 0.5- to 1.0-cm-long flowers occur singly or in clusters at the ends of the branches. Flowers are white, pink, or lavender. Each floral bract is tipped with a long, slender spine and is fringed with smaller spines.

**Fruits and seeds:** Seeds are oblong, blackish-brown with vertical brown to gray stripes, and about 3 mm long.

## Infestations

**Worst infested states:** Idaho, Montana, Oregon, and Washington.

**Habitat:** The plains and forested benchlands are typically infested environments in the eastern part of its geographical distribution. It is commonly distributed by human activity and can be found along roadsides, in abandoned areas, and in disturbed sites, where it has a tendency to grow in dense stands. Diffuse knapweed seems to prefer well-drained, light-textured soils and is intolerant of shade. It prefers drier conditions than spotted and meadow knapweed, but more moisture than squarrose knapweed. The plant occurs from sea level to 1,500 m and where annual precipitation ranges from 15 to 90 cm.

**Impacts:** Diffuse knapweed has become one of the most economically important rangeland weeds in the western United States. In 1979, it was estimated to infest 302,400 ha in Washington, 300,000 ha in Oregon, and 29,700 ha in Idaho. The main economic loss caused by diffuse knapweed is a reduction in high-quality forage, generally in areas where chemical use is cost-prohibitive.

## Comments

Several pathogens can be quite destructive to diffuse knapweed. Pathogens include two fungi, *Puccinia jaceae* var. *diffusae*, which attacks the leaves of diffuse knapweed, and *Sclerotinia sclerotiorum*, which attacks the crowns of both diffuse and spotted knapweed. However, these species are being studied and are not yet cleared as knapweed biocontrol agents or for movement across state lines. Livestock grazing of diffuse and other knapweed species in the spring can reduce seed production and cause plants to perennate. Grazing during flower bud production severely reduces the impact of the seed head biocontrol agents.

The seed head weevil *Larinus minutus* has caused remarkable reductions in diffuse knapweed density in some areas of Oregon and Washington with emerging success also reported in Colorado.

## Meadow knapweed *Centaurea pratensis (=C. jacea* x *nigra)*

Sunflower family—Asteraceae

## *E. M. Coombs, G. L. Piper, and C. Roché*

**Additional common names:** Protean knapweed, Bemis grass.

**Native range:** Europe.

**Entry into the United States:** Meadow knapweed was introduced into Oregon as a forage plant.

### Biology

**Life duration/habit:** Meadow knapweed is a highly variable perennial that often grows in thick stands.

**Reproduction:** Reproduction is by seeds, primarily, although some resprouting from roots and root crowns occurs.

**Roots:** Meadow knapweed has a taproot similar to other knapweeds.

**Stems and leaves:** Meadow knapweed grows from 0.6 to 1.3 m tall. Its many stems branch and have solitary flowers. The lower leaves are stalked and entire while the middle leaves are smaller and lobed. The upper leaves are even smaller and are not stalked and entire.

**Flowers:** Flowers are usually pink to purplish with bracts that are brown, paper-like, and fringed. Flowering occurs from July to October.

**Fruits and seeds:** Fruits may or may not have very small scales at the apex.

### Infestations

**Worst infested states:** Northwestern California and west of the Cascade Mountains of Oregon and

Meadow knapweed flower. (Photo credit: E. Coombs, Oregon Department of Agriculture)

Washington. It also grows in the northern Rockies and in some areas of the northeastern United States.

**Habitat:** It infests roadsides, moist pastures, ditch banks, vacant lots, and is becoming more common in clearcuts. This plant does not do well in the drier areas favored by the other knapweeds.

**Impacts:** Meadow knapweed forms thick stands, excluding most other vegetation, that are difficult to control even by grazing.

## Comments

Many consider meadow knapweed a fertile hybrid cross between brown knapweed (*C. jacea*) and black knapweed (*C. nigra*). At many sites, variants that resemble the parent species may be found. It has been grown as a forage species for livestock but has escaped cultivation and spread throughout much of western Oregon and Washington. A seed head fly, *Urophora quadrifasciata*, introduced against diffuse and spotted knapweed has become widely reassociated with meadow knapweed. The seed head moth *Metzneria paucipunctella* and the seed head weevils *Larinus minutus* and *L. obtusus* have been recovered at several sites in Oregon and Washington. *L. obtusus* is being shown to be an important seed predator. Meadow knapweed is being tested as a host for other approved knapweed biocontrol agents.

Meadow knapweed has been found to hybridize with yellow starthistle and diffuse knapweed. It is not known whether the offspring are fertile.

# Russian knapweed *Acroptilon repens (=Centaurea repens)*

Sunflower family—Asteraceae

*J. L. Littlefield*

**Additional common names:** Turkmenistan thistle, hardheads.

**Native range:** Central Asia to Turkey.

**Entry into the United States:** The plant was accidentally introduced about 1898, probably as a contaminant of alfalfa seed from Turkmenistan. DNA analyses indicate that U.S. populations of the weed may have originated in Uzbekistan and Kazakhstan.

## Biology

**Life duration/habit:** Russian knapweed is a perennial plant that spreads mainly by its extensive, lateral root system.

**Reproduction:** It reproduces mainly by creeping roots and, to a lesser extent, by seeds.

**Roots:** This species has an extensive branching root system, consisting of vertical and lateral roots.

**Stems and leaves:** The stems are erect, openly branched, 45 to 90 cm tall, and are covered with soft gray hairs. Lower leaves are deeply lobed and upper leaves are oblong and toothed.

201

**Flowers:** Flowers are numerous, rose or pink to purple, lavender, or blue in color, and are enclosed in solitary, flask-shaped heads on the ends of leafy branches. Flower heads are 5 to 10 mm in diameter. Flowering occurs from June to September.

**Fruits and seeds:** Seeds are gray to ivory, 2 to 3 mm long, and possess numerous whitish bristles. Seeds are produced from August through September. A single plant may produce more than 1,200 seeds.

## Infestations

**Worst infested states:** Arizona, California, Idaho, Montana, Nevada, New Mexico, Oregon, Utah, and Washington.

**Habitat:** The plant has become widespread in the United States, particularly in the arid west, and is a pest of grazing land, grain and other crops, waste places, roadsides, and irrigation ditches.

**Impacts:** This plant is difficult to control. It forms dense infestations in all types of agricultural and recreational lands. Prolonged ingestion of Russian knapweed is poisonous to horses, in which it has been known to cause the neurological disorder nigropallidal encephalomalacia (chewing disease). Consumption of 60 to 200% of plant material (dry or fresh) of animal body weight may result in death.

## Comments

Russian knapweed is relatively free of insects and pathogens in North America. Biological control of Russian knapweed in North America utilizing exotic organisms has been somewhat limited. Only the gall-inducing nematode *Subanguina picridis* has

202

Russian knapweed. (Photo credit: E. Coombs, Oregon Department of Agriculture)

Russian knapweed infestation. (Photo credit: E. Coombs, Oregon Department of Agriculture)

been introduced and established in North America. Although the nematode is effective in reducing plant biomass and flowering, infections are not consistent from year to year due to varying moisture conditions. The nematode does not move readily, thus it needs to be propagated and redistributed on a large scale, which is not cost-effective with present techniques. For these reasons, other organisms are being considered for biological control. Field surveys by various Russian, USDA, and more recently CABI scientists have identified other potential biocontrol organisms attacking the plant. As of 2003, the most promising of these (due to their reported host specificity and potential damage) are *Urophora kasachstanica* and *U. xanthippe* (flower gall flies that are under review by the TAG), *Aceria acroptiloni* (flower-infesting eriophyid mite), *Aulacidea acroptilonica* (stem gall wasp), *Dasineura* sp. (tip gall midge), *Napomyza* sp. near *lateralis* (root-boring fly), and *Cochylimorpha nomadana* (root-boring moth). Other insects are also under consideration. In North America, the seed head weevil *Eustenopus villosus*, a biological control agent of yellow starthistle, occasionally feeds on seed heads and causes them to abort.

Pathogens may also be of value for biological control. In North America, three fungi have been found on Russian knapweed. These are *Alternaria* sp. and *Puccinia acroptili*, which attack the leaves, and *Sclerotinia sclerotiorum*, which infests the roots.

# Spotted knapweed

*Centaurea stoebe* ssp. *micranthos*
*(=C. maculosa)*

## Sunflower family—Asteraceae

*J. M. Story, E. M. Coombs, and G. L. Piper*

**Additional common names:** None widely accepted.

**Native range:** Spotted knapweed is indigenous to southcentral and southeastern Europe and northwestern Asia.

**Entry into the United States:** The plant arrived in the United States as a contaminant in alfalfa seed.

## Biology

**Life duration/habit:** Spotted knapweed is an aggressive biennial or short-lived perennial that grows from 0.3 to 1 m tall. Seedlings emerge from spring through early fall. Plants bolt in early May and flower from late June to October.

**Reproduction:** This species reproduces by seed.

**Roots:** Spotted knapweed possesses a stout taproot.

**Stems and leaves:** Plants have alternately arranged, pale- or gray-green leaves measuring about 2.5 to 7.5 cm long. Rosette leaves are deeply divided into lobes; stem leaves have fewer lobes and become smaller toward the branch tips.

**Flowers:** Flower buds form at the ends of the main stem and upper branches. Flowers are pink to light purple and occasionally white. Spotted knapweed is usually

Spotted knapweed.
(Photo credit: N. Poritz, bio-control.com)

distinguished by black-tipped bracts on the flower heads with pinkish-purple flowers, although these characteristics can be seen in some diffuse knapweed (*C. diffusa*) plants. The primary difference between the two species is the presence of a distinct spine at the tip of diffuse knapweed bracts.

**Fruits and seeds:** Spotted knapweed is a prolific seed producer. Between five and 1,200 seed heads are produced by each plant; each seed head produces up to 30 seeds. The smooth seeds are brown to black with pale, longitudinal lines. Seeds remain viable in the soil for five or more years. Most seeds germinate in the spring, but those that germinate in the fall (often found in climates that lack summer rain) produce overwintering rosettes.

## Infestations

**Worst infested states:** Idaho, Montana, Oregon, and Washington, with lesser infestations throughout much of the United States.

Spotted knapweed infestation. (Photo credit: USDA-ARS)

**Habitat:** Dense infestations develop on well-drained, light soils receiving summer rainfall. Spotted knapweed prefers a moister environment than either diffuse or squarrose knapweed (*C. virgata* ssp. *squarrosa*). It occurs at elevations ranging from 100 to more than 3,040 m and in precipitation zones ranging from 20 to 200 cm annually.

**Impacts:** Spotted knapweed is the number one problem weed on western Montana rangelands. It allows increased soil erosion, decreases biodiversity, and reduces forage for wildlife and livestock.

## Comments

Spotted knapweed infests an estimated 3 million ha in the western United States. This plant has been reported in North America as *C. maculosa*, but should be called *C. stoebe* ssp. *micranthos* because it is perennial, polycarpic, and tetraploid, whereas *C. maculosa* is biennial, monocarpic, and diploid.

The fungus *Sclerotinia sclerotiorum* attacks the crowns of both diffuse and spotted knapweed. However, this fungus is being studied and is not cleared for use for biocontrol of these knapweeds or for transportation across state lines.

The root insects *Agapeta zoegana* and *Cyphocleonus achates* are having a significant impact on spotted knapweed. In some areas, the seed head weevils *Larinus minutus* and *L. obtusus* are beginning to exert some control.

205

# Squarrose knapweed

*Centaurea virgata* ssp. *squarrosa*

## Sunflower family—Asteraceae

*J. M. Story and D. M. Woods*

**Additional common names:** None widely accepted.

**Native range:** Bulgaria, Lebanon, Turkey, Transcaucasia, northern Iraq, Iran, Afghanistan, and Turkmenistan.

**Entry into the United States:** This species may have been introduced as seeds carried in wool, either on sheep or on woolen products.

## Biology

**Life duration/habit:** Squarrose knapweed is a taprooted perennial with a woody crown that produces many profusely branched stems 30 to 45 cm tall. The crown consists of one or more clusters of rosette leaves branching from the stout taproot.

**Reproduction:** This plant reproduces only by seed.

**Roots:** Squarrose knapweed possesses a taproot.

**Stems and leaves:** Basal leaves are deeply dissected and often wither by the time of flowering. Stem leaves lack petioles and have fewer lobes progressively up the stem, the uppermost ones being bract-like.

**Flowers:** Squarrose knapweed looks very similar to diffuse knapweed (*C. diffusa*) except for the recurved bract tips and its true perennial growth form. The flowering period is from June to August, although plants may be several years old before they flower. Rose-purple to pink flowers are borne within the slender, urn-shaped heads. Seed heads are closed and most are deciduous at fruiting time. The seed head bracts are pale or suffused with red or purple, especially near the tip, and are spiny, allowing the head to attach to almost anything (such as fleece, hair, and clothing) that brushes against it.

**Fruits and seeds:** Seeds are 3 to 3.8 mm long. There are usually two seeds per seed head.

Squarrose knapweed plant. (Photo credit: S. Dewey, Utah State University)

## Infestations

**Worst infested states:** California and Utah.

**Habitat:** Squarrose knapweed is aggressive and weedy, as are diffuse and spotted knapweed (*C. stoebe* ssp. *micranthos*), but is more adaptable to harsh growing conditions such as cold temperatures and drought. It appears to thrive in degraded rangeland.

**Impacts:** Its impact on rangeland is not understood at this time, but is suspected to be similar to that of diffuse and spotted knapweed. In Utah it crowds out most desirable vegetation.

Squarrose knapweed flower heads. (Photo credit: S. Dewey, Utah State University)

## Comments

Several biological control agents imported for diffuse or spotted knapweed have been successfully established on squarrose knapweed, such as the seed head weevils *Bangasternus fausti* and *Larinus minutus*, the seed head fly *Urophora quadrifasciata*, and the root beetle *Sphenoptera jugoslavica*.

Specialized organisms that attack squarrose knapweed in Turkey that may be good biological control agents include the rust *Puccinia jaceae*, the seed head flies *Chaetorellia* spp., a seed head wasp *Isoculus* sp., and an unidentified root-boring weevil.

## *Agapeta zoegana*

J. M. Story

**Common names:** Sulfur knapweed moth, yellow-winged knapweed root moth.

**Type of agent:** Insect: Moth (Lepidoptera: Cochylidae).

**Native distribution:** Primarily Europe and western Asia.

**Original sources:** Western Hungary and eastern Austria.

## Biology

**Generations per year:** One in the United States, two in warmer climates.

**Overwintering stage:** Larval (in the root).

**Egg stage:** Eggs are laid singly or in small groups on the surface of stems and leaves of knapweed and other vegetation. Eggs hatch in seven to 10 days, are whitish when laid

*A. zoegana* larva. (Photo credit: N. Poritz, bio-control.com)

but turn yellowish-red in three to four days. The eggs are somewhat flattened with a surface covered with a network of fine lines.

**Larval stage:** Newly hatched larvae migrate to the crown area where they mine the root. The exposed sides of lateral tunnels are covered with a silken web. There are six larval instars.

**Pupal stage:** Pupation occurs within the root.

**Adult stage:** Emergence of adults usually occurs over a 12-week period from mid-June to early September. Adults are about 11 mm long and have bright yellow wings with brown markings. Females mate within 24 hours of emergence and begin laying eggs one day later. The adults live for only several days in the field.

## Effect

**Destructive stage:** Larval.

**Plant species attacked:** Spotted knapweed and, to a lesser extent, diffuse knapweed.

**Site of attack:** Cortex of the root.

**Impact on the host:** Feeding by larvae reduces knapweed biomass and density. Larval feeding can result in death to small plants or plants being attacked by multiple larvae.

**Nontarget effects:** There have been no reports of attack on nontarget plants.

*A. zoegana* adult. (Photo credit: N. Poritz, bio-control.com)

## Releases

**First introduced into the United States:** 1984, Montana.

**Established in:** California, Colorado, Idaho, Minnesota, Montana, Oregon, Utah, Washington, and Wyoming.

**Habitat:** This insect prefers dry, well-drained, open sites with scattered (not dense) vegetation. The moth is known to survive in areas characterized by a moderately humid climate and in areas with arid, subcontinental climates.

**Availability:** This insect is becoming widespread, but collections are limited to sites with large populations.

**Stage to transfer:** Adult.

**Redistribution:** Collect the adults attracted to blacklights with a modified insect vacuum or rear them from infested roots.

## Comments

The moth has reduced knapweed biomass and density at several sites in western Montana. The insect was experimentally released on squarrose knapweed in Utah in 1994, but has not been recovered. Establishment of the moth at a site can be determined with the use of a blacklight suspended over a white sheet in the early evening in early August. However, most of the adults attracted to the light will be males, so the technique is not recommended for collection/redistribution.

# *Bangasternus fausti*

J. M. Story and E. M. Coombs

**Common name:** Broad-nosed seed head weevil.

**Type of agent:** Insect: Beetle, weevil (Coleoptera: Curculionidae).

**Native distribution:** Italy, Romania, Bulgaria, Greece, Turkey, Armenia, and Iran.

**Original source:** Northern Greece.

## Biology

**Generations per year:** One.

**Overwintering stage:** Adult (in the soil and litter surrounding the plant; may occur in seed heads in warmer climates).

**Egg stage:** Eggs are laid from May to mid-July on the underside of leaflets below the developing flower head or on the end of the stem and leaflets.

*B. fausti* adult. (Photo credit: N. Poritz, bio-control.com)

**Larval stage:** Depending upon the placement of the egg, the new larva mines directly into the midrib of the leaflet or into the stem before tunneling into the flower head.

**Pupal stage:** Pupation occurs within the seed head.

**Adult stage:** Adults are active from May until late July, while the plants are in the bud to mid-flowering stages. Adult weevils are grayish-black and 3.5 to 4.0 mm long. Adult weevils have a much shorter and more blunt snout than the *Larinus* weevils.

## Effect
**Destructive stage:** Larval.

**Plant species attacked:** Diffuse, spotted, and squarrose knapweed.

*B. fausti* egg below bud. (Photo credit: E. Coombs, Oregon Department of Agriculture)

**Site of attack:** Larvae consume up to 100% of the seeds in a flower head. The viability of unattacked seeds is reduced.

**Impact on the host:** In Greece, there is often up to 95% seed consumption in diffuse knapweed, less in spotted knapweed, probably because flower heads of the latter are larger. In California, this weevil destroys up to 100% of the seed in squarrose knapweed.

**Nontarget effects:** There have been no reports of attack on nontarget plants.

## Releases
**First introduced into the United States:** 1989, experimental release in Oregon.

**Established in:** California, Idaho, Oregon, Utah, and Washington.

**Habitat:** This weevil prefers hot, dry areas and does not do well at high elevations and in areas with prolonged rain.

**Availability:** The weevil is available for mass redistribution in Oregon and Washington.

**Stage to transfer:** Adult.

**Redistribution:** Collect adults with a sweep net in late spring to early summer when plants begin to bloom. Many areas also contain *Larinus minutus*; therefore, early season collecting favors *Bangasternus*. Adults can survive several days in shipment if kept cool and dry, and provided with ample food.

## Comments
The larvae will attack any other insects occupying the flower head.

# Chaetorellia acrolophi

J. M. Story and E. M. Coombs

**Common name:** Knapweed peacock fly.

**Type of agent:** Insect: Fly (Diptera: Tephritidae).

**Native distribution:** Spain and the European Alps to Greece and Turkmenistan; southern Austria and France, eastern and central Hungary, eastern Romania, southern Switzerland, and central Turkey.

**Original sources:** Austria, France, Hungary, Romania, and Switzerland.

## Biology

**Generations per year:** Two (in Europe).

**Overwintering stage:** Larval (in the seed head).

**Egg stage:** Each female produces about 70 eggs. Eggs are laid singly or in groups underneath the bracts of closed buds. The eggs measure 0.9 mm long and are shiny white, elongate, and have a long filament that is thick at one end. Eggs hatch in four to five days.

**Larval stage:** Upon emerging, the larva burrows horizontally through the bracts into the center of the flower head where it feeds in an immature floret. Older larvae feed on the seeds, florets, and receptacle tissue, often completely destroying the inner contents of the flower head. Larval development takes 10 to 15 days. The larvae and pupae of the first generation are white, while second generation larvae and pupae are yellow. There are three larval instars.

**Pupal stage:** Pupation takes place in a vertical position between the florets, in the upper area of the flower head. The larva forms a white puparium with florets or plant hairs affixed to the surface.

**Adult stage:** In Europe, adults appear in the field in May. Adults live up to four weeks in the laboratory. Females begin laying eggs one week after emergence. Adults are 4 to 5 mm long with light-brown banded wings.

*C. acrolophi* adult. (Photo credit: E. Coombs, Oregon Department of Agriculture)

## Effect

**Destructive stage:** Larval.

**Plant species attacked:** Spotted knapweed is the primary host, but it may also attack diffuse knapweed, purple starthistle, *C. leucophaea*, *C. vallesiaca*, and squarrose knapweed.

**Site of attack:** Flower head.

**Impact on the host:** The fly reduces seed production.

**Nontarget effects:** There have been no reports of attack on nontarget plants.

## Releases

**First introduced into the United States:** 1992, Montana.

**Established in:** Colorado, Oregon, Washington, and Wyoming.

**Habitat:** In Oregon, the fly does best where the seed head weevils are less abundant, generally at higher elevations, and in areas with higher rainfall.

**Availability:** This insect is available for limited redistribution.

**Stages to transfer:** Adult and larval.

**Redistribution:** Fly-infested seed heads can be collected in the fall and stored at 4 to 8° C, or collected in early spring and transported to release sites prior to adult emergence. To exclude other seed head insects and parasitoids, the adults should be recovered in a sleeve cage.

## Comments

There are two generations of the fly in Europe and the western United States. The fly appears to be having a difficult time establishing in some areas because of interspecific competition with the seed head weevils and moths.


## *Cyphocleonus achates*

J. M. Story

**Common name:** Knapweed root weevil.

**Type of agent:** Insect: Beetle, weevil (Coleoptera: Curculionidae).

**Native distribution:** Eastern and southern Europe and the eastern Mediterranean region.

**Original sources:** Austria, Greece, Hungary, and Romania.

## Biology

**Generations per year:** One.

**Overwintering stage:** Larval (in the root).

**Egg stage:** Eggs are laid singly in a notch excavated by the female on the root crown just below the soil surface. A female may deposit more than 100 eggs during her lifetime. Eggs are oval and white to pale yellow when laid, becoming more yellow with age. They hatch in 10 to 12 days.

**Larval stage:** Newly hatched larvae mine toward the cortex of the root. There are four larval instars. Third- and fourth-instar larvae often cause a gall-like enlargement of the root.

**Pupal stage:** Pupation occurs within the galled root. The pupal period lasts about two weeks.

**Adult stage:** Adults emerge from June to mid-September. The adult weevils feed on knapweed leaves. Adults are 14 to 15 mm long and generally live 8 to 15 weeks, but do not overwinter. A single female will mate several times during her lifetime.

## Effect

**Destructive stage:** Larval.

**Plant species attacked:** Spotted knapweed is the preferred host but the weevil also attacks diffuse knapweed.

*C. achates* adult. (Photo credit: N. Poritz, bio-control.com)

**Site of attack:** Larvae mine and gall the central vascular tissue of the root. Adults feed on the leaves.

**Impact on the host:** Feeding by larvae reduces knapweed biomass and density. Larval feeding can result in death to small plants or plants being attacked by multiple larvae.

**Nontarget effects:** There have been no reports of attack on nontarget plants.

## Releases

**First introduced into the United States:** 1988, Montana and Washington.

**Established in:** California, Colorado, Idaho, Montana, New Mexico, Oregon, Utah, Washington, and Wyoming.

**Habitat:** This weevil prefers hot, dry, well-drained sites with low, scattered vegetation in temperate areas.

**Availability:** Availability is limited; the weevil is only collectible at sites with large populations. Mass rearing was occurring in Colorado and Montana as of 2002.

**Stage to transfer:** Adult.

**Redistribution:** Adults can be collected from the field or reared from infested roots.

## Comments

The weevil is reducing knapweed biomass and density at several sites in western Montana.

# *Larinus minutus*

J. M. Story and E. M. Coombs

**Common name:** Lesser knapweed flower weevil.

**Type of agent:** Insect: Beetle, weevil (Coleoptera: Curculionidae).

**Native distribution:** Bulgaria, Greece, Israel, Romania, Turkey, Turkmenistan, and the Caucasus.

**Original source:** Greece.

## Biology

**Generations per year:** One.

**Overwintering stage:** Adult (in soil near the base of the plant).

**Egg stage:** Eggs are deposited between the pappus hairs in the flower head. Up to five eggs are laid throughout the florets. Females produce between 28 and 130 eggs. The elongate, yellow eggs are 1.5 mm long and hatch in three days.

**Larval stage:** Larvae feed within the flower head on the pappus hairs and developing seeds. There are three instars; larval development is completed in four weeks.

**Pupal stage:** The larva constructs a cocoon from seeds and pappus hairs that is vertically positioned in the receptacle of the flower head. Pupae are white but turn brown shortly before adult emergence.

**Adult stage:** Adults are active in the field from May through mid-September. In the laboratory, adults lived up to 14 weeks. Adults feed on the leaves and flowers prior to laying eggs. Adults are 4 to 5 mm long, mottled-brown, and have a large snout.

## Effect

**Destructive stages:** Larval and adult.

**Plant species attacked:** In Europe, diffuse knapweed is the preferred host plant, but the weevil also attacks meadow knapweed, spotted knapweed, squarrose knapweed, and plants associated with the subgenera Acrolophus (*C. arenaria*) and Calcitrapa (*C. calcitrapa* and *C. iberica*).

**Site of attack:** Larvae feed on the seeds in the flower heads. Adults feed on leaves.

**Impact on the host:** Knapweed defoliation by adults can be severe in sites with high weevil populations, resulting in the stunting and death of affected plants.

**Nontarget effects:** There have been no reports of attack on nontarget plants.

214

*L. minutus* adult. (Photo credit: E. Coombs, Oregon Department of Agriculture)

Diffuse knapweed infestation in Oregon in 1990. (Photo credit: E. Coombs, Oregon Department of Agriculture)

Same vista in 2001, seven years after releases of *L. minutus*. (Photo credit: E. Coombs, Oregon Department of Agriculture)

## Releases

**First introduced into the United States:** 1991, Montana, Washington, and Wyoming.

**Established in:** California, Colorado, Idaho, Indiana, Minnesota, Montana, Nevada, Oregon, Utah, Washington, and Wyoming.

**Habitat:** This weevil prefers hot, dry areas. It does not do well at higher elevations and in areas with prolonged rainfall.

**Availability:** This weevil can be collected from established populations in Montana, Oregon, and Washington.

**Stage to transfer:** Adult.

**Redistribution:** On diffuse knapweed, adults can be collected with an aspirator in the spring when they congregate in large numbers around the root crown of knapweed plants. On diffuse and spotted knapweed, adults can be collected with a sweep net during early to mid-bloom. Collection should be made at 50% bloom to minimize collection of *Bangasternus fausti*, which is often present. Using a passive sorter helps

215

reduce the chance of moving unwanted insects and seeds that often contaminate sweep net collections.

## Comments

The weevil has caused dramatic reductions of diffuse knapweed at sites in Montana, Oregon, and Washington. It is extremely difficult to distinguish between the adults of *L. minutus* and *L. obtusus* when they coexist at a site. *L. minutus* tends to be smaller, covered with more grayish hairs, and has more reddish tibia. Evidence suggests that these two species may be variants of a single species.

Livestock grazing while knapweed plants are bolting delays flowering which is very detrimental to maintaining strong populations of seed head agents. At some sites, this weevil is preyed on extensively during the pupal stage by mice.

# *Larinus obtusus*

J. M. Story and E. M. Coombs

**Common name:** Blunt knapweed flower weevil.

**Type of agent:** Insect: Beetle, weevil (Coleoptera: Curculionidae).

**Native distribution:** South, central, and eastern Europe, and the Middle East.

**Original sources:** Serbia and Romania.

## Biology

**Generations per year:** One.

**Overwintering stage:** Adult (in the soil litter).

**Egg stage:** Eggs are laid among the florets in newly opened flower heads. The eggs hatch in three days at 25° C. Eggs are yellowish, oval to round, and measure 1.3 mm long.

**Larval stage:** Larvae feed within the flower head on the pappus hairs and developing seeds. There are three instars; larval development is completed in 17 days.

*L. obtusus* adult. (Photo credit: N. Poritz, bio-control.com)

**Pupal stage:** Pupation occurs within a cocoon constructed from seeds and pappus hairs. The cocoon stands upright in the flower head receptacle. The pupal period lasts about nine days.

**Adult stage:** Adults are active in the field from May through August. Adults are 5 to 7 mm long, dark brown, and have a large snout.

## Effect

**Destructive stages:** Larval and to a lesser extent the adult.

**Plant species attacked:** Spotted knapweed, meadow knapweed, and occasionally diffuse knapweed.

**Site of attack:** Larvae feed on the seeds in the flower heads. Adults feed on leaves.

**Impact on the host:** This weevil destroys seeds.

**Nontarget effects:** There have been no reports of attack on nontarget plants.

## Releases

**First introduced into the United States:** 1992, Montana.

**Established in:** Montana, Oregon, Washington, and Wyoming.

**Habitat:** This weevil prefers dry, open sites.

**Availability:** Availability of this weevil is limited. It is collectable in Oregon and Washington.

**Stage to transfer:** Adult.

**Redistribution:** Adults can be collected with a sweep net during knapweed bloom.

Meadow knapweed damaged by adult *L. obtusus*. (Photo credit: E. Coombs, Oregon Department of Agriculture)

## Comments

Populations of this weevil on spotted knapweed increase much more slowly than populations of *L. minutus* on diffuse knapweed. It is extremely difficult to distinguish between the adults of *L. obtusus* and *L. minutus* when they coexist at a site. *L. obtusus* is larger, darker, and has dark reddish-black tibia. *L. minutus* tends to be smaller, covered with more grayish hairs, and has more reddish tibia. Also, the powdery covering on the sides of *L. obtusus* appears generally orange-red compared to yellow on *L. minutus*. Evidence suggests that these two species may be variants of a single species. Also, it seems to prefer larger-headed knapweeds.

# Metzneria paucipunctella

J. M. Story and E. M. Coombs

**Common name:** Spotted knapweed seed head moth.

**Type of agent:** Insect: Moth (Lepidoptera: Gelechiidae).

**Native distribution:** Austria, Bulgaria, France, Hungary, and Switzerland.

**Original source:** Switzerland.

## Biology

**Generations per year:** One.

**Overwintering stage:** Larval (in the seed head).

**Egg stage:** Each female lays an average of 80 eggs. The eggs are placed singly at the base of immature flower heads. Eggs are reddish-brown when laid, but turn yellowish as they mature. The eggs are elongate, oval, and measure about 0.75 mm long. The eggs hatch in about 10 days.

**Larval stage:** Upon hatching, the larvae tunnel into the flower head where they feed on florets and seeds. Only one larva survives per flower head. The larvae are white with dark brown head capsules and have distinct body segments. There are five larval instars. Larvae remain in the seed head until the next spring.

**Pupal stage:** Pupation occurs within the seed head receptacle during May.

**Adult stage:** Adults emerge in early June. The wings are light gray with numerous small, pepper-like spots and dark tips. Adults are about 8 mm long and characteristically rest with their wings folded over the back.

## Effect

**Destructive stage:** Larval.

**Plant species attacked:** Spotted knapweed is the preferred host, but the moth also attacks diffuse and meadow knapweed.

*M. paucipunctella* larva in a spotted knapweed seed head. (Photo credit: E. Coombs, Oregon Department of Agriculture)

Adult *M. paucipunctella* on a spotted knapweed seed head. (Photo credit: E. Coombs, Oregon Department of Agriculture)

**Impact on the host:** The moth attacks the seeds and complements the damage inflicted by *Urophora affinis, U. quadrifasciata,* and other flower head feeders. Older larvae bind uneaten seeds together, preventing their dispersal.

**Nontarget effects:** There have been no reports of attack on nontarget plants. Indirect effects occur when the larvae feed as facultative carnivores on other natural enemies in the seed heads. Because deer mice feed heavily on the moth larvae, the moth may contribute to an increase in deer mice numbers in knapweed-infested areas.

## Releases

**First introduced into the United States:** 1980, Montana.

**Established in:** Colorado, Idaho, Montana, Oregon, Virginia, and Washington.

**Habitat:** The moth does best at sites with winter snow cover.

**Availability:** The moth can be collected in Idaho, Montana, Oregon, and Washington.

**Stages to transfer:** Larval, pupal, and adult.

**Redistribution:** Moth-infested seed heads can be collected in the fall and stored at 4 to 8° C, or collected in early spring and transported to release sites prior to adult emergence. To exclude other seed head insects and parasitoids, the adults should be recovered in a sleeve cage. Also, caution should be used when rearing seed head insects as the seed heads are often infested with straw itch mites (*Pyemotes tritici*), which attack the emerging biocontrol agents and can cause severe itching when they attack humans.

## Comments

The larva will destroy all other biocontrol agents in the seed head. Although there is strong competition between the moth and the two gall flies, the greatest reduction in knapweed seed production occurs when all three species are present. In Oregon, deer mice harvest the infested seed heads throughout the winter, causing heavy mortality to the overwintering larvae. The moth is often heavily attacked by parasitoids.

# *Pelochrista medullana*

J. M. Story

**Common name:** Brown-winged root moth.

**Type of agent:** Insect: Moth (Lepidoptera: Tortricidae).

**Native distribution:** Eastern Austria, northern Hungary, eastern Romania, Turkey, and probably southeastern Russia.

**Original sources:** Eastern Romania and eastern Austria.

## Biology

**Generations per year:** One.

**Overwintering stage:** Larval (in the root).

**Egg stage:** Eggs are laid singly or in small groups on the leaves. The eggs are oval, flattened, and prominently ribbed. Eggs are whitish when laid but turn yellowish later. They hatch in seven to nine days.

**Larval stage:** Upon hatching, larvae move to the center of the root and mine the crown. There are six larval instars.

**Pupal stage:** Pupation occurs within the root.

**Adult stage:** Adults are about 8 mm long and tan to gray with mottled wings.

*P. medullana* larva in a root. (Photo credit: Montana State University)

*P. medullana* adult. (Photo credit: USDA-ARS)

## Effect

**Destructive stage:** Larval.

**Plant species attacked:** Spotted knapweed and diffuse knapweed.

**Site of attack:** Root cortex.

**Impact on the host:** Larval feeding reduces knapweed biomass.

**Nontarget effects:** There have been no reports of attack on nontarget plants.

## Releases

**First introduced into the United States:** 1984, Montana.

**Established in:** Montana.

**Habitat:** This moth prefers hot, dry sites.

**Availability:** This moth is not available as of 2003.

**Stages to transfer:** Adult.

**Redistribution:** Collect the adults with a modified insect vacuum or rear them from infested roots.

## Comments

Usually only one larva develops per root, apparently due to intraspecific competition. This moth has been very slow to establish for unknown reasons, so availability is very limited.

# *Pterolonche inspersa*

J. M. Story and E. M. Coombs

**Common name:** Grey-winged root moth.

**Type of agent:** Insect: Moth (Lepidoptera: Pterolonchidae).

**Native distribution:** Bulgaria, France, Greece, Hungary, Italy, Romania, Russia, Spain, and Turkey.

**Original source:** Austria, Greece, and Hungary.

## Biology

**Generations per year:** One.

**Overwintering stage:** Larval (in the root).

**Egg stage:** Eggs are deposited on the lower leaf surface singly or in small groups. Eggs are oval, black, and 0.04 mm long and 0.03 mm wide. The incubation period lasts about 12 days.

*P. inspersa* larva in a root. (Photo credit: E. Coombs, Oregon Department of Agriculture)

**Larval stage:** Larvae mine in either the cortex or the center of the root and cause galls to form. Larvae construct "chimney" silk tubes that extend from the galls to above the soil surface from which the adults will later emerge.

**Pupal stage:** Pupation occurs within the root.

**Adult stage:** In Europe, adults are present in late July. The white to brownish-gray adults are about 8.2 mm long.

*P. inspersa* adult. (Photo credit: Montana State University)

## Effect

**Destructive stage:** Larval.

**Plant species attacked:** Diffuse knapweed, spotted knapweed, and squarrose knapweed.

**Site of attack:** Central vascular tissue of the root.

**Impact on the host:** In Europe, larval feeding reduced knapweed biomass.

**Nontarget effects:** There have been no reports of attack on nontarget plants.

## Releases

**First introduced into the United States:** 1986, Idaho, Oregon, and Utah.

**Established in:** Oregon, but it has not been recovered since 2000.

**Habitat:** The moth apparently prefers hot, dry areas with limited vegetation. It does not do well at higher elevations and sites with rainfall.

**Availability:** This moth is not available as of 2003.

**Stage to transfer:** Adult.

**Redistribution:** Optimum redistribution techniques have not yet been established.

## Comments

This insect has been very difficult to establish in the United States. The moth can co-occur in the root below the galls created by *Sphenoptera jugoslavica* and at one time infested nearly 20% of the plants at a site in Oregon. The moth has dwindled to undetectable levels in Oregon because of the dramatic control of diffuse knapweed by the seed head weevils.

# Sphenoptera jugoslavica

G. L. Piper

**Common name:** Bronze knapweed root borer.

**Type of agent:** Insect: Beetle (Coleoptera: Buprestidae).

**Native distribution:** Bulgaria, Greece, Macedonia, Romania, Serbia, and Turkey.

**Original source:** Greece.

## Biology

**Generations per year:** One.

**Overwintering stage:** Larval (in the root).

**Egg stage:** Eggs are placed in the leaf axils of knapweed rosette leaves during July and August. The flat eggs are white when deposited, but turn bluish-gray after four or five days.

**Larval stage:** The first-instar larva feeds in the leaf axil. The second-instar larva tunnels into the root. After overwintering in the root, the larva resumes feeding. The whitish larvae are elongated with the anterior part of the body flat and wider than the rest of the body, and they have dark brown head capsules. The galls and tunnels are often packed with sawdust-like frass from the larvae.

**Pupal stage:** Pupation occurs within the upper root from June through early September. The pupation period lasts about nine days.

**Adult stage:** Peak adult emergence occurs in July, coinciding with the onset of flowering in diffuse knapweed. Adults are 8 to 10 mm long, somewhat flattened, and metallic, dark reddish-brown.

## Effect

**Destructive stages:** Larval; there is some minor leaf feeding by the adults.

**Plant species attacked:** Diffuse knapweed is the preferred host, although the moth also attacks spotted knapweed and squarrose knapweed.

**Site of attack:** Center of the root, which becomes swollen.

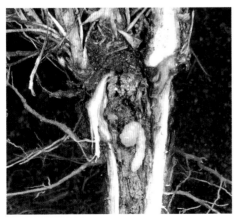

*S. jugoslavica* larva in a root. (Photo credit: N. Poritz, bio-control.com)

223

*S. jugoslavica* adult. (Photo credit: N. Poritz, bio-control.com)

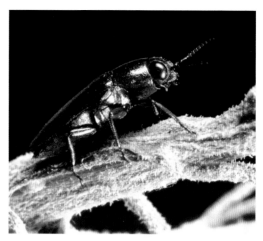

**Impact on the host:** The insect significantly reduces knapweed biomass, seed output, and density.

**Nontarget effects:** No nontarget effects have been reported.

## Releases

**First introduced into the United States:** 1980, California, Idaho, Oregon, and Washington.

**Established in:** California, Colorado, Idaho, Montana, Nevada, Oregon, Utah, Washington, and Wyoming.

**Habitat:** This beetle prefers hot, dry areas. Higher elevations and areas with higher, prolonged rainfall limit the insect's effectiveness.

**Availability:** This beetle is readily available in Oregon and Washington.

**Stage to transfer:** Adult.

**Redistribution:** Adults can be collected with a sweep net in mid-July in the early evening.

## Comments

Usually only a single larva develops within a root. Root aphids at some sites attract ants, which often kill larvae and pupae.

# *Subanguina picridis*

J. L. Littlefield and E. M. Coombs

**Common name:** Russian knapweed gall nematode.

**Type of agent:** Gall-forming nematode (Nematoda: Anguinidae).

**Native distribution:** Asia.

**Original sources:** Turkey and Uzbekistan.

## Biology

**Generations per year:** Multiple.

**Overwintering stage:** Second-stage larval (in disintegrating galls in the upper soil layers).

**Egg stage:** Females give birth to larval nematodes.

**Larval stage:** Larvae are long, slender, almost translucent, and microscopic. In early spring, the infective stage larvae are activated by moisture. They penetrate immature leaves and stems of the new, emerging shoots and galls eventually form at the infected sites. Nematodes multiply within these galls until August when the mature galls contain primarily second-stage larvae. These larvae disperse into the soil as the galls disintegrate and later become infective after at least a month in soil moisture.

**Adult stage:** Adults are present during the summer. This worm-like microscopic nematode is about 1.5 mm long.

Damage (galls) caused by *S. picridis* on Russian knapweed. (Photo credit: E. Coombs, Oregon Department of Agriculture)

## Effect

**Destructive stages:** Larval and adult.

**Plant species attacked:** Russian knapweed is a highly susceptible host and diffuse knapweed is a moderately susceptible host. Other closely related plant species were attacked during host specificity tests, but these failed to produce viable populations of the nematode.

**Site of attack:** Galls are formed on the stems, leaves, and root crowns. The galls are fuzzy and white and look like small tennis balls.

**Impact on the host:** Severely infested plants are often stunted and do not flower. The nematode is very effective in favorable climatic areas in Russia and adjacent countries, and is considered a valuable biocontrol agent there. In western North America, impact generally has been poor.

**Nontarget effects:** No nontarget effects have been reported.

## Releases

**First introduced into the United States:** 1984, Washington.

**Established in:** Colorado, Montana, Oregon, Utah, and Wyoming.

**Habitat:** This species does best in areas that are moist during winter or during the spring infection period. The nematode does not respond well in areas or years that are dry.

**Availability:** Availability is limited.

**Stage to transfer:** Larval, in dried galls.

225

**Redistribution:** Galls can be collected in the fall and placed upon the soil to permit the nematode larvae to emerge from disintegrating galls and penetrate the young knapweed shoots when they break through the soil in the spring. Nematodes may also be redistributed in contaminated soil, especially when the soil remains moist. In Uzbekistan, the nematode has been used as an augmentative biocontrol method. The nematodes are collected by soaking the galls in water, then sprayed over knapweed-infested fields. Nematodes can be stored for several years if kept in dry and frozen gall material.

## Comments

This was the first organism studied for biological control of Russian knapweed in the former USSR, Canada, and the United States and the first nematode to be introduced into North America for the biological control of a weed. Because it disperses so slowly, it has limited impact at most sites. At several sites in Oregon, native grasses have reappeared in small openings where the nematode has controlled the host. If the openings are too small, the weed will reinvade from the edges by the end of the growing season.

Releases have been made in Canada with limited success. Mass rearing and release of this nematode has not proved cost-effective as of 2003. This nematode is also referred to in the literature as *Paranguina picridis* or *Mesoanguina picridis*.

## *Terellia virens*

J. M. Story and E. M. Coombs

**Common names:** Green clearwing fly, verdant seed fly.

**Type of agent:** Insect: Fly (Diptera: Tephritidae).

**Native distribution:** Southern Austria, Czech Republic, southern France, Germany, Hungary, Israel, Italy, Jordan, Morocco, eastern Romania, Slovakia, southern Switzerland, and Turkey.

**Original source:** Austria.

### Biology

**Generations per year:** One to two.

**Overwintering stage:** Larval (in the seed head).

**Egg stage:** Each female produces about 80 eggs, which are inserted between the florets of newly opened flowers. Eggs are elongate, 1.0 mm long, shiny white, and hatch three to five days after deposition.

*T. virens* larva in a seed head. (Photo credit: Oregon Department of Agriculture)

**Larval stage:** Young larvae feed inside a single developing seed, but later attack other seeds and receptacle tissue. The larvae are white, but turn yellow-brown as they mature.

**Pupal stage:** First-generation larvae pupate upright above the receptacle in a loose cocoon of floral hairs. Second-generation larvae overwinter in a cocoon partially inserted into the receptacle. The existence of a second generation in North America has not yet been determined. The puparium is yellow-brown in color.

**Adult stage:** In Europe, adults occur from mid-May until late August, living from 22 to 48 days. Females begin laying eggs one week after emergence. Adult flies feed intensively on knapweed nectar. Adults are 4 to 5 mm long, greenish-gray, and have iridescent green eyes and clear wings.

Adult *T. virens*. (Photo credit: E. Coombs, Oregon Department of Agriculture)

## Effect

**Destructive stage:** Larval.

**Plant species attacked:** Spotted knapweed is the preferred host, but the fly also attacks diffuse knapweed.

**Site of attack:** Larvae feed on seeds in the flower head.

**Impact on the host:** Larval feeding reduces seed production.

**Nontarget effects:** There have been no reports of attack on nontarget plants.

## Releases

**First introduced into the United States:** 1992, Montana.

**Established in:** California and Oregon.

**Habitat:** The fly prefers cooler and wetter sites than the seed head weevils.

**Availability:** Limited numbers of this fly are available in Oregon.

**Stages to transfer:** Adult and larval.

**Redistribution:** Larvae-infested seed heads can be collected in the fall and stored at 4 to 8°C or collected in early spring and transported to release sites prior to adult emergence. To exclude other seed head insects and parasitoids, the adults should be recovered in a sleeve cage.

## Comments

The fly appears to be having a difficult time establishing at many sites because of competition with the *Urophora* spp. gall flies and seed head weevils.

# Urophora affinis

J. M. Story and E. M. Coombs

**Common name:** Banded gall fly.

**Type of agent:** Insect: Fly (Diptera: Tephritidae).

**Native distribution:** Europe and western Asia.

**Original sources:** Austria and France.

## Biology

**Generations per year:** One, and often a partial second generation.

**Overwintering stage:** Larval (in the seed head).

**Egg stage:** Eggs are deposited in June and July in immature knapweed flower heads and hatch in three to four days.

**Larval stage:** The developing larvae form hard, teardrop-shaped galls from receptacle tissues which may be misidentified by some as seeds. Larvae are white with distinct brown spiracular plates. There are three larval instars.

**Pupal stage:** Pupation occurs within the gall during late May for the first generation, and late August for the second generation. Pupation lasts about 14 days.

**Adult stage:** First-generation adults are present during June and July; second-generation adults appear in August and September. Adults are about 4 mm long, black, and have faint horizontal bands on the wings.

## Effect

**Destructive stage:** Larval.

**Plant species attacked:** Spotted knapweed, diffuse knapweed, and squarrose knapweed.

**Site of attack:** Developing flower heads.

*U. affinis* galls in a developing flower head. (Photo credit: N. Poritz, bio-control.com)

**Impact on the host:** The galls divert plant nutrients, resulting in reduced seed production. Each gall has been shown to reduce seed production in diffuse knapweed by 2.4 seeds per attacked head, and by 2.0 seeds per attacked head in spotted knapweed.

**Nontarget effects:** There have been no reports of attack on nontarget plants. Because deer mice feed heavily on the fly larvae, the flies may contribute to an increase in deer mice in knapweed-infested areas.

## Releases

**First introduced into the United States:** 1973, Montana and Oregon.

**Established in:** The fly is well established throughout most of the diffuse and spotted knapweed-infested areas of the United States, particularly the Northwest.

*U. affinis* adult. (Photo credit: N. Poritz, bio-control.com)

**Habitat:** The fly is distributed throughout most of the knapweed-infested range.

**Availability:** The fly is readily available throughout the northwestern United States.

**Stages to transfer:** Adult and larval.

**Redistribution:** Larvae-infested seed heads can be collected in early spring and transported to release sites prior to adult emergence, or collected in the fall and stored at 4 to 8° C. Adults can be reared from caged seed heads and then released in the field.

## Comments

*Urophora affinis* was the first insect released into the United States for knapweed control. It does not disperse as rapidly as the other introduced seed head gall fly, *U. quadrifasciata*, but has been the more persistent colonizer and is the dominant species at most North American sites where the two flies coexist. In areas where both flies are present, they are reducing spotted knapweed seed production by at least 50%. *Urophora* species are often destroyed by the seed head moth and weevils when they occur in the same seed head.

# Urophora quadrifasciata

J. M. Story and E. M. Coombs

**Common name:** UV knapweed seed head fly.

**Type of agent:** Insect: Fly (Diptera: Tephritidae).

**Native distribution:** Eastern Europe and western Asia.

**Original source:** Russia.

## Biology

**Generations per year:** Usually two.

**Overwintering stage:** Larval (in the seed head).

**Egg stage:** Eggs are deposited in June and July in immature knapweed flower heads and hatch in three to four days.

**Larval stage:** The developing larvae form papery galls from the developing seeds in the flower head. The larvae are white with distinct brown spiracular plates.

**Pupal stage:** Pupation occurs within the gall during late May for the first generation and late August for the second generation. The pupal period lasts about 14 days.

*U. quadrifasciata* adult. (Photo credit: N. Poritz, bio-control.com)

**Adult stage:** First-generation adults are present during June and July; second-generation adults appear in August and September. Adults are about 4 mm long, black, and have distinctive dark bands forming a "UV" pattern on each wing.

## Effect

**Destructive stage:** Larval.

**Plant species attacked:** Black, brown, diffuse, meadow, short-fringed (*C. nigrescens*), spotted, and squarrose knapweed, and cornflower.

**Site of attack:** Developing flower heads.

**Impact on the host:** The galls divert plant nutrients, resulting in reduced seed production in attacked seed heads.

**Nontarget effects:** There have been no reports of attack on nontarget plants. Because deer mice feed heavily on the fly larvae, the flies may contribute to an increase in deer mice in knapweed-infested areas.

## Releases

**First introduced into the United States:** The fly immigrated to Idaho, Montana, and Washington in 1981 from populations in Canada.

**Established in:** The fly is well established throughout most of the diffuse and spotted knapweed-infested areas of the United States, particularly the Northwest.

**Habitat:** The fly has established throughout most of the knapweed-infested range.

**Availability:** The fly is readily available throughout the northwestern United States.

**Stages to transfer:** Adult and larval.

**Redistribution:** Larvae-infested seed heads can be collected in early spring and transported to release sites prior to adult emergence, or collected in the fall and stored at 4 to 8˚ C. Adults can be reared from caged seed heads and then released in the field.

## Comments

*Urophora quadrifasciata* disperses more rapidly than *U. affinis*, but the latter insect has been the more persistent colonizer. In areas where the two flies coexist, they are reducing spotted knapweed seed production by at least 50%. Both *Urophora* species are often destroyed by the seed head moth and weevils when they occur in the same seed head.

## *References*

Callaway, R. M., T. H. DeLuca, and W. M. Belliveau. 1999. Biological control herbivores may increase competitive ability of the noxious weed *Centaurea maculosa*. Ecology 80: 1196-1201.

Groppe, K., and K. Marquardt. 1989. *Terellia virens* (Loew) (Diptera: Tephritidae), a suitable candidate for the biological control of diffuse and spotted knapweed in North America. Report. CAB, IIBC, European Station, Delémont, Switzerland.

Harris, P., and J. H. Myers. 1984. *Centaurea diffusa* Lam. and *C. maculosa* Lam. *s. lat.*, diffuse and spotted knapweed (Compositae). Pages 127-37 *in* J. S. Kelleher and M. A. Hulme, eds. Biological Control Programmes Against Insects and Weeds in Canada 1969-1980. CAB: Page Bros. (Norwich) Ltd.

Jordan, K. 1995. Host specificity of *Larinus minutus* Gyll. (Col., Curculionidae), an agent introduced for the biological control of diffuse and spotted knapweed in North America. J. Appl. Entomol. 119: 689-93.

Lang, R. F., J. M. Story, and G. L. Piper. 1996. Establishment of *Larinus minutus* Gyllenhal (Coleoptera: Curculionidae) for biological control of diffuse and spotted knapweed in the western United States. Pan-Pac. Entomol. 72: 209-12.

Maddox, D. M. 1982. Biological control of diffuse knapweed (*Centaurea diffusa*) and spotted knapweed (*C. maculosa*). Weed Sci. 30: 76-82.

Müller, H., D. Schroeder, and A. Gassmann. 1988. *Agapeta zoegana* (L.) (Lepidoptera: Cochylidae), a suitable prospect for biological control of spotted and diffuse knapweed, *Centaurea maculosa* Monnet de la Marck and *Centaurea diffusa* Monnet de la Marck (Compositae) in North America. Can. Entomol. 120: 109-24.

Ochsmann, J. 2001. On the taxonomy of spotted knapweed (*Centaurea stoebe* L.). Pages 33-41 *in* L. Smith, ed. Proc. First Int. Knapweed Symp. of the Twenty-First Century, 15-16 March 2001, Coeur d'Alene, ID. USDA-ARS/WRRC-EIW, Albany, CA.

Pearson, D. E., K. S. McKelvey, and L. F. Ruggiero. 2000. Nontarget effects of an introduced biological agent on deer mouse ecology. Oecologia 122: 121-28.

Piper, G. L., and S. S. Rosenthal. 1995. 64. Diffuse knapweed *Centaurea diffusa* Lamarck (Asteraceae). Pages 237-41 *in* J. R. Nechols, L .A. Andres, J. W. Beardsley, R. D. Goeden, and C. G. Jackson, eds. Biological Control in the Western United States: Accomplishments and Benefits of Regional Research Project W-84, 1964-1989. Univ. Calif. Div. Agric. Nat. Res. Pub. 3361. Oakland, CA.

Roché, C., and L. C. Burrill. 1992. Squarrose knapweed. Pacific Northwest Ext. Pub. PNW 422.

Roché, C. T., and B. F. Roché. 1989. Introductory notes on squarrose knapweed (*Centaurea virgata* Lam. ssp. *squarrosa* Gugl.). Northwest Sci. 63: 246-52.

Rosenthal, S. S., and G. L. Piper. 1995. 69. Russian knapweed, *Centaurea* (*Acroptilon*) *repens* L. (Asteraceae). Pages 256-57 *in* J. R. Nechols, L. A. Andres, J. W. Beardsley, R.D. Goeden, and C.G. Jackson, eds. Biological Control in the Western United States: Accomplishments and Benefits of Regional Research Project W-84, 1964-1989. Univ. Calif. Div. Agric. Nat. Res. Pub. 3361. Oakland, CA.

Rosenthal, S. S., E. M. Coombs, B. Haglan, and G. L. Piper. 1993. Use of critical nontarget plants by the Russian knapweed nematode, (*Subanguina picridis*). Weed Technol. 7: 759-62.

Rosenthal, S. S., G. Campobasso, L. Fornasari, R. Sobhian, and C.E. Turner. 1991. Biological control of *Centaurea* spp. Pages 292-302 *in* L.F. James, J.O. Evans, M.H. Ralphs, and R.D. Child, eds. Noxious Range Weeds. Westview Press, Boulder, CO.

Schirman, R. 1981. Seed production and spring seedling establishment of diffuse and spotted knapweed. J. Range Manage. 34: 45-47.

Sheley, R. L., J. S. Jacobs, and M. F. Carpinelli. 1998. Distribution, biology, and management of diffuse knapweed (*Centaurea diffusa*) and spotted knapweed (*Centaurea maculosa*). Weed Technol. 12: 353-62.

Smith, L., ed. 2001. Proc. First Int. Knapweed Symp. of the Twenty-First Century, 15-16 March 2001, Coeur d'Alene, ID. USDA-ARS/WRRC-EIW, Albany, CA.

Sobhian, R., G. Campobasso, and P. H. Dunn. 1992. A contribution to the biology of *Bangasternus fausti* (Col.: Curculionidae), a potential biological control agent of diffuse knapweed, *Centaurea diffusa*, and its effect on the host plant. Entomophaga 37: 171-79.

Stinson, C. S. A., D. Schroeder, and K. Marquardt. 1994. Investigations on *Cyphocleonus achates* (Fahr.) (Col. Curculionidae), a potential biological agent of spotted knapweed (*Centaurea maculosa* Lam.) and diffuse knapweed (*C. diffusa* Lam.) (Compositae) in North America. J. Appl. Entomol. 117: 35-50.

Story, J. M. 1995. 70. Spotted knapweed *Centaurea maculosa* Lamarck (Asteraceae). Pages 258-63 *in* J. R. Nechols, L. A. Andres, J. W. Beardsley, R. D. Goeden, and C. G. Jackson, eds. Biological Control in the Western United States: Accomplishments and Benefits of Regional Research Project W-84, 1964-1989. Univ. Calif. Div. Agric. Nat. Res. Pub. 3361. Oakland, CA.

Story, J. M. 2002. Spotted knapweed. Pages 169-80 *in* R. Van Driesche, S. Lyon, B. Blossey, M. Hoddle, and R. Reardon, eds. Biological Control of Invasive Plants in the Eastern United States. USDA Forest Serv. Pub. FHTET-2002-04. Morgantown, WV.

Story, J. M., and G. L. Piper. 2001. Status of biological control efforts against spotted and diffuse knapweed. Pages 11-17 *in* L. Smith, ed. Proc. First Int. Knapweed Symp. of the Twenty-First Century, 15-16 March 2001, Coeur d'Alene, ID. USDA-ARS/WRRC-EIW, Albany, CA.

Story, J. M., W. R. Good, and L. J. White. 2001. Response of the knapweed biocontrol agent, *Agapeta zoegana* L. (Lepidoptera: Cochylidae), to portable lights. Pan-Pac Entomol. 77: 219-25.

Story, J. M., W. R. Good, L. J. White, and L. Smith. 2000. Effects of the interaction of the biocontrol agent, *Agapeta zoegana* L. (Lepidoptera: Cochylidae), and grass competition on spotted knapweed. Biol. Control 17: 182-90.

Story, J. M., K. W. Boggs, W. R. Good, P. Harris, and R. M. Nowierski. 1991. *Metzneria paucipunctella* Zeller (Lepidoptera: Gelechiidae), a moth introduced against spotted knapweed: Its feeding strategy and impact on two introduced *Urophora* spp. (Diptera: Tephritidae). Can. Entomol. 123: 1001-7.

Story, J. M., K. W. Boggs, W. R. Good, L. J. White, and R. M. Nowierski. 1995. Cause and extent of predation on *Urophora* spp. larvae (Diptera: Tephritidae) in spotted knapweed capitula. Environ. Entomol. 24: 1467-72.

Watson, A. K., and A. J. Renney. 1974. The biology of Canadian weeds. 6. *Centaurea diffusa* and *C. maculosa*. Can. J. Plant Sci. 54: 687-701.

# Leafy spurge

*Euphorbia esula* (complex)

Spurge family—Euphorbiaceae

*R. W. Hansen, N. R. Spencer, L. Fornasari, P. C. Quimby, Jr., R. W. Pemberton, and R. M. Nowierski*

**Additional common name:** Faitour's grass.

**Native range:** Eurasia.

**Entry into the United States:** The plant was first reported in the United States in 1827.

## Biology

**Life duration/habit:** Leafy spurge is an aggressive, persistent, deep-rooted perennial, growing to a height of 1 m or taller. Vegetative stems manufacture sugars for root reserves while other stems produce flowers.

**Reproduction:** Leafy spurge reproduces by vegetative regrowth from spreading roots and by the production of large quantities of seeds that are distributed by birds, wildlife, humans, and in rivers and streams.

**Roots:** Leafy spurge roots are brown with pinkish buds. Plants are able to maintain high root reserves through an extensive root system, ranging from a massive network of small lateral roots near the soil surface (within 30.5 cm) to deep, penetrating taproots that may extend to depths of 3 to 7 m. This ability to maintain high root reserves permits the plant to recover quickly from physical and most herbicide damage.

**Stems and leaves:** The stems are thickly clustered and bear narrow, 2.5- to 10-cm-long leaves that are alternately arranged along the stems. When damaged, leaves and stems exude a milky latex.

**Flowers:** The small flowers are yellowish-green, arranged in clusters, and enclosed in yellow-green bracts.

**Fruits and seeds:** Seeds are oblong and gray to purple, and occur in clusters of three. When dry, the seed capsules shatter, scattering seeds away from the plant.

## Infestations

**Worst infested states:** Leafy spurge now occurs across southern Canada and the northern United States, and is approaching areas as far south as Texas. The largest infestations in the United States generally occur from

Leafy spurge. (Photo credit: N. Poritz, bio-control.com)

Leafy spurge infestation in Montana. (Photo credit: N. Poritz, bio-control.com)

Minnesota west to eastern Washington and Oregon and south to Colorado and Nebraska.

**Habitat:** Leafy spurge occurs primarily in grassland and rangeland habitats, but also infests forested and riparian areas and abandoned cropland.

**Impacts:** Leafy spurge is an aggressive competitor that displaces native and introduced forage plants, reducing the carrying capacity of infested rangeland for cattle. Its milky latex is poisonous to some animals and can cause blistering and skin irritation. The digestive tract is similarly affected when this plant is eaten by humans and some animals. In cattle it causes scours and weakness; when ingested in larger amounts it can cause death. Cattle usually refuse to eat leafy spurge unless it is given to them in dry, weedy hay or when better forage is not available. They generally avoid spurge-infested range, even when palatable plants are present.

A conservative 1979 estimated loss in the United States of $10.5 million annually was based on expenditures for controlling leafy spurge and loss of productivity. Although leafy spurge infestations are most severe on undisturbed lands, on cultivated cropland the weed can reduce crop yields from 10 to 100%. A 1990 study conducted by North Dakota State University estimated that the annual financial losses in Montana, North Dakota, South Dakota, and Wyoming exceed $130 million and are associated with the potential loss of 1,433 jobs.

## Comments

Leafy spurge is extremely difficult to control with herbicides and almost impossible to control by cultural or physical methods. It apparently has the ability to purge undesirable chemicals from the root system in approximately the top 45 cm of the soil, allowing the remaining portion of the root system to regenerate as soon as the effect of the chemical in the soil has dissipated.

Although leafy spurge causes problems with cattle that consume it, sheep generally can be taught to feed on it and goats will seek it out. Both sheep and goats are used in weed control programs to reduce flowering and seed production and to retard its spread.

People should handle the plant with caution because the latex can cause irritation, blotching, blisters, and swelling in sensitive individuals. The eyes should never be rubbed until after hands have been thoroughly washed. The dried latex is often very difficult to wash off; consider wearing lightweight latex gloves when handling the plant.

This chapter and many of the descriptions of biological control agents are largely based on information previously provided by Gaetano Campobasso, Robert Carlson, Paul Dunn, Luca Fornasari, Peter Harris, Eric Maw, Alec McClay, Robert Nowierski, Pasquale Pecora, Robert Pemberton, Chuck Quimby, Jr., Norman Rees, Robert Richard, and Neal Spencer in "Biological Control of Weeds of the West."

# Aphthona abdominalis

R. W. Hansen

**Common name:** Minute spurge flea beetle.

**Type of agent:** Insect: Beetle, flea beetle (Coleoptera: Chrysomelidae).

**Native distribution:** Found in northern and central Italy, Spain, France, southern Poland, Austria, eastern Europe, southwestern Asia, and northwestern Iran.

**Original source:** Europe.

## Biology

**Generations per year:** Up to four.

**Overwintering stages:** Larval and adult.

**Egg stage:** The eggs are laid singly or in clusters of two to six on plant stems near the soil surface or in the soil near the plants from April until October. Females lay as many as 100 eggs, which require three to six days of incubation during ideal conditions or up to 16 days under harsh conditions. The eggs require high relative humidity to survive.

**Larval stage:** There are three larval instars. The larvae feed on young roots, root buds, and subterranean shoots. These larvae are somewhat C-shaped and have prominent head capsules.

**Pupal stage:** Pupation occurs in the soil, the pupal stage lasting 10 to 11 days.

**Adult stage:** Adults live for 40 to 55 days, and those that are alive in December in Italy (generally fourth generation) begin diapause.

*A. abdominalis* adult. (Photo credit: R. Richard, USDA-APHIS)

The adults usually hide among plant duff and under stones and branches on the soil. This species is more gray to straw-colored than the other *Aphthona* species. The head, prothorax, and mesothorax are reddish-yellow and the abdomen and metathorax are black. This species is also relatively small, measuring an average of 2.0 mm long by 1.0 mm wide. The outer pair of wings is transparent and straw-colored.

## Effect

**Destructive stages:** Larval and adult.

**Additional plant species attacked:** None.

**Site of attack:** Adult beetles feed on the leaves and flowers while the larvae feed on root hairs and young roots.

**Impact on the host:** The larvae seriously damage shoots, shoot buds, and roots, and in heavy attacks plants are stressed and cannot produce new stems. Such injury reduces the plant's ability to take up moisture and nutrients, while the adults' feeding on the foliage, especially when it is concentrated on the youngest leaves at the tip of the plant, decreases the plant's sugar-making ability for root reserves. Larval root damage may also provide entry points for soil-inhabiting pathogenic fungi. Since this species feeds on the foliage, flowers, root hairs and young roots, it has great potential, under ideal conditions, to be extremely effective in reducing plant density. It is also possible that its four generations per year could give it an advantage over other flea beetle species in some locations. However, this beetle has apparently not become established after multiple releases in the United States.

**Nontarget effects:** No nontarget effects have been reported.

## Releases

**First introduced into the United States:** 1993, Montana and North Dakota.

**Established in:** Not established.

**Habitat:** European data suggest that favorable areas are more moist than for *Aphthona nigriscutis* but drier than for *A. czwalinae,* and that this species may not do well in heavy clay soils.

**Availability:** This species is unavailable as of 2004.

## Comments

*Aphthona abdominalis* is a grayish color and about one-third to one-fourth the size of most other introduced *Aphthona* spp. This species also has four generations per year and overwinters as larvae and/or adults, whereas the other approved species have only one generation per year and overwinter as larvae. A small native flea beetle in the genus *Glyptina* is often present in leafy spurge infestations in low numbers and may be confused with *A. abdominalis.*

# *Aphthona cyparissiae*

R. W. Hansen

**Common name:** Brown dot leafy spurge flea beetle.

**Type of agent:** Insect: Beetle, flea beetle (Coleoptera: Chrysomelidae).

**Native distribution:** Europe.

**Original sources:** Europe, Austria, Hungary, and Italy.

## Biology

**Generations per year:** One.

**Overwintering stages:** Mature larval or prepupal.

**Egg stage:** Eggs are generally laid on the lower stem next to the soil or on the soil next to the stem during July, August, and September. They hatch in about 13 days.

**Larval stage:** The larvae are active from August until early spring. There are three larval growth stages: the first lasts about eight days (under ideal conditions), the second lasts about 25 to 30 days, and the final stage requires about 45 days. A cold period is needed to cause the mature larvae to pupate. The head is well sclerotized and subcompressed and the body is whitish in color.

**Pupal stage:** Pupation occurs in the soil and lasts about 20 days.

**Adult stage:** Adult beetles are found on the leafy spurge plants from late June until about September, with many individuals surviving and laying eggs for three to four months. Adults are oval and brown and measure about 3.2 mm long.

## Effect

**Destructive stages:** Adult and larval.

**Additional plant species attacked:** Cypress spurge (*Euphorbia cyparissias*).

**Site of attack:** Adult beetles feed on leafy spurge leaves and flowers while larvae feed on or in root hairs and young roots.

**Impact on the host:** Adult feeding on foliage reduces the plant's photosynthetic production of sugars for the roots, while larval feeding on or in the root hairs and young roots reduces the plant's ability to take up moisture and nutrients. This decreases the height attained by the plant, delays the flowering period, and causes the plant to take its nourishment from the taproot. Over prolonged periods, continuous pressure by the beetles weakens the taproot, resulting in the death of the plant. Larval root damage

*A. cyparissiae* adult. (Photo credit: N. Poritz, bio-control.com)

237

may also provide entry points for soil-inhabiting pathogenic fungi. Generally, when *Aphthona* spp. successfully establish, definite impacts on leafy spurge and nontarget vegetation are observed three to five years after release. These include significant reductions in leafy spurge cover, density, and aboveground and root biomass, and significant increases in abundance, cover, and biomass of grasses and forbs.

**Nontarget effects:** No nontarget effects have been reported.

## Releases

**First introduced into the United States:** 1987, Montana.

**Established in:** Colorado, Idaho, Iowa, Minnesota, Montana, Nebraska, Nevada, New Mexico, New York, North Dakota, Oregon, Rhode Island, South Dakota, Utah, Washington, Wisconsin, and Wyoming.

**Habitat:** Canadian research suggests that green needle grass be present as an indicator, that flowering spurge stems be taller than 51 cm, that the density be between 50 and 125 stems/m$^2$, and that the soils be 40 to 60% sand. These requisites are often found on dry alluvial fans. *A. cyparissiae* may prefer warm, open, sunny areas and slightly more moist conditions than *A. nigriscutis*, but less moist than for *A. flava*. *A. cyparissiae* can survive subfreezing winter temperatures.

**Availability:** This insect is available in states with established populations.

**Stage to transfer:** Adult.

**Redistribution:** Collect the beetles with a sweep net. After they are sorted, they can be shipped or stored on leaf material for several days if kept cool, or for several weeks under cool temperatures with intermittent warm feeding periods. To release, sprinkle beetles among moderately dense leafy spurge plants. Areas of high ant activity should be avoided.

## Comments

After this species is released, leafy spurge plant density is greatly reduced at first. However, roots that are not attacked (including the taproot) are able to send up small new shoots to supply the sugars for root reserves. It is only through the persistence of the beetle over a long period of time in ecosystems that favor beetle development that the lateral roots of leafy spurge will be destroyed, the taproots weakened, and the plants eliminated from the area.

Host preference in Europe seems to be for *E. cyparissias, E. esula, E. seguieriana*, and *E. virgata*, in that order. Most collections have been made in mesic to dry areas with sandy to sand-gravel soils and sparse vegetation. Like many of the other flea beetle species that attack leafy spurge, *A. cyparissiae* tends to congregate for feeding, mating, and egg laying.

# Aphthona czwalinae

R. W. Hansen

**Common name:** Black leafy spurge flea beetle.

**Type of agent:** Insect: Beetle, flea beetle (Coleoptera: Chrysomelidae).

**Native distribution:** This species is found from central and eastern Europe to central Asia and eastern Siberia. Eastern Austria and northwestern Hungary are probably the southwestern limits of its range.

**Original source:** Hungary.

## Biology

**Generations per year:** One.

**Overwintering stage:** Mature larval (in or on leafy spurge roots).

**Egg stage:** Eggs are generally deposited in the ground next to leafy spurge stems during July, August, and September. The yellowish eggs are oval and measure 0.66 by 0.36 mm. They hatch in 16 to 17 days.

**Larval stage:** There are three larval instars. The larvae are slender, elongate, and whitish except for the head which is light brown. North Dakota studies suggest that *Aphthona* flea beetles require fine and lateral spurge roots within 5 cm of the soil surface to facilitate larval root feeding.

**Pupal stage:** Pupation occurs within a soil cell from late spring to early summer.

**Adult stage:** Adults can be located on leafy spurge plants from mid-June to August. Adults are black, while the front and middle femora (upper part of the legs) are yellow-brown. The hind femora are dark-colored. The males are about 2.9 mm long while the females measure about 3.1 mm. This species is slightly smaller than *A. cyparissiae* and *A. flava*.

## Effect

**Destructive stages:** Adult and larval.

**Plant species attacked:** *Euphorbia virgata* and *E. esula* (complex).

*A. czwalinae* adult. (Photo credit: N. Poritz, bio-control.com)

**239**

**Site of attack:** Adults feed on leaves and flowers and larvae feed in or on root hairs and young roots.

**Impact on the host:** Like the other flea beetle species, the adults feed on the leaves creating "shot-holes" that reduce the plant's photosynthetic production of sugars for the root. The larvae feed on the root hairs and young roots and reduce the plant's ability to take up moisture and nutrients. Moderate feeding reduces the plant's potential height and delays the flowering period. More intense feeding diminishes the number and vigor of aboveground plant stems. Larval root damage may also provide entry points for soil-inhabiting pathogenic fungi. Generally, when *Aphthona* spp. successfully establish, definite impacts on leafy spurge and nontarget vegetation are observed three to five years after release. These include significant reductions in leafy spurge cover, density, and aboveground and root biomass, and significant increases in abundance, cover, and biomass of grasses and forbs.

**Nontarget effects:** No nontarget effects have been reported.

### Releases

**First introduced into the United States:** 1987, Montana.

**Established in:** Colorado, Idaho, Iowa, Michigan, Minnesota, Montana, Nebraska, Nevada, New Hampshire, New Mexico, New York, North Dakota, Oregon, Rhode Island, South Dakota, Utah, Washington, Wisconsin, and Wyoming.

**Habitat:** This species does well in continental climates with warm and dry summers. Sites with relatively high humidity and mesic, loamy soils where the host plant is growing with other vegetation appear to be preferred. This species may also survive in well-drained, sandy or rocky, sun-exposed sites. Undesirable habitats include those with clay soils, open or dry sites, and areas with high ant populations. This species survives very cold, subfreezing winter temperatures.

**Availability:** This insect is readily available from populations mixed with *A. lacertosa*.

**Stage to transfer:** Adult.

**Redistribution:** Collect the beetles with a sweep net from mid-June through July. They can be shipped for several days with fresh leaves, and they can be stored under cool conditions for longer periods if they are provided with intermittent warm feeding periods. Maintaining them in cages at room temperature is possible if they are provided with fresh leafy spurge leaves, but the longer they remain in the cage, the fewer eggs will be deposited in the field. To release the adults, either sprinkle them on leafy spurge plants of moderate density, or quickly swing your arm with the open container to distribute them over a wider distance.

### Comments

This is one of two black flea beetles approved for release in the United States. It is generally established in mixed populations with *A. lacertosa*, although *A. lacertosa* is usually more abundant. Adults are about 3 mm long, black, and resemble *A. lacertosa*. The hind legs of *A. czwalinae* are dark brown or black next to the body, while the hind legs of *A. lacertosa* adults are yellow-brown in the similar position. However, these differences are difficult to detect with the naked eye. Like many of the other flea beetle species that attack leafy spurge, this species tends to congregate for feeding, mating, and egg laying.

After this species is released, leafy spurge plant density is greatly reduced at first. However, roots that are not attacked (including the taproot) are able to send up small new shoots to supply the sugars for root reserves. It is only through the persistence of the beetles over a long period of time in ecosystems that favor beetle survival that the lateral roots of leafy spurge will be destroyed, the taproots weakened, and the plants eliminated from the area.

# *Aphthona flava*

R. W. Hansen

**Common name:** Copper or amber leafy spurge flea beetle.

**Type of agent:** Insect: Beetle, flea beetle (Coleoptera: Chrysomelidae).

**Native distribution:** Europe.

**Original sources:** Italy and Hungary.

## Biology

**Generations per year:** One.

**Overwintering stage:** Larval (within young leafy spurge roots).

**Egg stage:** The eggs are deposited in June through early fall, generally on the plant stem at or below the soil surface, and sometimes on the soil but near the plant stem.

**Larval stage:** The larvae are active from July through early spring of the following year. The young larvae begin feeding in or on the root hairs; as they become older and larger, they migrate to the larger roots. They are difficult to observe except under a microscope. The more mature, whitish, worm-like larvae can be observed with the naked eye in freshly extracted roots. North Dakota studies suggest that *Aphthona* flea beetles require fine and lateral spurge roots within 5 cm of the soil surface to facilitate larval root feeding.

**Pupal stage:** The pupal stage occurs in late spring and early summer.

**Adult stage:** Adults emerge in June through early fall, depending on degree-days. This species is larger

*A. flava* adult. (Photo credit: N. Poritz, bio-control.com)

and more orange than *A. cyparissiae* and *A. nigriscutis.* It has the characteristic flea beetle appearance and jumps when disturbed. Adult males are about 3.4 mm long; females are about 3.6 mm long.

## Effect

**Destructive stages:** Adult (on the leaves) and larval (root hairs and young roots).

**Additional plant species attacked:** None.

**Site of attack:** Adult beetles feed on leaves and flowers; larvae feed in or on root hairs and young roots.

**Impact on the host:** Feeding on the foliage reduces photosynthesis, and flower consumption slightly reduces seed production. Feeding within the roots reduces the plant's ability to absorb moisture and nutrients. Larval root damage may also provide entry points for soil-inhabiting pathogenic fungi. Light populations reduce plant height and retard flowering, while high populations reduce plant density and cause

Damage to leafy spurge caused by *A. flava.* (Photo credit: E. Coombs, Oregon Department of Agriculture)

what is often referred to as "a hole in the spurge," which is characteristically grayish from the previous year's dead stems. Generally, when *Aphthona* spp. successfully establish, definite impacts on leafy spurge and nontarget vegetation are observed three to five years after release. These include significant reductions in leafy spurge cover, density, and aboveground and root biomass, and significant increases in abundance, cover, and biomass of grasses and forbs. At one research site this species reduced the aerial portion of leafy spurge in a 212 by 167 m area in six years from 57% to less than 2%.

**Nontarget effects:** No nontarget effects have been reported.

## Releases

**First introduced into the United States:** 1985, Montana.

**Established in:** Colorado, Idaho, Iowa, Michigan, Minnesota, Montana, Nebraska, New Hampshire, New Mexico, New York, North Dakota, Oregon, Rhode Island, South Dakota, Utah, Washington, Wisconsin, and Wyoming.

**Habitat:** Sunny locations are desirable. The beetles are hard to establish at sites with clay or acidic soils and in deeply shaded areas. They survive subfreezing winter temperatures.

242

**Availability:** The beetles are readily available in some locations.

**Stage to transfer:** Adult.

Leafy spurge infestation near Bozeman, MT. (Photo credit: USDA-ARS)

Same site several years after release of *A. flava*. (Photo credit: USDA-ARS)

**Redistribution:** Collect the beetles with a sweep net from late June through mid-August. Beetles can be kept several days at room temperature if given fresh leafy spurge leaves and confined in containers. The beetles can also be kept for several weeks at room temperature if kept in large cages and given fresh food, or for several weeks if kept cool and fed periodically at room temperature. However, the longer they are kept in captivity, the fewer eggs will remain for the field. To release, sprinkle beetles on moderately dense leafy spurge plants. Areas of high ant activity should be avoided for initial releases.

## Comments

This was the first leafy spurge flea beetle released in the United States. In one area near Bozeman, Montana, its effect has been spectacular. At many other sites, however, *A. flava* has persisted at fairly low levels with little noticeable impact on leafy spurge infestations. Like many of the other flea beetle species that attack leafy spurge, this species tends to congregate for feeding, mating, and egg laying.

After this species is released, leafy spurge plant density is greatly diminished at first. However, roots that were not attacked (including the taproot) are able to send up new shoots to supply the sugars for root reserves. It is only through the persistence of the beetles over a long period of time in ecosystems that favor beetle survival that the lateral roots of leafy spurge are destroyed, the taproots weakened, and the plants eliminated from the area.

243

# Aphthona lacertosa

R. W. Hansen

**Common name:** Brown-legged leafy spurge flea beetle.

**Type of agent:** Insect: Beetle, flea beetle (Coleoptera: Chrysomelidae).

**Native distribution:** Austria, Italy, and eastern Europe.

**Original source:** Eastern Europe.

## Biology

**Generations per year:** One.

**Overwintering stage:** Larval (within or on spurge roots).

**Egg stage:** The eggs are deposited in small batches underground near the root of their host over a period of several months during the summer.

**Larval stage:** Upon hatching, the larvae migrate to the root hairs and feed through the summer until cool temperatures cause them to depart the plant and enter the soil where they pupate. North Dakota studies suggest that *Aphthona* flea beetles require fine and lateral spurge roots within 5 cm of the soil surface to facilitate larval root feeding.

**Pupal stage:** Pupation occurs in the soil near the host plant.

**Adult stage:** The adults emerge in early summer and feed on the leaves of leafy spurge. Each female produces 200 to 300 eggs. Adults are about 3 mm long, black, and resemble *A. czwalinae*. The hind legs of *A. lacertosa* adults are yellow-brown next to the body, while the hind legs of *A. czwalinae* are dark brown or black in the similar position. However, these differences are difficult to detect with the naked eye.

## Effect

**Destructive stages:** Larval and adult.

**Additional plant species attacked:** Cypress spurge *(E. cyparissias)*, and *E. virgata*, with a lesser preference for *E. lucida* and *E. stepposa*. In laboratory studies, survival of this species was high on *E. myrsinites* and *E. seguieriana*. This species does not survive outside the subgenus *Esula*.

**244**

**Site of attack:** Adult beetles feed on leaves and flowers; larvae feed in or on root hairs and young roots.

*A. lacertosa* adults feeding on leafy spurge. (Photo credit: N. Poritz, bio-control.com)

**Impact on the host:** As with the other flea beetle species, the beetles reduce the plant's root reserves and diminish its ability to replace them. Larval root damage may also provide entry points for soil-inhabiting pathogenic fungi. Since the beetles are concentrated in the feeding areas, the effects are obvious. In low populations the affected plants are shorter and have delayed flowering periods. High concentrations of the beetles reduce plant density, causing what often is referred to as "a hole in the spurge." Generally, when *Aphthona* spp. successfully establish, definite impacts on leafy spurge and nontarget vegetation are observed three to five years after release. These include significant reductions in leafy spurge cover, density, and aboveground and root biomass, and significant increases in abundance, cover, and biomass of grasses and forbs.

**Nontarget effects:** No nontarget effects have been reported.

## Releases

**First introduced into the United States:** 1993, North Dakota.

**Established in:** Colorado, Idaho, Iowa, Michigan, Minnesota, Montana, Nevada, New Hampshire, New Mexico, New York, North Dakota, Oregon, Rhode Island, South Dakota, Utah, Washington, Wisconsin, and Wyoming.

**Habitat:** This species appears to do best at open, sunny, mesic to moderately dry sites in North America. It survives very cold, subfreezing winter temperatures.

**Availability:** This beetle is readily available from populations mixed with *A. czwalinae*.

**Stage to transfer:** Adult.

**Redistribution:** Collect the beetles from leafy spurge plants with a sweep net during the summer. The storage and shipping times are similar to those of other *Aphthona* spp. Sprinkle beetles in moderately dense leafy spurge. Areas of high ant activity should be avoided.

## Comments

*Aphthona lacertosa* and *A. czwalinae* are the only two black flea beetles released for leafy spurge control. Both species are present at most sites, although *A. lacertosa* is often considerably more abundant.

In host testing, *A. lacertosa* readily accepted Canadian leafy spurge as strongly as it did its native hosts. It has a strong preference for *E. virgata* and its host range is limited to a few species in the subgenus *Esula*. Judging from its distribution in Europe, this species is expected to establish in southern Canada and the northern part of the United States, but will probably not establish in the southern part of the United States. Like many of the other flea beetle species that attack leafy spurge, this species tends to congregate for feeding, mating, and egg laying.

After this species is released, leafy spurge plant density is greatly diminished at first. However, roots that are not attacked (including the taproot) are able to send up new shoots to supply the sugars for root reserves. It is only through the persistence of the beetles over a long period of time in ecosystems that favor beetle survival that the lateral roots of the leafy spurge will be destroyed, the taproots weakened, and the plants eliminated from the immediate area.

245

# *Aphthona nigriscutis*

R. W. Hansen

**Common name:** Black dot leafy spurge flea beetle.

**Type of agent:** Insect: Beetle, flea beetle (Coleoptera: Chrysomelidae).

**Native distribution:** Europe.

**Original sources:** Europe, including Hungary.

## Biology

**Generations per year:** One.

**Overwintering stage:** Larval (within the spurge roots).

**Egg stage:** The eggs are laid on the stem of the plant near or below the soil surface.

**Larval stage:** Larvae can be found from July to early spring of the following year. North Dakota studies suggest that *Aphthona* flea beetles require fine and lateral spurge roots within 5 cm of the soil surface to facilitate larval root feeding.

**Pupal stage:** Pupation occurs in the soil near the plant.

**Adult stage:** Adults are in the field in late June, July, and August. They are yellowish brown or brownish with a black dot on the back behind the thorax at the leading edge of the wings.

## Effect

**Destructive stages:** Adult and larval.

**Additional plant species attacked:** None.

**Site of attack:** Adult beetles feed on the leaves and flowers while larvae feed on the root hairs and young roots.

**Impact on the host:** Adult feeding on the foliage causes some injury, but larval feeding in and on the root hairs and young roots causes the greatest damage. The former reduces the plant's ability to make sugars for the root reserves, and the latter impairs the roots

*A. nigriscutis* adult. (Photo credit: N. Poritz, bio-control.com)

from taking up moisture and nutrients, thus reducing the potential plant height and retarding the flowering period. Larval root damage may also provide entry points for soil-inhabiting pathogenic fungi. Higher concentrations of the beetles often reduce plant density, causing what often is referred to as "a hole in the spurge." Generally, when *Aphthona* spp. successfully establish, definite impacts on leafy spurge and nontarget vegetation are observed three to five years after release. These include significant reductions in leafy spurge cover, density, and aboveground and root biomass, and significant increases in abundance, cover, and biomass of grasses and forbs.

**Nontarget effects:** Some feeding on the native Rocky Mountain spurge, *Euphorbia robusta*, has been observed at several locations in the western United States. It is not clear what, if any, impacts this feeding may have.

## Releases

**First introduced into the United States:** 1989, Montana.

**Established in:** Colorado, Idaho, Iowa, Michigan, Minnesota, Montana, Nebraska, Nevada, New Hampshire, New Mexico, New York, North Dakota, Oregon, Rhode Island, South Dakota, Utah, Washington, Wisconsin, and Wyoming.

**Habitat:** This agent is best suited to dry sites with maximum solar exposure. Canadian researchers recommend sites with needle and thread or porcupine grasses (*Stipa* spp.), flowering spurge stems less than 70 cm tall with fewer than 60 stems/m², and well-drained soils with less than 3% organic matter. It survives very cold, subfreezing winter temperatures.

**Availability:** This beetle is readily available in the aforementioned states.

**Stage to transfer:** Adult.

**Redistribution:** Collect the beetles with a sweep net from July through early August. The adult beetles can be shipped or stored for several days at cool temperatures if fed fresh leafy spurge leaves and confined in paperboard containers. They can also be kept at room temperature for several weeks in large cages with fresh food, or for several weeks in smaller cardboard can containers if kept cool and fed periodically under warmer conditions. To release, sprinkle beetles on moderately dense leafy spurge plants. Sites for first-time release that contain high ant or grasshopper populations should be avoided.

## Comments

*Aphthona nigriscutis* was first released in Canada in 1983 with spectacular results. It was the fourth flea beetle species cleared for release into the United States. Like many of the other flea beetle species that attack leafy spurge, this species tends to congregate for feeding, mating, and egg laying.

Leafy spurge plant density is greatly diminished at first. However, roots that are not attacked (including the taproot) are able to send up new shoots to supply the sugars for root reserves. It is only through the persistence of the beetles over a long period of time in ecosystems that favor beetle survival that the lateral roots of leafy spurge will be destroyed, the taproots weakened, and the plants eliminated from the area.

247

# *Chamaesphecia crassicornis*

R. W. Hansen

**Common name:** None widely accepted (leafy spurge clearwing).

**Type of agent:** Insect: Moth, clearwing (Lepidoptera: Sesiidae).

**Native distribution:** Eastern Austria, Romania, and southern Slovakia.

**Original source:** Romania.

## Biology

**Generations per year:** One, although some individuals may require two years to complete a generation.

**Overwintering stage:** Pupal.

**Egg stage:** Eggs are laid in mid-June in Romania. Eggs are generally deposited in groups of two to four along the stems. The incubation period is 11 to 16 days. Eggs are oval and flattened, dark brown when laid, becoming light brown prior to emergence of the larvae. The surface of the egg is covered with a network of slightly raised veins which form pentagonal and hexagonal shapes.

**Larval stage:** Newly hatched larvae either crawl down the stem from where the eggs were laid or drop to the ground before penetrating the plants. They feed just under the cortex below the crown of the plant. It is believed that there are five larval growth stages. Third-stage larvae begin to penetrate the central part of the root, and finally the last growth stage larvae make tunnels about 10 to 20 cm long in the root, which they fill with frass. By the time the larvae reach maturity, the tunneled section of the root is almost completely destroyed. Mature larvae move to the lower part of the stem where they chew exit holes, cover them with frass, and pupate.

*C. crassicornis* adult.
(Photo credit: R. Richard,
USDA-APHIS)

**Pupal stage:** The pupal period can last through the winter or through a season and a half, with the adult emerging the following year. During emergence, the pupa protrudes outside the stem for about three-fourths of its length and holds itself in place with its anal hooks. In this way, the emerging moth is able to free itself.

**Adult stage:** Adults are present from the end of May until the end of July. Males live about seven days while the females live an average of five days. Gestation lasts one to three days. Number of eggs per female is 29 to 33. Adults are dark brown with yellow and cream striping. Hyaline areas of the wing are devoid of scales. The moths are 10.4 to 11.6 mm long with a wingspan of 16 to 22 mm. The moth has a wasp-like appearance and may be confused with several native sesiid moth species.

## Effect

**Destructive stage:** Larval.

**Additional plant species attacked:** Leafy spurge (*Euphorbia virgata* complex).

**Site of attack:** The lower stems are slightly damaged but major damage occurs in the root system.

**Impact on the host:** The roots that contain larvae are nearly destroyed, the root reserves are reduced, and the plant's ability to replace these reserves is greatly decreased. The vigor of the plant is reduced as is the number of root buds. Larval feeding damage could provide entry points for pathogenic fungi.

**Nontarget effects:** No nontarget effects have been reported.

## Releases

**First introduced into the United States:** 1994, Montana.

**Established in:** This insect has not yet been established in the field, although researchers have successfully raised this species in laboratory and greenhouse cultures. Rearing efforts are ongoing in Canada.

**Habitat:** Undetermined.

**Availability:** This species is unavailable as of 2004.

**Stages to transfer:** Adult or egg.

**Redistribution:** If proper equipment and understanding is available, fertile eggs can be collected in the laboratory and applied to the plants in the field.

## Comments

This species has a narrow host range, only utilizing hosts in the *E. virgata* complex, and it appears to have a very narrow microclimate and habitat niche.

Two additional *Chamaesphecia* species were unsuccessfully introduced into the United States in the 1970s. The European clearwing moths *C. empiformis* and *C. tenthridiniformis* were introduced into Idaho and Montana in 1975, but never established viable populations. Both feed on roots as larvae. *C. empiformis* targeted cypress spurge and has a very specific host range. *C. tenthridiniformis* is extremely restricted to a specific biotype of leafy spurge and was not able to accept the host plants offered it and, therefore, did not survive. It is highly unlikely that these species will be reintroduced.

249

# Chamaesphecia hungarica

R. W. Hansen

**Common name:** None widely accepted (leafy spurge clearwing moth).

**Type of agent:** Insect: Moth, clearwing (Lepidoptera: Sesiidae).

**Native distribution:** Southeastern Austria and eastern Europe.

**Original sources:** Hungary, Bosnia, and Serbia.

## Biology

**Generations per year:** One.

**Overwintering stage:** Larval (in the root).

**Egg stage:** Eggs are laid shortly after mating. Each female produces an average of 205 eggs that are usually laid singly. During the spurge flowering period, most eggs are laid on the bracts (modified leaves around the flower). After the plants have flowered, eggs are laid on the leaves and stems. The oval, flattened, light brown eggs measure 0.75 mm by 0.50 mm. The eggshell is divided into distinct polygonal structures and covered with minute papillae.

**Larval stage:** The larvae emerge about 17 days after the eggs are deposited and penetrate into the shoot a few centimeters above the ground. Larvae from eggs deposited earlier around the flower apparently have a better survival rate than those laid later on the leaves and stems. The young larvae mine the stem just below the epidermis for a few centimeters before they move into the pith and down into the root where most of the larval feeding occurs. There are seven larval stages; the sixth and seventh occur before winter. In the spring the larvae mine up to the stem base and prepare an emergence hole a few centimeters above the ground.

**Pupal stage:** Pupation occurs in the spring within the stem of the host plant a few centimeters above the ground. The empty pupal case is left protruding from the exit hole after emergence.

*C. hungarica* adult. (Photo credit: R. Richard, USDA-APHIS)

**Adult stage:** The adults emerge between mid-May and the end of June in eastern Europe; at higher elevations they may emerge until the end of July. Adult females use a pheromone to attract the males. The abdomen is black and contains whitish bands on segments two, four, and six, with numerous greenish scales on the back. The male also has whitish bands on segment seven. Their bodies are 10 to 14 mm long. The dorsal side of the antennae is black while the ventral side is brown. The outermost area of the 7 to 10 mm-long forewing is black with yellow scales between some of the veins. A black spot is located at the tip of the wing. Three transparent areas of the forewing are clearly visible. The moth has a wasp-like appearance and may be confused with several native sesiid moth species.

## Effect

**Destructive stage:** Larval.

**Additional plant species attacked:** *Euphorbia lucida, E. palustris*, leafy spurge (*E. esula-virgata* complex), and possibly *E. lathyris* as a marginal host.

**Site of attack:** Roots.

**Impact on the host:** By feeding within the roots of the host, larvae deplete root reserves and impair the plant's ability to replace those reserves, causing loss of plant vigor and often death.

**Nontarget effects:** In host testing, it was determined that *C. hungarica* only utilized Eurasian spurges in the *E. esula-virgata* complex, with *E. lathyris* as a marginal host. The *Euphorbia* subgenera *Chamaesyche, Agaloma*, and *Poinsettia* did not support any larval feeding. No rare and endangered native North American spurges are at risk, nor is the economically important *E. pulcherrima*. Other plants of concern, such as *E. robusta*, which belongs to the subgenus *Esula*, live in dry climates outside the range of this species. No *Chamaesphecia* species have been recorded on annual spurges. In its native range, *C. hungarica* has only been reported on *E. lucida* and *E. palustris*, although the latter has not been confirmed.

## Releases

**First introduced into the United States:** 1993, Montana.

**Established in:** This insect has not yet been established in the United States.

**Habitat:** In its native range, this species is generally found in plants growing in moist loamy soils and in partly shaded habitats such as river banks, swampy areas, and ditches. It apparently does not do well in dry sunny areas.

**Availability:** This insect is unavailable as of 2004.

**Stage to transfer:** Adult.

**Redistribution:** Dig leafy spurge plants of *E. esula-virgata* after this agent has become established. Infested roots can be stored at cool temperatures of 4 to 12° C for three or four months if the plants are dug in the fall. Rear the larvae and release adults. Adults can be collected with a sweep net, but only a few sweeps should be made per series so the captured moths are not damaged.

## Comments

This moth may be potentially valuable as a biocontrol agent of certain leafy spurge biotypes in moist areas while presenting minimal risk to native spurges in the United States.

# *Dasineura* sp. nr. *capsulae*

R. W. Hansen

**Common name:** None widely accepted.

**Type of agent:** Insect: Fly, gall midge (Diptera: Cecidomyiidae).

**Native distribution:** Italy.

**Original source:** Italy.

## Biology

**Generations per year:** One.

**Overwintering stages:** Larval and pupal (in the soil).

**Egg stage:** Eggs are generally deposited within 24 hours of mating, usually in the inflorescence between the bracts and the cyathium in groups of about 35. Total number of eggs deposited per female is about 89. The incubation period is three to five days. Freshly laid eggs are white, slightly elongated with rounded ends. They have a soft, smooth, translucent covering. Eggs are about 0.27 mm long.

**Larval stage:** The larvae are first located in the inner part of the bracts that cover the cyathium or inside the cyathium where the galls start to form. Larvae develop for about five weeks before they exit the gall and enter the soil to hibernate until the next

*Dasineura* sp. nr. *capsulae* adult.
(Photo credit: USDA-ARS)

spring. Mature larvae are yellowish and about 3.1 mm long with distinct and well-formed sternal plates.

**Pupal stage:** Pupation occurs in early April after mature larvae move to the surface. The pupal stage lasts two to four days. Pupae are about 2 mm long and are a light red except for the reddish-brown wing pads and leg appendages.

**Adult stage:** In Italy, adults emerge between mid-April and mid-May and mate shortly thereafter. Adult females live about three days while adult males live for only 2.4 days. The adult body is reddish-yellow with brown, hardened parts. The female is 2.3 mm long; the male is 1.7 mm long.

## Effect

**Destructive stage:** Larval.

**Additional plant species attacked:** None.

**Site of attack:** Fly larvae attack the inner part of the bracts that cover the cyathium or develop inside the cyathium. Galls are produced mainly by the enlargement and distortion of the cyathium and also by the deformation of the bracts that cover the cyathium or a deformation of the leaves of the growing tips.

**Impact on the host:** This agent attacks the seed-producing portion of the plant, thus reducing seed output.

**Nontarget effects:** No nontarget effects have been reported.

## Releases

**First introduced into the United States:** This species has yet to be released. It was permitted for release in 1991; however, no live adults from foreign collections have been reared in quarantine for release in the United States because of very high parasitism rates.

**Established in:** This species has not yet been established in the United States.

**Habitat:** Undetermined.

**Availability:** This species is not available.

**Stages to transfer:** Egg, larval, pupal (while on or in the host plant), and adult (but adults are very short-lived and therefore can only be transferred short distances).

**Redistribution:** Collect galls that contain mature larvae and pupae. Let mature larvae leave the spurge galls by placing field-collected galls into plastic bags, then refrigerate at 8° C to force the larvae to leave the galls. Place the larvae in a mixture of sphagnum moss and sand so that they can pupate. This material can be kept up to six months in cool storage to mimic winter conditions. The midge pupae can be reared and the adults released. The adults are fragile flies that need to be released in the early morning or late evening on calm days when possible. It is during these periods in their short adult lives that they mate and lay eggs.

## Comments

This insect has not yet been described and named. It is very similar in appearance to *Spurgia esulae* and attacks the plant at nearly the same location. However, the galls of the two species are different: *S. esulae* causes the plant to enclose the larvae in a protective encasement of leaves while the larvae of *D.* sp. nr. *capsulae* burrow into the

253

ovary and cause the plant to create a gall of swollen plant tissue. A second difference is that *D.* sp. nr. *capsulae* has only one generation per year. However, both attack the growing tip of the plant and reduce seed production.

## Hyles euphorbiae

R. W. Hansen

**Common name:** Leafy spurge hawk moth.

**Type of agent:** Insect: Moth (Lepidoptera: Sphingidae).

**Native distribution:** Southern and central Europe, northern India, and central Asia.

**Original sources:** France, Germany, Hungary, and Switzerland.

### Biology

**Generations per year:** Two.

**Overwintering stage:** Pupal (in the soil).

**Egg stage:** Each female deposits between 70 and 150 eggs, generally in batches, on leafy spurge leaves and bracts in June and August. The eggs are round and a green fluid can be seen at first through the transparent egg covering in younger eggs; the developing larvae can be seen in the older eggs.

**Larval stage:** The larvae generally hatch during June and early August. There are five larval growth stages that require two to three weeks for completion. Larvae at each growth stage have their own distinctive and conspicuous color pattern. The first-stage larvae are a dark black or blackish green, have six legs and 10 prolegs, and spin very thin threads of silk that adhere to the plant and keep the larvae from falling off. Later the coloration changes to longitudinal yellow and dark brown stripes that change to green with white spots, and by two weeks of age, the larva will have increased its

*H. euphorbiae larva.* (Photo credit: N. Poritz, bio-control.com)

*H. euphorbiae* adult. (Photo credit: N. Poritz, bio-control.com)

weight to about 2 to 3 gm. The last stage larva is green, black, and red with yellow spots. Well-fed larvae are about 11 cm long. The color changes from the bright green, red, black, and yellow to a darker combination of these colors and the body length contracts during the prepupal stage.

**Pupal stage:** Mature larvae excavate a hole in the soil or litter 2.5 to 8 cm deep and cement soil particles and other loose material together to form the watertight pupal chambers. If the insect does not enter diapause, the pupal stage lasts 15 to 20 days. Otherwise, it will remain a pupa all winter. Young pupae are greenish-white and soft, while mature pupae are dark brown. Pupae are about 4 to 5 cm long.

**Adult stage:** The adults begin to appear in the field from late June to July and again in late August and September. They are fast fliers and hover at flowers. Moths have a wingspan of about 5 cm and have distinctive markings, although they can sometimes be confused with other native hawk moths. There is no simple visual way to distinguish *H. euphorbiae* adults from other species. Generally, *H. euphorbiae* moths are active during the day and are usually found in or near leafy spurge stands, while many other hawk moths are active in the evening and at night.

## Effect

**Destructive stage:** Larval.

**Additional plant species attacked:** None.

**Site of attack:** The larvae feed on leaves and bracts of leafy spurge.

**Impact on the host:** Larval consumption of leafy spurge is apparent in dense patches. However, this feeding does not result in plant mortality. By itself, this species is ineffective as a biological control agent.

**Nontarget effects:** No nontarget effects have been reported.

255

## Releases

**First introduced into the United States:** 1966, Montana.

**Established in:** Idaho, Minnesota, Montana, Nebraska, North Dakota, South Dakota, and Wyoming.

**Habitat:** This species prefers open areas near trees with an abundance of spurge. Mortality may be high in areas with many ground squirrels, birds, and other small animals that eat the pupae. It survives subfreezing winter temperatures.

**Availability:** Once established, this species may periodically become quite abundant for several years and cause noticeable defoliation of leafy spurge stems. In intervening years, however, *H. euphorbiae* is usually rare and inconspicuous.

**Stages to transfer:** Larval and pupal. Movement at these stages has been traditional, although if one can transfer mated females, eggs will be deposited less conspicuously and might be less subject to predation.

**Redistribution:** Collect the larvae from plants in July and September and keep them supplied with fresh leafy spurge foliage. Larvae can be shipped if the moths have fresh food, the container is kept cool, and the duration of the travel is only two or three days. Adults can be collected with sweep nets or black lights in late June to July and again in late August and September, although care should be taken not to damage the wings. Adults should be moved in large cardboard containers that are kept cool, well ventilated, and provided with spurge branches to which the moths can cling. It is preferable and less destructive to the insects to hand-carry the containers with the adults to new locations rather than ship them commercially. The best method of redistribution is to rear them to the adult stage, allow them to mate, and then release the mated moths. The adult females will deposit the eggs where they have the best chance of survival.

## Comments

A generation is completed in 45 to 60 days. The larval skin contains toxins of the leafy spurge plant, which tends to deter predators. This protection, however, is lacking in the pupal skin.

There may be two strains of *H. euphorbiae*, each with slightly different host preferences. This biocontrol agent is very visible and its feeding damage quite noticeable. However, it is not considered to be a good agent by itself since defoliation alone does not reduce the leafy spurge infestation. Also, an insect virus may reduce caterpillar populations.

# *Oberea erythrocephala*

R. W. Hansen

**Common name:** Red-headed leafy spurge stem borer.

**Type of agent:** Insect: Beetle (Coleoptera: Cerambycidae).

**Native distribution:** Europe.

**Original sources:** Europe, Italy, and Switzerland.

## Biology

**Generations per year:** One, with some generations lasting two years in colder climates.

**Overwintering stage:** Larval (within the stem or crown).

**Egg stage:** Eggs are deposited from the end of June to mid-July. The female adult often girdles the upper part of the stem one to four times (usually twice) by cutting grooves completely or partly around it. She then gnaws a hole into the stem above the girdle marks and deposits an egg into it. The hole becomes covered with latex, which eventually dries. Usually only one egg is deposited in each shoot. Each female can produce about 60 eggs during her lifetime. The eggs are 1.8 to 2.0 mm long, pale yellow at first but changing to pinkish-white or pink shortly before the larvae hatch.

**Larval stage:** The larvae hatch seven to 10 days after oviposition and begin to feed immediately on the pith. In thicker stems, they begin to tunnel downward, while in thinner stems they tunnel upward first and then down. In Montana, stem diameters of at least 3.0 mm are usually selected by the females for egg-laying. The first three to four larval stages consume all the pith of the stem leaving only the cortical tissue, and fill their galleries with fibrous frass. The mined stems wilt and dry up by the end of July and do not produce flowers and seeds. Eventually, the larvae mine the root crown.

Regardless of the length of time required to complete development, the larvae remain in the crown of the plant during the winter. Only a single larva develops per stem; if several larvae occupy the same stem, the more mature larva will survive at the expense of the younger. In well-developed roots of large plants with several attacked stems, several larvae can complete their development. The whitish larvae are long and slender, and have sclerotized heads.

**Pupal stage:** Pupation occurs in cells formed within the root crown during May.

**Adult stage:** Males emerge several days before females. Both sexes are sexually immature for

*O. erythrocephala* adult. (Photo credit: N. Poritz, bio-control.com)

257

Damage to leafy spurge stems caused by *O. erythrocephala*. (Photo credit: N. Poritz, bio-control.com)

two weeks. The hind wings are grayish-black, the head is red with black eyes, and the body is very slender. Antennae are nearly as long as the body; the female's antennae are a little shorter than the male's.

## Effect

**Destructive stages:** Adult feeding on the leaves and stems does not greatly affect plant survival. However, girdling by the adult with subsequent egg-laying generally results in shoot death. The larvae in the stem also cause the stem to die, and destructive feeding in the crown and root reduces the plant's root reserves. Larval galleries may allow pathogenic fungi to enter leafy spurge roots.

**Additional plant species attacked:** None.

**Site of attack:** Larvae live and feed in the stem and crown of plants with stem diameters in excess of 3.0 mm.

**Impact on the host:** Although this agent has the potential to greatly depress leafy spurge populations, it attacks only specific biotypes of leafy spurge and, therefore, has not yet increased its population sufficiently in many areas to produce noticeable impacts.

**Nontarget effects:** No nontarget effects have been reported.

## Releases

**First introduced into the United States:** 1982, Montana.

**Established in:** Colorado, Minnesota, Montana, North Dakota, Oregon, South Dakota, and Wyoming.

**Habitat:** This species seems to prefer fairly mesic areas with trees and may do well in riparian areas. It survives subfreezing winter temperatures.

**Availability:** This insect has limited availability in several states.

**Stage to transfer:** Adult.

**Redistribution:** Use a sweep net or hand-collect the adults. These can be stored up to several weeks if kept cool and allowed to warm up and feed for two-hour periods, three times per week. They can be shipped in a cool environment with plant stems and leaves for food. However, shipments should take no longer than six days.

## Comments

This species can be confused with a flower beetle that has the same general coloration and appearance, except the flower beetle's head is slightly larger and black and the abdomen is larger and much more flattened. There are also a number of native *Oberea* species whose adults are superficially similar to those of *O. erythrocephala*. However, these native beetles attack various trees and shrubs and will not be found on leafy spurge plants.

This was the second insect species introduced to control leafy spurge. There is some indication that in some Montana locations the life cycle tends to require two years rather than just one, probably because of the cooler temperatures. It appears that this species is very host-specific and apparently prefers certain leafy spurge biotypes over others.

# *Spurgia esulae*

R. W. Hansen

**Common name:** Leafy spurge tip gall midge.

**Type of agent:** Insect: Fly, gall midge (Diptera: Cecidomyiidae).

**Native distribution:** Italy.

**Original source:** Italy.

## Biology

**Generations per year:** Two in Montana with a partial third, but three to five in warmer areas; emergence coincides with the availability of new shoots.

**Overwintering stage:** Mature larval.

**Egg stage:** The females lay eggs on the external and internal leaves of the growing tip. Newly laid eggs measure about 0.35 mm and are light red at first, darkening with age. The egg is cigar-shaped and slightly curved with a smooth, soft covering.

*S. esula* larvae. (Photo credit: N. Poritz, bio-contrrol.com)

259

**Larval stage:** There are three larval growth stages. Larvae are generally concentrated within the gall; the newly hatched larvae move to the internal part of the growing tips to feed on young leaves. Mature larvae are orange and possess sternal plates on the lower side of the first thoracic segment.

**Pupal stage:** Pupation of all generations, except for the overwintering generation, occurs in the gall. The overwintering generation pupates in the soil. Pupae measure about 1.84 mm long and are light red, except for the legs and wings, which tend to be reddish-brown. Pupae are enclosed within a white, thin, silk cocoon.

**Adult stage:** Adults emerge in the spring. This very delicate fly is short lived, its life being measured in hours rather than days. Mating and egg laying generally occur in the cool, calm periods around twilight and dusk. During the heat of the day they seek shady areas and move very little. The males are about 1.86 mm long and possess forceps on the end of the abdomen, whereas females are about 1.90 mm long and have a tapered abdomen with an exposed ovipositor.

## Effect

**Destructive stage:** Larval.

**Additional plant species attacked:** None.

**Site of attack:** Growing points of the plant.

**Impact on the host:** Each generation attacks the growing tips of leafy spurge plants, destroying the shoots' ability to flower and produce seeds. The tips eventually die, and the plants then produce new shoots from below the attacked areas. These shoots are then attacked by the next generation of midges.

**Nontarget effects:** No nontarget effects have been reported.

## Releases

**First introduced into the United States:** 1985, Montana and North Dakota.

**Established in:** Colorado, Idaho, Michigan, Minnesota, Montana, Nebraska, New York, North Dakota, Rhode Island, South Dakota, and Wyoming.

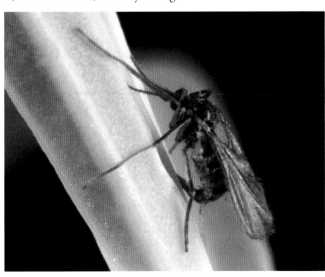

*S. esula* adult.
(Photo credit: N. Poritz,
bio-control.com)

**Habitat:** This agent prefers fairly dense leafy spurge populations growing on south-facing slopes in cooler climates. It appears to tolerate some shading.

**Availability:** Limited availability in Montana, North Dakota, and several other states.

**Stages to transfer:** Larval and pupal (in galls).

**Redistribution:** To collect, clip galls containing mature larvae. (Sweeping with a net will damage the very fragile adults.) Clipped stems should be bunched and the bottoms wrapped in damp towels or damp cotton. They should be taken to the field as quickly as possible and placed upright in a wire frame or other device so that the larvae will not be found by ants and other predatory insects.

## Comments

This species was tested and introduced into the United States under the name of *Bayeria capitigena*. Established populations may be attacked by native predators and parasitoids of native cecidomyiid fly species. Once established, adults can be spread over long distances by wind and initiate new populations.

## *References*

Anderson, G. L., E. S. Delfosse, N. R. Spencer, C. W. Prosser, and R. D. Richard. 2003. Lessons in developing successful invasive weed control programs. J. Range Manage. 56: 2-12.

Bangsund, D. A., F. L. Leistritz, and J. A. Leitch. 1999. Assessing economic impacts of biological control of weeds: the case of leafy spurge in the northern Great Plains of the United States. J. Environ. Manage. 56: 35-43.

Bangsund, D. A., D. J. Nudell, R. S Sell, and F. L. Leistritz. 2001. Economic analysis of using sheep to control leafy spurge. J. Range Manage. 54: 322-29.

Batra, S. W. T. 1983. Establishment of *Hyles euphorbiae* (L.) (Lepidoptera: Sphingidae) in the United States for control of two weedy spurges *Euphorbia esula* L. and *E. cyparissias* L. J. NY Entomol. Soc. 91: 304-11.

Belcher, J. W., and S. D. Wilson. 1989. Leafy spurge and the species composition of a mixed-grass prairie. J. Range Manage. 42: 172-75.

Caesar, A. J., G. Campobasso, and G. Terragitti. 1999. Effects of European and U.S. strains of *Fusarium* spp. pathogenic to leafy spurge on North American grasses and cultivated species. Biol. Control 15: 130-36.

Fellows. D. P., and W. E. Newton. 1999. Prescribed fire effects on biological control of leafy spurge. J. Range Manage. 52: 489-93.

Fornasari, L., and P. Pecora. 1995. Host specificity of *Aphthona abdominalis* Duftschmid (Coleoptera: Chrysomelidae), a biological control agent for *Euphorbia esula* L. (leafy spurge, Euphorbiaceae) in North America. Biol. Control 5: 353-60.

Gassmann, A., and D. Schroeder. 1995. The search for effective biological control agents in Europe: history and lessons from leafy spurge (*Euphorbia esula* L.) and cypress spurge (*Euphorbia cyparissias* L.). Biol. Control 5: 466-77.

Gassmann, A., and I. Tosevski. 1994. Biology and host specificity of *Chamaesphecia hungarica* and *C. astatiformis* (Lep.: Sesiidae), two candidates for the biological control of leafy spurge, *Euphorbia esula* (Euphorbiaceae) in North America. Entomophaga 39: 237-45.

Hansen, R. W., R. D. Richard, P. E. Parker, and L. E. Wendel. 1997. Distribution of biological control agents of leafy spurge (*Euphorbia esula* L.) in the United States: 1988-1996. Biol. Control 10: 129-42.

Harris, P., P. H. Dunn, D. Schroeder, and R. Vonmoos. 1985. Biological control of leafy spurge in North America. Pages 79-92 *in* A.K. Watson, ed. Leafy Spurge. Weed Sci. Soc. Monogr. Ser. No. 3. Weed Sci. Soc. Am., Champaign, IL.

Jacobs, J. S., R. L. Sheley, N. S. Spencer, and G. Anderson. 2001. Relationships among edaphic, climatic, and vegetation conditions at releases sites and *Aphthona nigriscutis* population density. Biol. Control 22: 46-50.

Jonsen, I. D., R. S. Bourchier, and J. Roland. 2001. The influence of matrix habitat on *Aphthona* flea beetle immigration to leafy spurge patches. Oecologia 127: 287-94.

LeSage, L., and P. Paquin. 1996. Identification keys for *Aphthona* flea beetles (Coleoptera: Chrysomelidae) introduced in Canada for the control of spurge (*Euphorbia* spp., Euphorbiaceae). Can. Entomol. 128: 593-603.

Lym, R. G. 1991. Economic impact, classification, distribution, and ecology of leafy spurge. Pages 169-81 *in* L. F. James, J. O. Evans, M. H. Ralph, and R. D. Child, eds. Noxious Range Weeds. Westview Press, Boulder, CO.

Lym, R. G., and J. A. Nelson. 2000. Biological control of leafy spurge (*Euphorbia esula*) with *Aphthona* spp. along railroad right-of-ways. Weed Technol. 14: 642-46.

Lym, R. G., S. J. Nissen, M. L. Rowe, D. L. Lee, and R. A. Masters. 1996. Leafy spurge (*Euphorbia esula*) genotype affects gall midge (*Spurgia esulae*) establishment. Weed Sci. 44: 629-33.

Masters, R. A., and S. J. Nissen. 1998. Revegetating leafy spurge (*Euphorbia esula*)-infested rangeland with native tallgrasses. Weed Technol. 12: 381-90.

Nowierski, R. M., and R. W. Pemberton. 2002. Leafy Spurge. Pages 181-94 *in* R. Van Driesche, S. Lyon, B. Blossey, M. Hoddle, and R. Reardon, eds. Biological Control of Invasive Plants in the Eastern United States. USDA Forest Serv. Pub. FHTET-2002-04. Morgantown, WV.

Nowierski, R. M., Z. Zeng, D. Schroeder, A. Gassmann, B. C. Fitzgerald, and M. Cristofaro. 2002. Habitat associations of *Euphorbia* and *Aphthona* species from Europe: development of predictive models for natural enemy release with ordination analysis. Biol. Control 23: 1-17.

Pecora, P., M. Cristofaro, and M. Stazi. 1989. *Dasineura* sp. near *capsulae* (Diptera: Cecidomyiidae), a candidate for biological control of *Euphorbia esula* complex in North America. Ann. Entomol. Soc. Am. 82: 693-700.

Pecora, P., R. W. Pemberton, M. Stazi, and G. R. Johnson. 1991. Host specificity of *Spurgia esulae* Gagne (Diptera: Cecidomyiidae), a gall midge introduced into the United States for control of leafy spurge (*Euphorbia esula* L. "complex"). Environ. Entomol. 20: 282-87.

Pemberton, R. W. 1995. 76. Leafy spurge, *Euphorbia esula* L. (Euphorbiaceae). Pages 289-95 *in* J. R. Nechols, L. A. Andres, J. W. Beardsley, R. D. Goeden, and C. G. Jackson, eds. Biological Control in the Western United States: Accomplishments and Benefits of Regional Research Project W-84, 1964-1989. Univ. Calif. Div. Agric. Nat. Res. Pub. 3361. Oakland, CA.

Radcliffe-Smith, A. 1985. Taxonomy of North American leafy spurges. Pages 14-25 *in* A. K. Watson, ed. Leafy Spurge. Monogr. Ser. No. 3. Weed. Sci. Soc. Am., Champaign, IL.

Rees, N. E., and N. R. Spencer. 1991. Biological control of leafy spurge. Pages 182-92 *in* L. F. James, J. O. Evans, M. H. Ralphs, and R. D. Child, eds. Noxious Range Weeds. Westview Press, Boulder, CO.

Rowe, M. L., D. J. Lee, S. J. Nissen, B. M. Bowditch, and R. A. Masters. 1997. Genetic variation in North America leafy spurge (*Euphorbia esula*) determined by DNA markers. Weed Sci. 45: 446-54.

Tosevski, I., A. Gassmann, and D. Schroeder. 1996. Description of European *Chamaesphecia* spp. (Lepidoptera: Sesiidae) feeding on *Euphorbia* (Euphorbiaceae), and their potential for biological control of leafy spurge (*Euphorbia esula*) in North America. Bull. Entomol. Res. 86: 703-14.

U.S. Department of Agriculture. 2000. Purge Spurge: Leafy Spurge Database v. 4.0 CD-ROM. USDA-APHIS and USDA-ARS, Sidney, MT.

# Mediterranean sage                               *Salvia aethiopis*

Mint family—Lamiaceae

*E. M. Coombs and L. M. Wilson*

**Additional common names:** None widely accepted.

**Native range:** Southern Eurasia.

**Entry into the United States:** The plant was probably introduced as a contaminant of alfalfa seed. It was first collected in 1882 in California.

## Biology

**Life duration/habit:** Mediterranean sage is a woolly biennial or short-lived perennial.

**Reproduction:** Reproduction is entirely by seeds.

**Roots:** A primary taproot is attended by numerous smaller fibrous roots.

**Stems and leaves:** During the plant's first year of growth, a rosette of gray-green woolly leaves with coarse, irregularly toothed margins is produced. Leaves are opposite on the square flowering stems and are mostly clustered at the base. The upper leaves are smaller and clasp the stem. Stem leaves are reduced to purple-tinged bracts with long, tapering points. The leaves have a strong, pungent odor that some have described as a mixture of mint and diesel. The stems are many-branched and stiff, giving the plant a rounded shape.

**Flowers:** The snapdragon-like flowers are whitish to pale yellow with an arched upper lip about twice the length of the lower lip. They are borne in clusters of three on profusely branched stems. Flowering takes place from May to August.

**Fruits and seeds:** The seeds are dark, smooth, egg-shaped, and 2 to 3 mm long. Each flower produces four seeds. The stiff plants break off at a natural abscission layer about 8 cm from the base and tumble in the wind, dispersing seeds as they tumble. Seeds need fall rains to germinate because they have a mucilaginous coating to prevent desiccation and enhance germination.

Mediterranean sage rosette. (Photo credit: E. Coombs, Oregon Department of Agriculture)

## Infestations

**Worst infested states:** California, Colorado, Idaho, Oregon, and Washington.

**Habitat:** Mediterranean sage invades rangelands and pastures, infesting habitats ranging from salt desert scrub to the upper sagebrush steppes. The most susceptible communities are those dominated by annual grasses and forbs and those heavily disturbed by overgrazing, fire, and roadsides.

**Impacts:** Mediterranean sage degrades rangeland and pastures and competes with desirable forage. It is unpalatable to livestock, but nontoxic. Because of its tumbleweed method of seed dispersal, whole plants can clog culverts and streams and become lodged in large masses along fencerows and hedges.

## Comments

Biological control has been successful in some areas, particularly where perennial grasses are well managed. Areas with annual

Mediterranean sage. (Photo credit: E. Coombs, Oregon Department of Agriculture)

grasses and salt desert shrub communities have continued to have heavy infestations.

Two species of weevils were introduced against Mediterranean sage—*Phrydiuchus tau* which is well established, and *P. spilmani*, which failed to establish.

## *Phrydiuchus tau*

E. M. Coombs and L. M. Wilson

**Common name:** Mediterranean sage root weevil.

**Type of agent:** Insect: Beetle, weevil (Coleoptera: Curculionidae).

**Native distribution:** Southern Eurasia.

**Original Source:** Turkey.

## Biology

**Generations per year:** One.

**Overwintering stages:** Egg, larval (inside plant), and adult (outside plant).

**Egg stage:** Eggs are laid in the leaf axils and petioles, and on the undersides of rosette leaves, singly or in groups of

*P. tau* larva. (Photo credit: E. Coombs, Oregon Department of Agriculture)

two to four. The eggs are white, oval, and 0.8 mm long by 0.5 mm wide. Eggs hatch 24 to 28 days after deposition.

**Larval stage:** Hatched larvae tunnel through the leaves into the root crown where most feeding and further development occurs. The immature stages complete development in the spring; mature larvae exit the plant to pupate in the surrounding soil. The larvae are white with brown head capsules. They pass through three larval instars.

**Pupal stage:** The mature larvae pupate from early May to July in earthen cells 2 to 3 cm beneath the soil surface.

*P. tau* adult. (Photo credit: E. Coombs, Oregon Department of Agriculture)

Mediterranean sage infestation near Lakeview, Oregon, in 1988. (Photo credit: E. Coombs, Oregon Department of Agriculture)

Same vista in 1995. (Photo credit: E. Coombs, Oregon Department of Agriculture)

265

**Adult stage:** Adults emerge in the late spring or early summer, feed on foliage, then rest in summer aestivation until the onset of fall rains and fall growth of the plant. The adults are dark-colored with a white "T" (Greek letter tau) on the back. The body is about 4 to 5 mm long, not including the snout.

## Effect

**Destructive stages:** Larval and adult (minor defoliation).

**Additional plant species attacked:** Clary sage (*Salvia sclarea*).

**Site of attack:** Larvae feed inside root crowns; adults feed externally on foliage and flowering shoots.

**Impact on the host:** Larval feeding damages flower shoot buds and root crowns, damaging or killing plants and reducing or preventing bolting. Small plants can be directly killed by larval feeding. Adults can cause minor defoliation of rosette leaves. One to several larvae can be contained in a single root crown, depending on root size.

**Nontarget effects:** No nontarget effects have been reported.

## Releases

**First introduced into the United States:** 1971, Oregon.

**Established in:** California, Colorado, Idaho, Oregon, and Washington.

**Habitat:** The agent does best at warm, dry sites such as south-facing slopes; it does not do well at higher elevations that receive snow before mid-November.

**Availability:** This weevil is readily available in the Lakeview, Oregon, area and along the Salmon River in Idaho.

**Redistribution:** Adults can be collected with a sweep net or by knocking them into a container with a badminton racquet or stick in the late spring and early summer when flowers are in 25% bloom. It is best to handpick or aspirate adults from rosettes in late fall during the mating season. Ship in cool paperboard containers with sufficient food.

## Comments

The life cycle of the weevil is synchronized to that of the host plant: periods of active adult feeding occur during periods of active rosette growth during the spring and fall. The weevil appears to have successfully controlled the weed at sites that have a strong perennial component and are not heavily grazed. The weevil seems to have little effect in salt-desert scrub communities and sites dominated by annuals.

*Phrydiuchus tau* is the *Phrydiuchus* species referred to in the Andres (1966) paper below.

## References

Andres, L. A. 1966. Host specificity studies of *Phrydiuchus topiarius* and *Phrydiuchus* sp. J. Econ. Entomol. 59: 69-76.

Andres, L. A., E. M. Coombs, and J. P. McCaffrey. 1995. 77. Mediterranean sage. Pages 296-98 *in* J. R. Nechols, L. A. Andres, J. W. Beardsley, R. D. Goeden, and C. G. Jackson, eds. Biological Control in the Western United States: Accomplishments and Benefits of Regional Research Project W-84, 1964-1989. Univ. Calif. Div. Agric. Nat. Res. Pub. 3361. Oakland, CA.

Roché, C. T., and L. M. Wilson. 1999. Mediterranean sage. Pages 261-70 *in* R. L. Sheley and J. K. Petroff, eds. Biology and Management of Noxious Rangeland Weeds. Oregon State Univ. Press, Corvallis.

Wilson, L. M., and J. P. McCaffrey. 1993. Bionomics of *Phrydiuchus tau* (Coleoptera: Curculionidae) associated with Mediterranean sage in Idaho. Environ. Entomol. 22: 704-8.

Wilson, L. M., J. P. McCaffrey, and E. M. Coombs. 1994. Biological control of Mediterranean sage. Pacific Northwest Ext. Pub. 473. Univ. Idaho Coop. Ext., Moscow.

# Melaleuca

*Melaleuca quinquenervia*

Myrtle family—Myrtaceae

*P. D. Pratt, T. D. Center, M. B. Rayamajhi, and T. K. Van*

**Additional common names:** Broad-leaved paperbark tree, cajeput, punk tree, white bottlebrush tree.

**Native range:** Eastern Australia, New Guinea, and New Caledonia.

**Entry into the United States:** Multiple introductions into Florida in the early 1900s as an ornamental, a timber source, to inhibit soil erosion, and otherwise drain wetlands. Fifty years later melaleuca began to invade the marshes and prairies of the Everglades.

## Biology

**Life duration/habit:** Melaleuca is a broad-leaved evergreen tree growing 25 to 30 m tall with a soft, layered, peeling bark.

**Reproduction:** Seeds.

**Roots:** Melaleuca roots are well adapted to fluctuating water levels. An abundance of vertical sinker roots extend at least to the water table's lowest annual level. Portions of the trunk below the high-water level also produce thread-like adventitious roots that dry and brown during dry periods but regenerate each flood season.

**Stems and leaves:** Leaves are simple, narrowly lance-shaped, alternate, grayish-green and emit a camphor-like odor when crushed. Woody branches occur at irregular intervals along the trunk.

**Flowers:** Flowers consist of a creamy white spike or "bottle brush" inflorescence approximately 8 cm long. In Florida, flowering occurs primarily in fall with a minor secondary bloom in summer, yet some flowers can be observed all year. Melaleuca trees can become reproductive within a year of germination.

Melaleuca "bottlebrush" flowers. (Photo credit: P. Pratt, USDA-ARS)

**Fruits and seeds:** Capsular fruits are persistent, arranged in a series of clusters, and may remain attached to the trunks, branches, or twigs for several years. In Florida, a flower spike can produce 30 to 70 sessile capsules; more than seven linearly occurring capsule clusters (each separated by series of leaves) have been recorded from melaleuca branches. Capsules contain 200 to 350 small seeds that are released after vascular connections are disrupted by increased bark thickness or stresses such as fire, frost, mechanical damage, herbicide treatments, or self-pruning of branches. Overall, 15% of the canopy-held seeds are embryonic and of these only 62% are viable. Less than 1% of the seed rain remains viable for longer than two years when buried in topsoil.

## Infestations

**Worst infested states:** Florida primarily, but also Hawaii to a lesser degree.

**Habitat:** Melaleuca has invaded essentially every existing plant community in south Florida, including those that possess healthy, vigorously growing native vegetation. Habitats most threatened by the invasive weed include sawgrass prairies, freshwater marshes, cypress swamps, pine flatwoods, hardwood bottomlands, and mangrove swamps. Melaleuca occurs abundantly within zones 9a to 10b of the U.S. Department of Agriculture's plant-hardiness zone map. Its distribution may be limited more by suitable habitat and hydroperiods than by climate.

**Impacts:** Invasion of freshwater herbaceous marsh communities in Florida results in a 60 to 80% loss in biodiversity, including a reduction in native wetland plant species, small mammals, and prey for birds. Melaleuca pollen is also a mild respiratory allergen, to which as much as 20% of the population may suffer allergic reactions. When considering economic impacts to the region, the South Florida Water Management District alone spent nearly $11 million to control melaleuca from 1991 to 1997, and estimates of losses to the local economy ranged as high as $168.6 million per year.

## Comments

Melaleuca is mainly present in south Florida below Lake Okeechobee. Major infestations occur along the eastern and western coasts of the southern peninsula and

A stand of melaleuca trees. (Photo credit: P. Pratt, USDA-ARS)

southern edge of Lake Okeechobee. Recent estimates (2002) suggest that melaleuca infests 202,000 ha of wetlands in south Florida. Melaleuca also occurs in Puerto Rico, Cuba, Belize, and the Virgin Islands.

In Australia, melaleuca occurs in a 40-km-wide zone along the eastern coast of Queensland and northern New South Wales. In contrast to Florida, most melaleuca habitats are threatened by development in Australia because they are located in highly desirable coastal areas of low topography, high rainfall, and mild climate.

## *Oxyops vitiosa*

P. D. Pratt, T. D. Center, M. B. Rayamajhi, T. K. Van, and S. Wineriter

**Common name:** Melaleuca snout beetle.

**Type of agent:** Insect: Beetle, weevil (Coleoptera: Curculionidae).

**Native distribution:** Eastern Australia.

**Original source:** Australia.

### Biology

**Generations per year:** Continuously brooded, overlapping generations when immature foliage is available.

**Overwintering stage:** Adult.

**Egg stage:** Eggs are laid singly on young leaves and expanding buds, often near the leaf apex; however, they may also occur on newly developed twigs. Eggs are approximately 1.0 mm long and individually coated by the female with a black or brown substance derived from feces and possibly a glandular secretion.

**Larval stage:** In Florida, larval densities are highest during late fall and early winter in conjunction with susceptible phenological stages of the host plant. Larvae are absent

*O. vitiosa* larva. (Photo credit: P. Pratt, USDA-ARS)

*O. vitiosa* adult. (Photo credit: P. Pratt, USDA-ARS)

or uncommon from April to August unless there is regrowth from damaged trees. Larvae have four instars, appear slug-like, and are covered with an oily exudate that has been shown to have anti-predator effects. Larvae excrete a coiled fecal filament that frequently becomes attached to the exudate. Near the end of the last growth stage, larvae stop feeding, drop from the host plant, wander, and ultimately burrow into the soil.

**Pupal stage:** Pupation occurs in a pupal cell constructed of soil, feces, and the oily exudate. Pupae are creamy white and lack the larval exudate. Failure to establish the biological control agent in permanently flooded habitats is attributed primarily to drowning of larvae when they drop to search for pupation sites.

**Adult stage:** Adults are long-lived (more than one year) and occur on melaleuca year-round. Shortly after eclosion adults are light reddish to brown, turning darker gray with age. They measure 9 mm long.

## Effect

**Destructive stages:** *Oxyops vitiosa* larvae are specialized flush feeders, consuming the seasonal flush of newly developed, expanding leaves at branch apices. Larval feeding occurs on one side of a leaf through to the cuticle on the opposite side, which results in a window-like feeding scar. This damage may persist for months, ultimately resulting in leaf drop. Adults feed superficially on young leaves and fully expanded, tough leaves resulting in narrow scars along the leaf surface. Adults may also feed on stems of the seasonal flush.

**Additional plant species attacked:** None.

**Site of attack:** Tender, expanding buds and leaves are destroyed by larvae.

**Impact on the host:** Severe larval feeding results in tip dieback and defoliation. Repeated damage of the growing tips removes apical dominance and induces branching through release of axillary buds. The new growth acts as a nutrient sink and sustains continual adult and larval populations. Foliar damage and the subsequent diversion of photosynthetic resources to the development of new foliage appears to limit reproductive performance of melaleuca. In preliminary studies, for instance, a 50% reduction in flowering was observed among severely damaged *M. quinquenervia* trees as compared to a similar undamaged group.

271

Undamaged plant (left); *O. vitiosa* larval feeding damage (right) on melaleuca. (Photo credit: P. Grebb, USDA-ARS)

**Nontarget effects:** Extensive field-based assessments of species closely related to melaleuca or identified as suboptimal hosts in quarantine testing have been monitored for nontarget damage. As predicted in host specificity testing, minor feeding by larvae has been observed on the Australian natives *Callistemon rigidis* and *C. viminalis*.

## Releases

**First introduced into the United States:** 1997, Florida.

**Established in:** Southern Florida.

**Habitat:** Habitats with short hydroperiods, dry winter conditions, and abundant young foliage enhance growth and development of *O. vitiosa* colonies. The weevil populations did not establish in permanently aquatic sites.

**Availability:** Adults can be collected all year but are most abundant in winter and spring.

**Stages to transfer:** Adults, although larvae can also be collected and released if suitable foliage is available at the release site.

**Redistribution:** Because of this weevil's slow rate of dispersal (about 1 km/year), coordinated collection and redistribution efforts have been implemented in Florida. Field-based mass-rearing sites may be developed by cutting trees and regularly mowing regrowth to maximize resource availability. The biological control agent has been redistributed to more than 150 locations in south Florida.

## Comments

Weevils used in quarantine testing and ultimately released into south Florida were collected within 320 km of Brisbane, Queensland, Australia.

An artificial diet has been developed for *O. vitiosa* which may facilitate mass production and release of the agent.

# *Boreioglycaspis melaleucae*

P. D. Pratt, S. Wineriter, T. D. Center, M. B. Rayamajhi, and T. K. Van

**Common name:** Melaleuca psyllid.

**Type of agent:** Insect: Psyllid (Homoptera: Psyllidae).

**Native distribution:** Australia.

**Original source:** Australia.

## Biology

**Generations per year:** Because of its recent (2002) introduction into south Florida, the number of generations completed per year has yet to be quantified.

**Overwintering stages:** Adult and nymphal.

**Egg stage:** Each egg has a spine-like projection near one end which the female inserts into the plant tissue. Eggs begin hatching approximately 18 days after they are laid; there is no evidence of an egg diapause.

**Nymphal stage:** The melaleuca psyllid has five nymphal instars. Early instars crawl about the leaves, but later stages are less active unless disturbed. Nymphs produce copious amounts of honeydew and also exude waxy filaments from glands located on their dorsum. These filaments form a dense, woolly mass that may partially cover later instars and facilitate locating colonies in the field.

**Adult stage:** When viewed with the naked eye, adult psyllids are unadorned, small insects about 3 mm long and pale yellow-orange to white. Adults can be observed jumping between leaves and plants. Females lay an average of 80 eggs during their lifetime.

## Effect

**Destructive stages:** Both adults and nymphs feed on melaleuca, but most damage is attributed to the nymphs.

**Additional plant species attacked:** None.

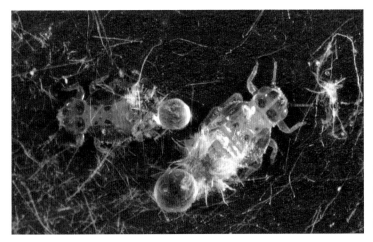

*B. melaleucae* nymphs. (Photo credit: S. Wineriter, USDA-ARS)

273

*B. melaleucae* adult female. (Photo credit: S. Wineriter, USDA-ARS)

Damage to melaleuca flowers and capsule clusters caused by *B. melaleucae* feeding. (Photo credit: P. Pratt, USDA-ARS)

**Site of attack:** Tender, expanding buds and leaves as well as mature older leaves are destroyed by nymphs. When populations are large, damage may extend to somewhat woody stems.

**Impact on the host:** Insufficient time has elapsed since the recent release of the psyllid to accurately quantify impacts on melaleuca in the field.

**Nontarget effects:** No nontarget impacts have been observed.

## Releases

**First introduced into the United States:** 2002, Florida.

**Established in:** Florida.

**Habitat:** This agent has successfully established over the wide range of habitats invaded by its host plant, including permanently flooded wetlands as well as upland pine flat woods. Climatic limits are under investigation.

**Availability:** Limited numbers are available from initial release colonies in Broward County, Florida.

**Stages to transfer:** Adult and nymphal.

**Redistribution:** Collect stems colonized by the psyllid, transport under cool conditions, and liberate as soon as possible.

## Comments

In regular field surveys of eastern Australia, the melaleuca psyllid was collected in north and southeastern Queensland and northern New South Wales, although collection records exist for Western Australia and the Northern Territory. Psyllids used in quarantine testing and ultimately released into south Florida were collected within 362 km of Brisbane, Queensland, Australia.

## *References*

Balciunas, J. K., D. W. Burrows, and M. F. Purcell. 1994. Field and laboratory host ranges of the Australian weevil, *Oxyops vitiosa*, a potential biological control agent of the paperbark tree, *Melaleuca quinquenervia*. Biol. Control 4: 351-60.

Bodle, J. J., A. P. Ferriter, and D. D. Thayer. 1994. The biology, distribution, and ecological consequences of *Melaleuca quinquenervia* in the Everglades. Pages 341-55 *in* S. M. Davis and J. C. Ogden, eds. Everglades: The Ecosystem and Its Restoration. St. Lucie Press, Delray Beach, FL.

Buckingham, G. R. 2001. Quarantine host range studies with *Lophyrotoma zonalis*, an Australian sawfly of interest for biological control of melaleuca, *Melaleuca quinquenervia*, in Florida, USA. Biocontrol 46: 363-86.

Center, T. D., T. K. Van, M. Rayachhetry, G. R. Buckingham, F. A. Dray, S. Wineriter, M. F. Purcell, and P. D. Pratt. 2000. Field colonization of the melaleuca snout beetle (*Oxyops vitiosa*) in south Florida. Biol. Control 19: 112-23.

Purcell, M. F., and J. K. Balciunas. 1994. Life history and distribution of the Australian weevil *Oxyops vitiosa*, a potential biological control agent for *Melaleuca quinquenervia*. Ann. Entomol. Soc. Am. 87: 867-73.

Rayachhetry, M. B., T. K. Van, and T. D. Center. 1998. Regeneration potential of the canopy-held seeds of *Melaleuca quinquenervia* in south Florida. Int. J. Plant Sci. 159: 648-54.

Rayachhetry, M. B., T. K. Van, T. D. Center, and M. L. Elliot. 2001. Host range of *Puccinia psidii*, a potential biological control agent of *Melaleuca quinquenervia* in Florida. Biol. Control 22: 38-45.

Rayamajhi, M. B., M. F. Purcell, T. K. Van, T. D. Center, P. D. Pratt, and G. R. Buckingham. 2002. Australian Paperbark Tree (Melaleuca). Pages 117-30 *in* R. Van Driesche, S. Lyon, B. Blossey, M. Hoddle, and R. Reardon, eds. Biological Control of Invasive Plants in the Eastern United States. USDA Forest Serv. Pub. FHTET-2002-04. Morgantown, WV.

Turner, C. E., T. D. Center, D. W. Burrows, and G. R. Buckingham. 1998. Ecology and management of *Melaleuca quinquenervia*, an invader of wetlands in Florida, U.S.A. Wetlands Ecol. Manage. 5: 165-78.

Van, T. K., M. B. Rayachhetry, T. D. Center, and P. D. Pratt. 2000. Litter dynamics and phenology of *Melaleuca quinquenervia* in south Florida. J. Aquat. Plant Manage. 38: 62-67.

Wheeler, G. S., and J. Zahniser. 2001. Artificial diet and rearing methods for the *Melaleuca quinquenervia* biological control agent *Oxyops vitiosa*. Florida Entomol. 84: 439-41.

Wineriter, S. A., and G. R. Buckingham. 1997. Love at first bite—introducing the Australian melaleuca weevil. Aquatics 19: 10-12.

Wood, M. 1997. Aussie weevil opens attack on rampant melaleuca. Agric. Res. 45: 4-7.

# Puncturevine

*Tribulus terrestris*

Caltrop family—Zygophyllaceae

## B. Villegas

**Additional common names:** Bull's head, caltrop, goat head, ground bur nut.

**Native range:** Africa and Eurasia.

**Entry into the United States:** The seed pods of the plant probably contaminated the wool of sheep imported from the Mediterranean region into the midwestern United States. Puncturevine was first reported in California in 1903.

## Biology

**Life duration/habit:** The plant is a prostrate, herbaceous annual.

**Reproduction:** Seeds.

**Roots:** The root system of puncturevine consists of a simple taproot branching into a network of fine rootlets.

**Stems and leaves:** The plant produces prostrate stems that radiate out from the root crown to form a mat. The stems often grow to 2 m long, are green to reddish or brownish in color, and are hairy. The leaves are pinnately compound, opposite, and hairy.

**Flowers:** Flowering occurs from June to September. The small yellow flowers are produced in leaf axils.

**Fruits and seeds:** The spiny fruits are made up of five burs that break apart at maturity. Each bur has two stout spines and contains two to four seeds.

## Infestations

**Worst infested states:** Puncturevine is widespread; the worst infestations are in Arizona, California, New Mexico, Nevada, and Texas. It is also reported in Colorado, Hawaii, Idaho, Kansas, Nevada, Oklahoma, Oregon, Utah, and Washington.

Puncturevine. (Photo credit: E. Coombs, Oregon Department of Agriculture)

Puncturevine infestation. (Photo credit: E. Coombs, Oregon Department of Agriculture)

**Habitat:** This plant is found most often in croplands, pastures, and corrals, along transportation rights-of-way, and in urban areas.

**Impacts:** The spiny burs can cause injury to the mouths and digestive tracts of livestock, are a nuisance to people, and diminish the value of alfalfa hay and wool.

## Comments

The plant has been controlled with biological control agents in areas without cold winters.

## *Microlarinus lareynii*

B. Villegas

**Common name:** Puncturevine seed weevil.

**Type of agent:** Insect: Beetle, weevil (Coleoptera: Curculionidae).

**Native distribution:** Africa and Eurasia.

**Original source:** Italy.

### Biology

**Generations per year:** The number of generations apparently depends on the climate. Multiple generations may be produced in warm climates.

**Overwintering stage:** Adult (in surface litter).

**Egg stage:** Eggs are deposited in pits chewed into the immature fruits. A female can deposit up to 324 eggs. The egg is oval, about 0.5 mm long, and is dark yellow to pale amber.

**Larval stage:** The larvae develop inside the fruits where they feed on the seeds. They are C-shaped and a dirty yellow color. There are four larval growth stages.

**Pupal stage:** Pupation occurs within the damaged fruit in an open cell. The pupae are surrounded by loosely packed frass. The pupal period lasts four to five days. The pupae are creamy white to pale yellow and are about 4.5 mm long.

277

**Adult stage:** Adults emerge from plants through exit holes in the fruits and begin to feed upon the stems, leaves, flowers, buds, and fruits. The adults are 4 to 5 mm long (not including the snout), brown, and covered with gray, erect hairs that give them a shaggy appearance. They differ from *Microlarinus lypriformis* in that the body is slightly larger and tear-shaped.

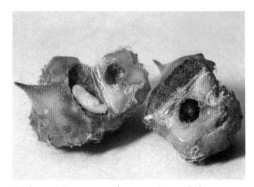

*M. lareynii* pupa and emerging adult. (Photo credit: E Coombs, Oregon Department of Agriculture)

## Effect

**Destructive stages:** Larval, though adults can cause minor defoliation.

**Additional plant species attacked:** Jamaica feverplant (*Tribulus cistoides*) and some *Kallstroemia* spp.

**Site of attack:** Fruits and seeds are infested by the larvae. Stems, leaves, buds, flowers, and fruits are eaten by the adults.

**Impact on the host:** Larval feeding destroys the seeds. Together with *M. lypriformis*, the weevil has provided a good level of control in areas with warm climates.

**Nontarget effects:** No significant damage to nontarget plants has been reported.

## Releases

**First introduced into the United States:** 1961, California.

**Established in:** Arizona, California, Colorado, Hawaii, Kansas, Nevada, New Mexico, Oklahoma, Oregon, Texas, and Utah.

**Habitat:** The distribution and abundance of the beetle is limited by cold winter temperatures.

**Availability:** The weevil is readily available.

**Stage to transfer:** Adult.

**Redistribution:** Adults can be collected from the soil litter beneath plants with a motorized vacuum apparatus or an aspirator. Also, infested plants and soil litter can be collected in paper bags or cardboard boxes and briefly left in the sun. Adults can be collected as they climb the walls of the bag or box.

## Comments

Together with *M. lypriformis*, *M. lareynii* has provided good control of puncturevine in warm climates on the mainland, and excellent control of puncturevine and *T. cistoides* in Hawaii. Both weevils are sensitive to cold winter temperatures.

# Microlarinus lypriformis

B. Villegas

**Common name:** Puncturevine stem weevil.

**Type of agent:** Insect: Beetle, weevil (Coleoptera: Curculionidae).

**Native distribution:** Africa and Eurasia.

**Original source:** Italy.

## Biology

**Generations per year:** The number of generations apparently depends on the climate. Multiple generations may be produced in warm climates.

**Overwintering stage:** Adult (in surface litter near the plant).

**Egg stage:** Eggs are deposited singly in pits chewed in the root crowns and in the undersides of stems. The pale yellow eggs are oval and about 0.5 mm long.

**Larval stage:** The larvae mine the stems and root crowns of the plants. Larvae of this species look similar to the C-shaped larvae of *Microlarinus lareynii*. There are four larval growth stages.

**Pupal stage:** Pupation occurs in the damaged stems. The pupal stage lasts from four to five days. Pupae are about 4 mm long and creamy white to pale yellow.

**Adult stage:** Adults feed on stems and leaves and emerge from the plant through exit holes in the stems and root crowns. The adults are about 4 to 5 mm long (not including the snout), brown, and differ from *M. lareynii* in that they are slightly smaller and the body is not tear-shaped.

## Effect

**Destructive stages:** Larval, though adults can cause minor defoliation.

**Additional plant species attacked:** Jamaica feverplant (*T. cistoides*) and some *Kallstroemia* spp.

**Site of attack:** Larvae tunnel stems and root crowns while adults feed on leaves and the undersurface of the stems.

**Impact on the host:** Larval feeding inside the stems and root crowns damages plant vascular tissue, and heavy feeding can lead to stem breakage. Together with *M. lareynii*, this weevil has provided good control in areas with warm climates.

**Nontarget effects:** No significant damage to nontarget plants has been reported.

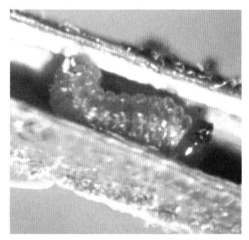

*M. lypriformis* larva in puncturevine. (Photo credit: USDA-ARS)

279

## Releases

**First introduced into the United States:** 1961, California.

**Established in:** Arizona, California, Florida, Hawaii, Kansas, Nevada, New Mexico, Oklahoma, Oregon, Texas, and Utah.

**Habitat:** The weevil's distribution is limited by cold winter temperatures.

**Availability:** The weevil is readily available.

**Stage to transfer:** Adult.

*M. lypriformis* adult. (Photo credit: USDA-ARS)

**Redistribution:** Adults can be collected from the soil litter beneath plants with a motorized vacuum apparatus or an aspirator. Also, infested plants and soil litter can be collected in paper bags or cardboard boxes and briefly left in the sun. Adults can be collected as they climb the walls of the bag or box.

## Comments

Together with *M. lareynii*, *M. lypriformis* has provided good control of puncturevine in warm climates on the mainland, and excellent control of puncturevine and *T. cistoides* in Hawaii. Both weevils are sensitive to cold winter temperatures.

## References

Andres, L. A., and G. W. Angelet. 1963. Notes on the ecology and host specificity of *Microlarinus lareynii* and *M. lypriformis* (Coleoptera: Curculionidae) and the biological control of puncturevine, *Tribulus terrestris*. J. Econ. Entomol. 56: 333-40.

Andres, L. A., and R. D. Goeden. 1995. 83. Puncturevine. Pages 318-21 *in* J. R. Nechols, L. A. Andres, J. W. Beardsley, R. D. Goeden, and C. G. Jackson, eds. Biological Control in the Western United States: Accomplishments and Benefits of Regional Research Project W-84, 1964-1989. Univ. Calif. Div. Agric. Nat. Res. Pub. 3361. Oakland, CA.

Huffaker, C. B., J. Hamai, and R. M. Nowierski. 1983. Biological control of puncturevine, *Tribulus terrestris*, in California after twenty years of activity of introduced weevils. Entomophaga 28: 387-400.

Kirkland, R. L., and R. D. Goeden. 1978. Biology of *Microlarinus lareynii* (Col.: Curculionidae) on puncturevine in southern California. Ann. Entomol. Soc. Am. 71: 13-18.

Kirkland, R. L., and R. D. Goeden. 1978. Biology of *Microlarinus lypriformis* (Col.: Curculionidae) on puncturevine in southern California. Ann. Entomol. Soc. Am. 71: 65-69.

# Purple loosestrife

*Lythrum salicaria*

Loosestrife family—Lythraceae

## *G. L. Piper, E. M. Coombs, B. Blossey, P. B. McEvoy, and S. S. Schooler*

**Additional common names:** Purple lythrum.

**Native range:** Asia, northern Africa, and Europe.

**Entry into the United States:** This plant was probably introduced via seeds in European soil used as ship ballast during the early 1800s. There is also evidence that seeds arrived inadvertently on raw wool or sheep imported from Europe.

## Biology

**Life duration/habit:** The plant is an herbaceous perennial that often grows up to 3.6 m tall.

**Reproduction:** Reproduction is primarily by seeds, but the plant can spread by sprouting from cut stems and generating from fragmented roots.

**Roots:** This plant possesses an abbreviated, robust, woody taproot.

**Stems and leaves:** Multiple, semi-woody, four- to six-sided stems bear lance-shaped leaves with smooth margins arranged either in whorls or opposite patterns.

**Flowers:** The flowers are showy rose-purple with five to six petals and arranged in long, spike-shaped inflorescences.

**Fruits and seeds:** The oval, 1-mm-long seeds develop in capsules that burst when mature. More than two million seeds can be produced by an average-sized plant.

## Infestations

**Worst infested states:** The plant is found throughout the northeastern, midwestern, and western United States.

**Habitat:** Aquatic sites such as marshy areas, streams and stream banks, ponds, irrigation canals, freshwater tidal flats, and ditches support the worst infestations of purple loosestrife.

**Impacts:** The weed displaces the resident plant community, reduces and degrades available wildlife habitat, decreases flow in irrigation canals and ditches, degrades hunting and fishing areas, and reduces the quality of wetland pastures.

Purple loosestrife flower stalk. (Photo credit: E. Coombs, Oregon Department of Agriculture)

281

## Comments

Because of the aquatic nature of the plant, chemical treatment is often inadvisable or impossible. Also, because plant fragments can root and produce new plants, mechanical control is difficult. The purple loosestrife biocontrol project is one of the most widely implemented projects in the United States. In several states, infestations have been reduced by 90% in 10 years.

The seed-consuming weevil *Nanophyes brevis* was approved for release in the United States. However, a parasitic nematode on *N. brevis* was included with its original shipment from Europe in 1994. Field release was postponed and it is unlikely that this insect will be released in the United States.

Purple loosestrife planted as an ornamental. (Photo credit: S. Dewey, Utah State University)

# *Galerucella calmariensis*

G. L. Piper, E. M. Coombs, B. Blossey, P. B. McEvoy, and S. S. Schooler

**Common name:** Black-margined loosestrife beetle.

**Type of agent:** Insect: Beetle (Coleoptera: Chrysomelidae).

**Native distribution:** Asia and Europe.

**Original source:** Northern Germany.

## Biology

**Generations per year:** One, with a partial second generation in warmer climates.

**Overwintering stage:** Adult. Adults may overwinter in the soil and vegetation surrounding wetlands, thus it is important to restrict soil disturbance within 50 m from the edge of the wetland in the spring. Once the adults have moved onto the loosestrife, soil cultivation can continue.

**Egg stage:** Egg laying occurs from May to June and from August to September. Females lay up to 10 eggs per day in groups of three to six on the stems, leaves, and leaf axils. Females can produce 300 to 400 eggs. The incubation period is 12 days. The eggs are 0.5 mm in diameter, barrel-shaped, tan- to cream-colored, and are often decorated with a line of frass on top.

**Larval stage:** There are three larval growth stages. The larvae initially feed upon and often kill young buds and leaves located at the shoot tips. The larvae later feed upon the developing leaves. Feeding occurs for about 14 days. The developing larvae are light green to orange with dark brown, sclerotized heads and have rows of black spots down their backs.

**Pupal stage:** Pupation occurs in the soil beneath the plant and in the spongy tissue of stems in standing water. The pupal period lasts about seven days. The pupae are yellowish to light brown and about 4 mm long.

**Adult stage:** The peak dispersal of overwintered adults occurs during the first several weeks of plant development. Overwintering adults emerge from hibernation, mate, and lay eggs from April through June. First-generation adults emerge and disperse to new locations during July and August; a few beetles may lay eggs before hibernation. The 3- to 5-mm-long adults are orange-brown with dark bands along the wing margins and have a dark, triangular marking behind the head.

## Effect

**Destructive stages:** Adult and larval.

**Additional plant species attacked:** None.

**Site of attack:** Buds and leaves.

**Impact on the host:** Adult and larval feeding upon the buds results in stunted plants and reduced seed production. After emerging from soil litter or from off site in the early spring, adults feed on exposed shoots that are about 5 to 10 cm long. With heavy defoliation, the host plant becomes skeletonized and turns brown. Heavily defoliated plants may die or produce fewer shoots the following year.

*G. calmariensis* larva. (Photo credit: N. Poritz, bio-control.com)

**Nontarget effects:** The beetle can feed on two native plants (*Decodon verticillatus* and *Lythrum alatum*) and two introduced plants (*L. hyssopifolia* and *Lagerstroemia indica*), but do not reproduce on these hosts.

*G. calmariensis* adult. (Photo credit: N. Poritz, bio-control.com)

283

Early season damage to purple loosestrife plants caused by adult *G. calmariensis*. (Photo credit: N. Poritz, bio-control.com)

## Releases

**First introduced into the United States:** 1992, Idaho, Maryland, Minnesota, New York, Oregon, Pennsylvania, Virginia, and Washington.

**Established in:** More than 30 states across the northern United States.

**Habitat:** Continuously flooded habitats are not suitable for beetle survival unless drier overwintering sites are available nearby. Cold winter weather does not negatively impact beetle survival.

**Availability:** The beetle is readily available in the northern United States.

**Stages to transfer:** Adult and larval.

**Redistribution:** Use a sweep net or handpick the adults and transfer them to other infestations. Insects can be kept two weeks if provided with fresh food. Releases are generally made with 250 to 500 adults. It is best to use freshly overwintered adults once egg-laying has begun in the early spring (late April or early May). Experiments using late-stage larvae-infested stems have had varying results.

## Comments

This agent has established well at release sites. *Galerucella calmariensis* adults are highly mobile and readily able to find host plants. This beetle and its sister species (*G. pusilla*) are difficult to tell apart, especially in the adult stage during the first week after pupation.

The biomass at several purple loosestrife stands in Oregon and Washington has been reduced by 90%. With adequate moisture, larger plants may regrow and flower after attack by the leaf beetles, but after several years of intensive attack, plant size decreases and mortality increases. After population outbreaks of adults, the beetles may disperse from the site resulting in a temporary increase of plant biomass the following year. However, beetle populations usually build up again and dampen the resurgence. At some locations, one or the other species seems to dominate after several years for unknown reasons.

# *Galerucella pusilla*

G. L. Piper, E. M. Coombs, B. Blossey, P. B. McEvoy, and S. S. Schooler

**Common name:** Golden loosestrife beetle.

**Type of agent:** Insect: Beetle (Coleoptera: Chrysomelidae).

**Native distribution:** Asia and Europe.

**Original source:** Northern Germany.

## Biology

**Generations per year:** One, with a partial second generation in warmer climates.

**Overwintering stage:** Adult (in the soil and vegetation near the host plant).

**Egg stage:** Eggs are laid from May to June and from August to September. Females lay up to 10 eggs per day, often in groups of three to six or more, although occasionally single eggs are found. The eggs are 0.5 mm in diameter, barrel-shaped, tan-cream colored, and often possess a line of frass on top.

**Larval stage:** Feeding and development of larvae are similar to *Galerucella calmariensis*. In the field, larvae may feed on plants for five to six weeks, but the development time of an individual larva from eclosion to pupa is generally two weeks, depending upon temperature and food quality and quantity. The larvae may reach 5 mm long after three instars and are light green with black spots down their backs and have darker, sclerotized heads.

**Pupal stage:** Larvae pupate in the soil litter or elevated root mass under the plants. However, in standing water they will pupate in the soft tissue surrounding the plant stem. Pupae are yellowish to pale brown and about 4 mm long.

**Adult stage:** Overwintering adults emerge from hibernation, begin feeding, mate, and lay eggs from late April through June. New adults emerge between July and August and may lay eggs before hibernation. Generally, females emerging before mid-July will lay eggs prior to hibernation. Adults are a light golden color and average 4 mm long. Gravid females are slightly larger than males.

*G. pusilla* adult. (Photo credit: E. Coombs, Oregon Department of Agriculture)

Infestation of purple loosestrife, near Ontario, Oregon, 1994. (Photo credit: E. Coombs, Oregon Department of Agriculture)

Same location in 2000. Note gray area of dead plants caused by *Galerucella* beetles. (Photo credit: E. Coombs, Oregon Department of Agriculture)

## Effect

**Destructive stages:** Adult and larval.

**Additional plant species attacked:** None.

**Site of attack:** Leaves and buds.

**Impact on the host:** Adult and larval feeding on the buds results in stunted plants and reduced seed production. After emerging from soil litter or from off site in the early spring, adults feed on exposed shoots that are about 5 to 10 cm long. With heavy defoliation, the host plant becomes skeletonized and turns brown. Heavily defoliated plants may die or produce fewer shoots the following year.

**Nontarget effects:** The beetle can feed on two native plants (*Decodon verticillatus* and *Lythrum alatum*) and two introduced plants (*L. hyssopifolia* and *Lagerstroemia indica*), but do not reproduce on these hosts.

## Releases

**First introduced into the United States:** 1992, Idaho, Maryland, Minnesota, New York, Oregon, Pennsylvania, Virginia, and Washington.

**Established in:** More than 30 states across the northern United States.

**Habitat:** The beetle readily establishes in purple loosestrife-infested areas that do not remain flooded unless drier overwintering sites are available nearby. Cold winter weather does not negatively impact beetle survival.

**Availability:** The beetle is readily available throughout the northern United States.

**Stages to transfer:** Adult and larval.

Purple loosestrife before release of biocontrol agents. (Photo credit: E. Coombs, Oregon Department of Agriculture)

Same site five years after release of *G. pusilla* and *G. calmariensis* showing heavy defoliation by larvae. (Photo credit: E. Coombs, Oregon Department of Agriculture)

**Redistribution:** Use a sweep net or handpick the adults and transfer them to other infestations. Insects can be kept two weeks if provided with fresh food. Releases are generally made with 250 to 500 adults. It is best to use freshly overwintered adults once egg-laying has begun in the early spring (late April or early May). Experiments using late-stage larvae-infested stems have had varying results.

## Comments

This natural enemy has readily established at most release sites. About 70 to 80% of the original mixed *Galerucella* releases made were composed of *G. pusilla*. The beetle skeletonizes and defoliates its host plant so severely that the plant turns brown. At some locations, one or the other species seems to dominate after several years for unknown reasons.

# *Hylobius transversovittatus*

G. L. Piper, E. M. Coombs, B. Blossey, P. B. McEvoy, and S. S. Schooler

**Common name:** Loosestrife root weevil.

**Type of agent:** Insect: Beetle, weevil (Coleoptera: Curculionidae).

**Native distribution:** Europe.

**Original source:** Germany.

## Biology

**Generations per year:** One, or one generation over two years.

**Overwintering stages:** Egg, larval, pupal, or adult.

**Egg stage:** Eggs are laid from June through August. Only one or two eggs are laid daily with a peak of about three eggs per day in June and early July. Most eggs are deposited in the soil while some are inserted into the stems just above the soil surface. Females lay about 300 eggs each over a two-year period. Eggs hatch in 11 days. The pale yellow eggs are oval and about 2 mm long.

**Larval stage:** The larvae are normally present from August through June of the next year, although the larval stage may last for two years. The larvae mine the roots and feeding tunnels are packed with light brown frass. Larvae are cream-colored with dark brown head capsules, somewhat crescent-shaped, and 8 to 10 mm long.

**Pupal stage:** Pupation occurs within the damaged root crown during late spring to early summer. Pupae are cream-colored.

**Adult stage:** The adults emerge in mid- to late summer and may live for up to three years. The adults are robust, 8- to 12-mm-long, reddish-brown weevils with two rows of dots comprised of white tufted hairs across the back.

## Effect

**Destructive stages:** Adult and larval.

**Additional plant species attacked:** Wing-angled loosestrife (*L. alatum*).

*H. transversovittatus* larva in a purple loosestrife root. (Photo credit: Oregon Department of Agriculture)

**Site of attack:** Larvae live in the roots while the adults feed on the foliage.

**Impact on the host:** Effects of larval feeding are dependent upon root size, attack intensity, and duration. Small roots can be destroyed within two years if infested by several larvae. Larger roots may die after several consecutive years of infestation.

**Nontarget effects:** The native *L. alatum* and swamp loosestrife (*Decodon verticillatus*) were utilized as hosts by the adults and/or larvae in host specificity tests, but it has not been determined whether such feeding has occurred in the field.

*H. transversovittatus* adult. (Photo credit: N. Poritz, bio-control.com)

## Releases

**First introduced into the United States:** 1992, Maryland, Minnesota, New York, Oregon, Pennsylvania, Virginia, and Washington.

**Established in:** Eight states and released in more than 20 others in the northern United States.

**Habitat:** Sites without prolonged flooding are favored for weevil development, but this species tolerates a wide range of environmental conditions.

**Availability:** This weevil is available for extensive redistribution within the United States. Adults can be reared on semi-artificial diets.

**Stages to transfer:** Adult, egg, and first-stage larval.

**Redistribution:** Adults can be hand-collected at night, reared from infested roots, or reared on semi-artificial diets. Weevils can easily be kept in captivity for two months as long as they are cool and supplied with new plants for feeding and egg laying. Eggs can be inserted in holes in the base of stems, but inoculation of plants is a laborious undertaking. Eggs can also be placed on the surface of exposed roots then gently covered with moist debris.

## Comments

This species increases and spreads more slowly than the leaf beetles. However, since during the growing season it feeds continuously on the root storage reserves of the plants, it is an important agent in the control of purple loosestrife. In stands of large, healthy plants, the leaf beetles may produce temporary severe defoliation, but the plants may recover after the beetles enter diapause in midsummer. By reducing root storage reserves, the weevil limits the plant's ability to recuperate after defoliation. The combined impact of both agents is enough to cause plants to die, leaving brown "flags" that indicate their presence within the purple loosestrife stand.

289

Since adults are nocturnal and larvae feed in the roots, the weevil is difficult to find. However, the dark green inky droppings and damage to the lower leaves (uniform removal of leaf tissue along the leaf edge, often leaving a notch near the petiole) indicate adult presence. Larval presence is associated with drooping upper plant stems and can be verified by digging roots and breaking them apart. Often the root will break at the weakest part, which is where the larva is feeding.

# Nanophyes marmoratus

G. L. Piper, E. M. Coombs, B. Blossey, P. B. McEvoy, and S. S. Schooler

**Common name:** Loosestrife seed weevil.

**Type of agent:** Insect: Beetle, weevil (Coleoptera: Brentidae [= Apionidae]).

**Native distribution:** Throughout Europe and western Siberia.

**Original sources:** France and Germany.

## Biology

**Generations per year:** One.

**Overwintering stage:** Adult.

**Egg stage:** Eggs are laid from June to September inside immature flower buds found on the upper flower spike. Females each produce 60 to 100 eggs.

**Larval stage:** Larvae consume the stamens, petals, and ovaries of unopened floral buds. Infested buds fail to open and sometimes prematurely drop from the plant. Only one larva is found per bud.

**Pupal stage:** Pupation occurs within a chamber formed inside the damaged bud.

*N. marmoratus* adult. (Photo credit: E. Coombs, Oregon Department of Agriculture)

**Adult stage:** Overwintered adults appear on the plant during mid-May and feed on young leaves near the shoot tips. Feeding usually produces a number of small "shot holes" on the upper leaves of the plant. The adults move to the spike and feed on flower buds when they begin to form. These weevils are 2.0 to 2.5 mm long, reddish-brown with light-colored shoulder patches, and have a long snout and a blunt abdomen. Because they are quite small, these weevils are difficult to find. However, when the flower spike is knocked onto a sheet of light-colored material, the seed weevils fall out and are easily seen.

## Effect

**Destructive stages:** Adult and larval.

**Additional plant species attacked:** None.

**Site of attack:** Unopened flower buds by larvae and developing leaves by adults.

**Impact on the host:** Flower buds that are fed upon by either adults or larvae usually abort and fail to produce seeds.

**Nontarget effects:** No nontarget effects have been reported.

## Releases

**First introduced into the United States:** 1994, Minnesota, New York, and Oregon.

**Established in:** California, Colorado, Idaho, Minnesota, Montana, New Jersey, New York, Oregon, and Washington.

**Habitat:** Sites without prolonged flooding and deficient in *Galerucella* spp. are favored for beetle development. Heavy *Galerucella* species populations severely defoliate the weed and reduce inflorescence availability for *Nanophyes marmoratus* oviposition.

**Availability:** This insect is available for extensive redistribution within the United States.

**Stage to transfer:** Adult.

**Redistribution:** Heavy-duty sweep nets or a beating tray and stout beat stick can be used to dislodge adults from flower spikes. The adults can then be collected with an aspirator. A release of 100 to 200 adults at a site is recommended for establishment.

## Comments

*Nanophyes marmoratus* tolerates a wide range of environmental conditions and possesses an excellent host-finding ability. It has successfully overwintered on exposed islands in an estuary with high tidal exchange where multiple releases of the leaf beetles have failed. The weevils can also persist where plants are scattered at low densities. Their impact is currently being overshadowed by the dramatic defoliation and plant death caused by the leaf beetles and the root weevil; however they may play an important role after loosestrife abundance declines and the other agents become less effective. This weevil has been recently placed in the family Brentidae.

# References

Blossey, B. 2002. Purple loosestrife. Pages 149-57 *in* R. Van Driesche, S. Lyon, B. Blossey, M. Hoddle, and R. Reardon, eds. Biological Control of Invasive Plants in the Eastern United States. USDA Forest Service Pub. FHTET-2002-4. Morgantown, WV.

Blossey, B., D. Eberts, E. Morrison, and T. R. Hunt. 2000. Mass rearing the weevil *Hylobius transversovittatus* (Coleoptera: Curculionidae), a biological control agent of *Lythrum salicaria*, on semiartificial diet. J. Econ. Entomol. 93: 1644-56.

Hight, S. D., B. Blossey, J. Laing, and R. DeClerck-Floate. 1995. Establishment of insect biological control agents from Europe against *Lythrum salicaria* in North America. Biol. Control 24: 967-77.

Kaufman, L. N., and D. A. Landis. 2000. Host specificity testing of *Galerucella calmariensis* L. (Coleoptera: Chrysomelidae) on wild and ornamental plant species. Biol. Control 18: 157-64.

Kok, L. T., T. J. McAvoy, R. A. Malecki, S. D. Hight, J. J. Drea, and J. R. Coulson. 1992. Host specificity tests of *Galerucella calmariensis* (L.) and *G. pusilla* (Duft.) (Coleoptera: Chrysomelidae), potential biological control agents of purple loosestrife, *Lythrum salicaria* L. (Lythraceae). Biol. Control 2: 282-90.

Kok, L. T., T. J. McAvoy, R. A. Malecki, S. D. Hight, J. J. Drea, and J. R. Coulson. 1992. Host specificity tests of *Hylobius transversovittatus* Goeze (Coleoptera: Curculionidae), a potential biological control agent of purple loosestrife, *Lythrum salicaria* L. (Lythraceae). Biol. Control 2: 1-8.

Malecki, R. A., B. Blossey, S. D. Hight, D. Schroeder, L. T. Kok, and J. R. Coulson. 1993. Biological control of purple loosestrife. BioScience 43: 680-86.

Mullin, B. 1999. Purple loosestrife. Pages 298-307 *in* R. L. Sheley and J. K. Petroff, eds. Biology and Management of Noxious Rangeland Weeds. Oregon State Univ. Press, Corvallis.

Schooler, S. C., E. M. Coombs, and P. B. McEvoy. 2003. Nontarget effects on crepe myrtle by *Galerucella pusilla* and *G. calmariensis* (Chrysomelidae), used for biocontrol of purple loosestrife. Weed Sci. 51: 449-55.

# Rush skeletonweed

*Chondrilla juncea*

Sunflower family—Asteraceae

## *G. L. Piper, E. M. Coombs, G. P. Markin, and D. B. Joley*

**Additional common names:** Devil's-grass, gum succory, hog-bite, naked weed.

**Native range:** Western Europe and north Africa to west central Asia.

**Entry into the United States:** Rush skeletonweed was apparently introduced in contaminated plant material, contaminated seed, or with animal fodder or bedding.

## Biology

**Life duration/habit:** Rush skeletonweed is a long-lived herbaceous perennial in some locations and a biennial or short-lived perennial in its native area.

**Reproduction:** It reproduces by seed and by vegetative regrowth. Flowers are self-fertile. Seed is produced over a relatively long period of the summer, the number of seeds produced per multi-stemmed plant exceeds 15,000 under stress conditions and 20,000 under more ideal conditions. The pappused seeds are adapted to wind dispersal, but may also be spread by water, animals, and humans.

**Roots:** The slender taproot often penetrates 2.5 m or more into the soil and branches at various depths. When undisturbed, buds at the root crown and along the major lateral roots give rise to daughter rosettes, enabling a single plant to become a colony. New plants also can develop from root fragments.

**Stems and leaves:** Rush skeletonweed produces a rosette of lance-shaped, deeply lobed leaves that are 5 to 12.5 cm long. Each rosette produces one or more stems 50 to 150 cm long with multiple spreading or ascending light-green branches. The slender, nearly leafless stems and branches are smooth except for the lowermost 5 to 8 cm which are densely covered with stiff, downward-directed hairs.

**Flowers:** Flower heads are produced along or at the ends of stems, either individually or in groups of two to three. Each flower head contains 10 to 12 bright yellow florets.

**Fruits and seeds:** The seeds are about 3 mm long, pale to dark brown, and vertically ribbed with scales and tooth-like projections at the tufted end.

Rush skeletonweed flower. (Photo credit: E. Coombs, Oregon Department of Agriculture)

293

## Infestations

**Worst infested states:** California, Idaho, Oregon, and Washington. The species also occurs in several states on the eastern seaboard but is not considered a troublesome plant there.

Rush skeletonweed infestation. (Photo credit: N. Poritz, bio-control.com)

**Habitat:** Rush skeletonweed is primarily a weed of cultivation and of open, waste areas with disturbed soils. It also grows along roadsides and in unimproved pastures that have been weakened by drought or overgrazing. The weed thrives under a wide range of climatic conditions, although semiarid and Mediterranean-type climates appear to provide the optimum conditions of cool winters and warm summers without severe drought and with a predominance of winter and spring rainfall. Rush skeletonweed attains its greatest development on sandy, sandy loam, and sandy clay soils.

**Impacts:** The weed is established on more than 1.4 million ha in 12 northern and southwestern Idaho counties, 404,000 ha in 15 eastern Washington counties, and 73,000 ha in one western Oregon county. Sizeable infestations also occur in California. Rush skeletonweed infestation levels in the eastern United States are not known.

The weed's extensive root system enables it to compete effectively with crop plants for moisture and nutrients, especially nitrogen. The tall, wiry, latex-producing stems also hinder the operation of crop harvest machinery. Rangeland infestations displace desirable forage plants.

## Comments

This plant has some beneficial attributes. It is a drought-tolerant forage plant whose rosette leaves and pre-flowering stems are palatable and nutritious. Continuous sheep grazing keeps rush skeletonweed in the rosette stage during the summer months when only small amounts of other forage plants may be available.

# Bradyrrhoa gilveolella

G. L. Piper, E. M. Coombs, G. P. Markin, and D. B. Joley

**Common name:** Skeletonweed root moth.

**Type of agent:** Insect: Moth (Lepidoptera: Pyralidae).

**Native distribution:** Bosnia, Bulgaria, Greece, Serbia, and Turkey.

**Original source:** Northern Greece.

## Biology

**Generations per year:** One in the interior Pacific Northwest; two or more possible in the warmer coastal area of Oregon, interior northern California, and southern coastal California.

**Overwintering stages:** Late instar larval or pupal, in a tube attached to the root.

**Egg stage:** Eggs are laid in the rosette crown or possibly in the soil. A female can produce nearly 300 eggs. The incubation period lasts six to 10 days. The spherical, slightly flattened, creamy white eggs are 0.65 to 0.80 mm long and 0.45 mm wide. The egg's exterior surface is reticulated.

**Larval stage:** Upon hatching, the caterpillars penetrate the soil to a depth of 5 to 10 mm and begin to feed externally on the roots. They spin elongated silken tubes affixed to the roots in which they live and continue to feed both internally and externally on root tissues. These protective, subterranean tubes are enlarged over time and are composed of coagulated latex, root fragments, frass, and sand grains. The tubes vary in length from 30 to 60 mm and are 5 to 7 mm wide. Larvae complete development in 45 to 60 days. Mature larvae are 20 to 26 mm long and are a dirty white color with brown head capsules.

**Pupal stage:** Fully grown larvae extend their shelter tubes to the soil surface to form exit chimneys comprised of silk and frass. The larvae then close the tube entrances with a thin layer of silk and pupate inside. The pupae are light brown and 11 to 14 mm long. The pupal period lasts seven to 10 days. Adults emerge by forcing their way out of the capped tubes.

**Adult stage:** In Europe, winter generation adults emerge during May and June; summer generation adults appear from late August to early October. Adults are 11 to 13 mm long with a wing span of 25 to 28 mm. They are creamy buff in color and have three distinct horizontal bands on the front wings.

Adult *Bradyrrhoa gilveolla*. (Photo credit: G. Markin, U.S. Forest Service)

Effect of *B. gilveolella* on a rush skeletonweed root. (Photo credit: G. Markin, U.S. Forest Service)

*B. gilveolella* larva on a rush skeletonweed root. (Photo credit: G. Markin, U.S. Forest Service)

## Effect

**Destructive stage:** Larval.

**Additional plant species attacked:** None.

**Site of attack:** Roots.

**Impact on the host:** Concurrent feeding by several larvae can destroy the cortical and vascular tissues of roots, resulting in death of aboveground plant parts, but new plants may be regenerated from root buds below the point of attack. Larval attack also can destroy regenerative root buds and diminish root carbohydrate reserves, thus adversely impacting plant vigor and overwintering capability. Soilborne plant pathogens may invade and develop on insect-damaged root tissues.

**Nontarget effects:** No nontarget effects have been reported.

## Releases

**First introduced into the United States:** 2002, Idaho.

**Established in:** Establishment has not been confirmed.

**Habitats:** Plants growing in sandy, granitic, or loose-textured soils are more readily attacked by the moth.

**Availability:** This moth is unavailable as of 2004.

**Stage to transfer:** Adult.

**Redistribution:** Collect adults with a sweep net during the evening. An alternative approach would be to transplant infested pots of rush skeletonweed into new areas.

# Cystiphora schmidti

G. L. Piper, E. M. Coombs, G. P. Markin, and D. B. Joley

**Common name:** Rush skeletonweed gall midge.

**Type of agent:** Insect: Fly, gall midge (Diptera: Cecidomyiidae).

**Native distribution:** Bosnia, Greece, Iran, and Poland.

**Original source:** Greece.

## Biology

**Generations per year:** Four to five.

**Overwintering stages:** Mature larval, prepupal or pupal (in stem or rosette leaf galls or in the soil).

**Egg stage:** Females insert eggs within rosette and stem leaves and stems. A female can produce about 100 eggs. The incubation period averages nine days. The eggs are elongate oval, 0.01 mm wide by 0.02 mm long, and pale white with a green or yellow tinge.

**Larval stage:** The larvae feed on the leaf or stem tissues at the egg deposition site. Feeding activity initiates gall formation that is characterized by a swelling and yellowish to maroon discoloration of affected tissue. Leaf galls are circular, about 3.0 mm in diameter, and slightly raised, whereas stem galls are elongated and usually more elevated. The larvae complete development in four to seven days. The larvae are flattened, 1 to 2.5 mm long, and are pink or orange.

**Pupal stage:** The mature larvae sometimes leave the galls and drop to the soil or surface litter where they pupate. More often, however, they pupate inside a silken cocoon within the gall. The pupae rupture the gall tissue with their "horns" to facilitate adult emergence. The pupal exuviae can often be seen partially protruding from the exit hole after the adults have departed. The pupal period lasts four to six days.

**Adult stage:** Adults are found in the field from April to October; males live one to two days and females three to four days. First-generation females lay eggs in the rosette leaves. Succeeding generations attack the flowering stems. The adults are small, light to medium brown, and lightly sclerotized. Females are 0.82 to 1.64 mm long; males are 1.03 to 1.56 mm long. The female's abdomen is swollen and terminates in a bulbous enlargement. The male's abdomen is slender with the genitalia readily visible.

*C. schmidti* larvae and unopened galls. (Photo credit: N. Poritz, bio-control.com)

297

## Effect

**Destructive stage:** Larval.

**Additional plant species attacked:** None.

**Site of attack:** Leaves and stems.

**Impact on the host:** The midge damages both the rosette and flowering stems, reducing the quantity of photosynthate available for plant growth and maintenance. Leaf and stem tissues are injured or destroyed, causing premature yellowing, desiccation, and death. Rosettes may die. Infested plants have fewer branches and flower heads than do uninfested plants. Seeds exhibit decreased weight and reduced viability.

**Nontarget effects:** No nontarget effects have been reported.

## Releases

**First introduced into the United States:** 1975, California.

**Established in:** California, Idaho, Oregon, and Washington.

**Habitats:** The insect is most abundant in areas where the yearly average temperature exceeds 17° C and precipitation is less than 400 mm. Plants subject to the heaviest attack grow in open locations in well-drained soil.

**Availability:** The insect is available for mass collection in California, Idaho, and Washington.

**Stages to transfer:** Larval and pupal within stem and leaf galls.

**Redistribution:** The midge is best collected by harvesting galled stems from early July to late September. Any seed heads and flowers on the stems of infected plants should be removed and the stems then tied into bundles to form "tepees." Tepees should be placed among the uninfested plants at the release site. Many of the immature midges will complete their development within the galls, emerge, and attack the new plants.

## Comments

Native parasitoids have decimated this agent in California. In Washington *Mesopolobus* sp. diminishes the effects of this agent on rush skeletonweed. The midge attacks all rush skeletonweed biotypes in the western United States.

# *Eriophyes chondrillae (= Aceria chondrillae)*

G. L. Piper, E. M. Coombs, G. P. Markin, and D. B. Joley

**Common name:** Rush skeletonweed gall mite.

**Type of agent:** Mite: Gall mite (Acari: Eriophyidae).

**Native distribution:** From Portugal through central and Mediterranean Europe, north to Germany, and east through Turkmenistan, Uzbekistan, and Kazakhstan.

**Original source:** Italy.

## Biology

**Generations per year:** Multiple.

**Overwintering stage:** Adult (females on the rosette bud).

**Egg stage:** Each female can deposit between 60 and 100 eggs within the gall it occupies. Eggs are spherical, about 0.04 mm in diameter, and translucent upon deposition but turn pale orange as the embryo develops.

**Nymphal stage:** Several hundred nymphs feed within a gall. Nymphal development can be completed in 10 days during the summer. The pale yellow-orange first- and second-stage nymphs look somewhat humpbacked, have four legs, and lack a genital opening. Body length of the first-stage nymph is 0.08 to 0.10 mm; second-stage nymph length is 0.10 to 0.17 mm.

**Adult stage:** Overwintered adults invade shoot buds when the weed bolts in the spring. Feeding by *Eriophyes chondrillae* transforms the buds into contorted, leaf-like galls that may reach a diameter of 5 cm. Mites increase and galls form until floral shoots cease growing in the fall. Adults may live three to four weeks. Adults are worm-like, soft-bodied, pale yellow-orange, and possess two pairs of legs. Males range in length from 0.16 to 0.18 mm and females from 0.19 to 0.26 mm.

## Effect

**Destructive stages:** Nymphal and adult.

**Additional plant species attacked:** None.

**Site of attack:** Axillary and terminal buds.

**Impact on the host:** Infestation of the buds by the mite decreases plant vigor by reducing root carbohydrate reserves, hinders rosette formation from established roots, stunts the plant and reduces the number of vegetative shoots produced, decreases or completely prevents seed production, and commonly results in death of seedlings or first-year satellite plants.

**Nontarget effects:** No nontarget effects have been reported.

## Releases

**First introduced into the United States:** 1977, California, Idaho, and Oregon; 1979, Washington.

**Established in:** California, Idaho, Oregon, and Washington.

**Habitats:** The mite rapidly colonizes plants growing in undisturbed, well-drained soils on south- or west-facing slopes. Mite populations do not persist in sites subjected to repetitive soil disturbance, such as cropland. The mite is very tolerant of climatic conditions and can

*E. chondrillae* in bud gall. (Photo credit: E. Coombs, Oregon Department of Agriculture)

299

withstand temperatures below 0° C. Inside the gall, mites can endure summer temperatures of 30 to 35° C without ill effects.

**Availability:** The mite is readily available in all states where it is established.

**Stages to transfer:** Adult and nymphal (with stem galls).

**Redistribution:** Galled stems can be gathered from July to mid-October. Galled shoots should be placed in direct contact with uninfested stems. Once the galls begin to dry up, the mites will exit and infest the buds of uninfested plants. Mite-galled stems retained within plastic garbage bags can be stored at 5 to 10° C for several weeks without any adverse effect.

## Comments

Despite its small size, *E. chondrillae* is an excellent disperser and will quickly colonize stands of all biotypes of rush skeletonweed found in the western United States. It has been the most effective natural enemy released to date against the weed in the region. However, in California its impact has been diminished because of mortality by predaceous mites. This mite has also been known as *Aceria chondrillae*.

Bud galls on rush skeletonweed caused by *E. chondrillae*. (Photo credit: N. Poritz, bio-control.com)

# *Puccinia chondrillina*

G. L. Piper, E. M. Coombs, G. P. Markin, and D. B. Joley

**Common name:** Rush skeletonweed rust fungus.

**Type of agent:** Rust fungus (Uredinales: Pucciniaceae).

**Native distribution:** Bosnia, Morocco, Spain, Portugal, France, Italy, Greece, Slovenia, Turkey, Iran, and central Asia.

**Original source:** Italy (pathotypes PC-1 and PC-16).

## Biology

**Generations per year:** Multiple.

**Overwintering stages:** Urediospores and/or teliospores.

**Life cycle of the fungus:** From spring to fall, cinnamon brown, circular, eruptive pustules (uredia) develop on all aboveground plant parts and release infective spores. Lesions (telia) form at the bases of flowering shoots in the fall. The lesions produce spores that remain dormant until spring. Spores germinate on the rosette leaves and form clusters of yellowish pycnia which soon yield pycniospores. These produce aecia and aeciospores on the leaves. Aeciospores germinate to produce brown pustules, thus completing the life cycle.

## Effect

**Destructive stages:** Uredia and telia.

**Additional plant species attacked:** None.

**Site of attack:** Leaves, stems, buds, and flowers.

**Impact on the host:** Fall and spring rosette infection often results in premature death of plants, especially seedlings. Open lesions cause desiccation, reduce photosynthetic area, and decrease plant vigor. Fungus-infected stems are stunted, and deformed, and produce few branches and floral buds. Seed yield, weight, and viability are reduced in rusted plants. Rush skeletonweed's ability to regenerate from root buds is also diminished as a consequence of *Puccinia chondrillina* infection.

**Nontarget effects:** No nontarget effects have been reported.

## Releases

**First introduced into the United States:** Intentionally released 1976, California; 1977, Idaho and Oregon; 1978, Washington. The rust was apparently introduced (year unknown) along with rush skeletonweed in the eastern United States and now occurs from Maryland south to Virginia.

**Established in:** California, Idaho, Maryland, Oregon, Virginia, and Washington.

**Habitats:** The rust survives under a variety of moisture regimes but its development is best in mesic habitats. It is less damaging in hot and dry sites.

**Availability:** The rust fungus is readily available for redistribution within California, Idaho, Oregon, and Washington.

**Stage to transfer:** Urediospores.

**Redistribution:** Rosettes infected with uredia can be dug and transplanted among uninfested plants during the spring and fall. During the summer, rusted floral stems can be harvested and placed among nonrusted rush skeletonweed to initiate infection. To ensure spore germination and

Rush skeletonweed rosette infected with *P. chondrillina*. (Photo credit: E. Coombs, Oregon Department of Agriculture)

301

subsequent infection, urediospore-laden material should be released in the evening when temperatures are cooler and when an extended dew period is anticipated. It is also worthwhile to spray the uninfested plants with water prior to pathogen dissemination to enhance the infectivity rate. Rust-infected stems can be stored at 5 to 10° C for several weeks without hurting spore viability.

## Comments

*Puccinia chondrillina* is the first exotic plant pathogen successfully employed for the classical biological control of a weed in North America. The fungus is well established at many rush skeletonweed-infested sites in California and the Pacific Northwest. In California, some weed management practitioners consider it to be a more effective biocontrol agent than either the midge or mite. The time required for the pathogen to reduce a rush skeletonweed infestation to a tolerable level depends on the amount and distribution of the initial inoculum, plant population density and age, and the environmental characteristics of the release site. Usually, the effects of the rust fungus are readily apparent the fourth year after release.

Pathotype PC-16 will not impact the early-flowering rush skeletonweed biotype found in parts of northeastern Washington and northern Idaho or a late-flowering biotype that occurs in southwestern Oregon. European plant pathologists are attempting to find *P. chondrillina* pathotypes infective to these resistant plant biotypes.

Damage to rush skeletonweed stem caused by the rust fungus *P. chondrillina*. (Photo credit: N. Poritz, bio-control.com)

## *References*

Caresche, L. A., and A. J. Wapshere. 1974. Biology and host specificity of the *Chondrilla* gall mite *Aceria chondrillae* (G. Can.) (Acarina: Eriophyidae). Bull. Entomol. Res. 64: 183-92.

Caresche, L. A., and A. J. Wapshere. 1975. The *Chondrilla* gall midge *Cystiphora schmidti* (Rübsaamen) (Diptera: Cecidomyiidae). II. Biology and host specificity. Bull. Entomol. Res. 65: 55-64.

Caresche, L. A., and A. J. Wapshere. 1975. Biology and host specificity of the *Chondrilla* root moth *Bradyrrhoa gilveolella* (Treitschke) (Lepidoptera: Phycitidae). Bull. Entomol. Res. 65: 171-85.

Dodd, J., and F. D. Panetta. 1987. Seed production by skeleton weed (*Chondrilla juncea* L.) in western Australia in relation to summer drought. Aust. J. Agric. Res. 38: 689-705.

Hasan, S., and A. J. Wapshere. 1973. The biology of *Puccinia chondrillina*, a potential biological control agent of skeleton weed. Ann. Appl. Biol. 74: 325-32.

Lee, G. A. 1986. Integrated control of rush skeletonweed (*Chondrilla juncea*) in the western U.S. Weed Sci. 34 (Suppl. 1): 2-6.

McVean, D. N. 1966. Ecology of *Chondrilla juncea* L. in south-eastern Australia. J. Ecol. 54: 345-65.

Panetta, F. D., and J. Dodd. 1987. The biology of Australian weeds. 16. *Chondrilla juncea* L. J. Aust. Inst. Agric. Sci. 53: 83-95.

Piper, G. L. 1990. Biological control efforts with rush skeletonweed. Pages 77-80 *in* B. F. Roché, Jr. and C. T. Roché, eds. Range Weeds Revisited. Symp. Proc. 1989 Pacific Northwest Range Management Short Course, Jan. 24-26, 1989, Spokane, WA. Washington State Univ. Coop. Ext. MISC 0143.

Sheley, R. L., J. M. Hudak, and R. T. Grubb. 1999. Rush skeletonweed. Pages 308-14 *in* R. L. Sheley and J. K. Petroff, eds. Biology and Management of Noxious Rangeland Weeds. Oregon State Univ. Press, Corvallis.

Sobhian, R., and L. A. Andres. 1978. The response of the skeletonweed gall midge, *Cystiphora schmidti* (Diptera: Cecidomyiidae), and gall mite, *Aceria chondrillae* (Eriophyidae), to North American strains of rush skeletonweed (*Chondrilla juncea*). Environ. Entomol. 7: 506-08.

Supkoff, D. M., D. B. Joley, and J. J. Marois. 1988. Effect of introduced biological control organisms on the density of *Chondrilla juncea* in California. J. Appl. Ecol. 25: 1089- 95.

Wehling, W. F., and G. L. Piper. 1988. Efficacy diminution of the rush skeletonweed gall midge, *Cystiphora schmidti* (Diptera: Cecidomyiidae), by an indigenous parasitoid. Pan-Pac. Entomol. 64: 83-85.

# Russian thistle

*Salsola tragus (=S. iberica, S. kali, S. pestifer)*

Goosefoot family—Chenopodiaceae

## M. J. Pitcairn

**Additional common names:** Russian tumbleweed, salwort, tumbleweed, windwitch.

**Native range:** Southern interior regions of eastern Europe and central Asia, including Pakistan, Ukraine, Russia, Turkmenistan, and Uzbekistan.

**Entry into the United States:** First introduced as a contaminant of flaxseed in South Dakota in 1873, but was likely reintroduced several times as a contaminant of grain seed in several states.

## Biology

**Life duration/habit:** Russian thistle is a bushy summer annual with rigid branches and reduced, stiff, prickly upper stem leaves at maturity. It grows 1 to 1.5 m tall with height more or less equal to width.

**Reproduction:** Reproduction occurs solely by seed. At death, the plant separates from the root crown and tumbles across the landscape dropping seeds along its route.

**Roots:** Plants form a single taproot up to 1.5 m deep with laterals spreading 1.8 m.

**Stems and leaves:** Stems are rigid and typically curve upward. Lower leaves are fleshy, 10 to 50 mm long, 0.5 to 1 mm wide, and have sharp tips. Upper leaves are reduced to awl-shaped bracts, the tips of which are very sharp.

**Flowers:** Minute flowers are solitary and occur in leaf axils. Petals are lacking but there are four to five membranous sepals. Five yellow stamens extend beyond the sepals. Flowers are wind-pollinated.

**Fruits and seeds:** Fruits are one-seeded and enclosed by calyces. Seeds are round and slightly flattened, 1.5 to 2 mm in diameter with a thin gray to brown seed coat. At

Russian thistle stem and leaves. (Photo credit: J.DiTomaso, University of California)

maturity, seeds are completely filled with a small, coiled embryo. One plant can produce thousands of seeds, but the seeds appear to be short-lived as viability during storage is usually one to two years.

## Infestations

**Worst infested states:** Occurs in almost every state in the continental United States, but is most abundant in Arizona, California, Idaho, eastern Montana, eastern Oregon, and eastern Washington.

**Habitat:** Russian thistle occurs throughout the semiarid regions of the United States and is heavily favored by disturbance. It infests sandy soils on disturbed sites, waste places, roadsides, cultivated and abandoned fields, and disturbed natural plant communities. Russian thistle can persist in dryland cropping systems and overgrazed rangeland.

**Impacts:** In the Pacific Northwest, Russian thistle infests spring wheat fields and reduces yield. In California, it is a host for the beet leafhopper which vectors curly top virus, a devastating pathogen of tomatoes, sugar beets, green beans, melons, cucurbits, and peppers. It also is a host for several polyphagous crop pests including lygus and stink bugs. Tumbling plants can surprise drivers and cause traffic accidents. Windblown tumbleweeds fill irrigation canals and catchments and pile against fences and dwellings. Immature plants can provide extra forage for livestock on arid rangelands. However, when the plants are exposed to heavy nitrogen fertilizer application, nitrates or oxalates can accumulate to levels poisonous to sheep.

## Comments

Recent research shows that Russian thistle consists of three variants or types (designated A, B, and C) in California. Types A and B differ in chromosome number (A: 2n=36; B: 2n=18), spinescence, fruit size, calyx characteristics, fruit weight, and pubescence. Type C has 2n=54 chromosomes and some characteristics of both types.

A renewed biological control effort was begun against Russian thistle in 1999 and several new natural enemies have been discovered. The most promising agents are the blister mite (*Aceria salsolae*) that galls the new growing tips and a moth (*Gymnacella canella*) that consumes seeds.

Windblown Russian thistle plants tumble and pile against fencerows. (Photo credit: J. DiTomaso, University of California)

# Coleophora klimeschiella

M. J. Pitcairn

**Common names:** Russian thistle casebearer, tumbleweed casebearer.

**Type of agent:** Insect: Moth (Lepidoptera: Coleophoridae).

**Native distribution:** Middle East and central Asia, including Pakistan, Russia, Turkmenistan, and Uzbekistan.

**Original source:** Pakistan and Turkey.

## Biology

**Generations per year:** Three in California, one to two in more northern areas. In laboratory cultures, egg-to-adult development took 40 to 45 days.

**Overwintering stage:** Mature larval in cases on dried host plants.

**Egg stage:** The yellowish-brown, ovoid eggs are laid singly on leaves or in leaf axils.

**Larval stage:** Larvae complete five growth stages. The first two growth stages feed as leaf miners, the later stages are casebearers feeding externally on the leaves. Cases are made from a hollowed-out leaf and only the head and forelegs protude from the case. The casebearing larvae are quite mobile and move from leaf to leaf and even to other plants. They feed by attaching themselves to leaves, chewing through the epidermis at the point of attachment, and hollowing out approximately 1 cm of the leaf base. A single larva can destroy 15 to 20 leaves during its development. Larvae are orange and 17 mm long by 3 mm wide at maturity.

**Pupal stage:** Mature larvae attach their case to a leaf and pupate within the case.

C. klimeschiella pupal case attached to a Russian thistle leaf. (Photo credit: E. Coombs, Oregon Department of Agriculture)

Russian thistle leaves mined by *C. klimeschiella* larvae. (Photo credit: J. Clark, University of California)

**Adult stage:** Adults usually live 11 to 13 days. The moths are brownish-gray and 5 to 6 mm long. When at rest on the plant, they fold their antennae back over their bodies (compared to *Coleophora parthenica*, which extends the antennae forward, in front of the body).

## Effect

**Destructive stage:** Larval.

**Additional plant species attacked:** None.

**Site of attack:** Leaves.

**Impact on the host:** Young plants heavily infested with *C. klimeschiella* usually die. Impact on older plants is not known, but attacked plants presumably produce fewer seeds than those not infested.

**Nontarget effects:** No nontarget effects have been observed.

## Releases

**First introduced into the United States:** 1977. This insect has been released in California, Colorado, Idaho, Kansas, Minnesota, Montana, Nebraska, Nevada, Oregon, South Dakota, Texas, Utah, Washington, and Wyoming.

**Established in:** California, Colorado, Idaho, Nevada, Oregon, Texas, Utah, and Washington.

**Habitat:** The insect's preferred habitat is not specified in the literature.

**Availability:** This insect is readily available in all states where it has established.

**Stages to transfer:** Overwintering mature larval or newly emerged adult.

**Redistribution:** Two distribution methods have been used for this species. First, it can be easily moved by relocating infested plant material during the winter months when Russian thistle can be gathered with pitchforks. The infested plant material should be tied down to prevent it from tumbling away but still allow the moths to emerge naturally in the new area. Be aware, however, that this method also moves several larval parasitoids known to attack this species and is recommended only in areas

307

where the moth is free of native parasitoids. Second, branches with overwintering casebearing larvae can be hand-collected and held in sleeve cages to allow moth emergence in the laboratory. The emerged moths can be aspirated into collecting tubes and transported to the release sites. It is recommended that at least 50 adults be released to ensure colonization. Colonization can be determined through surveys approximately two months after field release.

## Comments

This insect is not considered to be an effective biological control agent because it is often heavily attacked by native larval parasitoids and predators. Parasitization and predation rates exceeding 65% of overwintering larvae prevent this insect from building up high population densities needed to control Russian thistle. In California and Utah, *C. klimeschiella* was released on halogeton (*Halogeton glomeratus)* but it failed to establish.

# *Coleophora parthenica*

M. J. Pitcairn

**Common names:** Russian thistle stem-mining moth, tumbleweed stem moth.

**Type of agent:** Insect: Moth (Lepidoptera: Coleophoridae).

**Native distribution:** North Africa, the Middle East, and central Asia, including Pakistan, Russia, Turkmenistan, and Uzbekistan.

**Original source:** Egypt, Pakistan, and Turkey.

## Biology

**Generations per year:** Three in Arizona, California, and Texas; one to two in more northern areas. During summer, egg-to-adult development takes 42 to 60 days.

**Overwintering stage:** Mature larval, in stems of dried plants.

**Egg stage:** The yellowish-brown, ovoid eggs are 0.36 mm in diameter and are covered with punctures and ridges. Females lay eggs singly on leaves near branch tips.

**Larval stage:** Upon hatching, the larvae burrow into leaves and down into stems. They feed mainly on the central pith tissue of the larger stems of the plant. The larvae complete three growth stages. Larvae are orange and about 17 mm long by 3 mm wide at maturity.

**Pupal stage:** Before pupation, the larvae prepare exits by chewing a hole in the stem but leaving a thin layer or "window" of epidermis as a cover. After this, the larvae retreat into the stems and pupate. Upon emergence, the adults force their way out of the stems through the windows. Pupae are about 10 mm long by 2 mm wide.

**Adult stage:** Adults usually live only three to six days. The moths are creamy white and 10 mm long. When at rest on the plant, their antennae are extended forward in front of the body (compared to *Coleophora klimeschiella,* which fold their antennae back over the body).

## Effect

**Destructive stage:** Larval.

**Additional plant species attacked:** Host tests suggest that *Coleophora parthenica* may attack several species of *Salsola* and *Halogeton.*

**Site of attack:** Inside the stem; feeding is limited to the central pith.

**Impact on the host:** Minimal; damage to the pith appears to have little effect on growth and seed production. Exit windows may damage vascular and mechanical support tissues resulting in some broken branches.

**Nontarget effects:** No nontarget effects have been observed.

## Releases

**First introduced into the United States:** 1973. This insect has been released in Arizona, California, Colorado, Idaho, Minnesota, Montana, Nebraska, Nevada, Oregon, South Dakota, Texas, Utah, Washington, and Wyoming.

*C. parthenica* larva exposed in Russian thistle stem (above); emergence hole (below). (Photo credit: J. Clark, University of California)

**Established in:** Arizona, California, Colorado, Idaho, Nevada, Oregon, Texas, Utah, and Washington.

**Habitat:** This insect is widespread throughout the western United States. Its preferred habitat is not specified in the literature.

**Availability:** This insect is readily available in all states where it has established.

**Stages to transfer:** Overwintering mature larval or newly emerged adult.

**Redistribution:** Two distribution methods have been used for this species. First, infested plant material can be relocated to new areas during the winter months when Russian thistle can be gathered with pitchforks. The infested plant material should be tied down to prevent it from tumbling away but still allow the moths to emerge naturally in the new area. Be aware, however, that this method also moves several larval parasitoids known to attack this species and it is recommended that collections of infested plant material occur only in areas where the moth is free of native parasitoids. Second, infested stems can be held in sleeve cages to allow moth emergence in the laboratory. The emerged moths can be aspirated into collecting tubes and transported to release sites. It is recommended that at least 50 adults be released to ensure colonization. Colonization can be determined through surveys approximately two months after field release.

*C. parthenica* adult. (Photo credit: USDA-ARS)

## Comments

This insect is considered ineffective as a biological control agent. Feeding damage appears to have little effect on growth and reproduction of its host plant. In addition, *C. parthenica* larvae and pupae are attacked by several hymenopterous parasitoids and by small rodents that feed on overwintering larvae. Several generalist predators such as spiders feed on the adults. The resulting high mortality severely limits its population growth.

*Coleophora parthenica* was released on halogeton (*Halogeton glomeratus*) in California, Idaho, Nevada, and Utah, but it failed to establish.

## *References*

Goeden, R. D., and D. W. Ricker. 1979. Field analyses of *Coleophora parthenica* (Lepidoptera: Coleophoridae) as an imported natural enemy of Russian thistle, *Salsola iberica*, in the Coachella Valley of Southern California. Environ. Entomol. 8: 1099-1101.

Hafez, M., Y. H. Fayad, and A. A. Sarhan. 1978. *Coleophora parthenica* Meyrick (Lepidoptera, Coleophoridae) in Egypt, a potential agent for the biological control of the noxious thistle, *Salsola kali* L. (Chenopodiaceae). Prot. Ecol. 1: 33-44.

Hawkes, R. B., and A. Mayfield. 1978. *Coleophora klimeschiella*, biological control agent for Russian thistle: host specificity testing. Environ. Entomol. 7: 257-61.

Khan, A. G., and G. M. Baloch. 1976. *Coleophora klimeschiella* (Lep: Coleophoridae), a promising biocontrol agent for Russian thistles, *Salsola* spp. Entomophaga 21: 425-28.

Müller, H., G. S. Nuessly, and R. D. Goeden. 1990. Natural enemies and host-plant asynchrony contributing to the failure of the introduced moth, *Coleophora parthenica* Meyrick (Lepidoptera: Coleophoridae), to control Russian thistle. Agric. Ecosyst. Environ. 32: 133-42.

Pemberton, R. W. 1986. The impact of a stem-boring insect on the tissue, physiology and reproduction of Russian thistle. Entomol. Exp. Appl. 42: 169-77.

# Saltcedar

*Tamarix ramosissima, T. chinensis, T. parviflora, T. canariensis, T. gallica,* and hybrids

Saltcedar family—Tamaricaceae

## C. J. DeLoach and R. I. Carruthers

**Additional common names:** Tamarix, tamarisk.

**Native range:** Central Asia to China, Mongolia, India, to northern Africa and southern Europe; the Canary Islands.

**Entry into the United States:** Introduced about 1823 as ornamentals and for stream bank stabilization, saltcedar escaped cultivation in the early 1900s and spread rapidly after the late 1920s.

## Biology

**Life duration/habit:** Saltcedars are deciduous shrubs or small trees 1 to 10 m tall (average 3 to 7 m) with spineless, willowy foliage. They are tolerant of saline soils to about 30,000 ppm salt, as well as of fire, drought, floods, and cold temperatures. They are facultative phreatophytes that grow mostly in riparian areas.

**Reproduction:** Saltcedars reproduce from profuse seeds produced from first-year seedlings and older plants. Saltcedars also may reproduce vegetatively from root and stem fragments.

311

Saltcedar in flower in California. (Photo credit: USDA-ARS)

Infestation of saltcedar, displacing native riparian species. (Photo credit: USDA-ARS)

**Roots:** Saltcedars are very deep-rooted if in deep soil, or roots extend down to the water table.

**Stems and leaves:** Leaves are small, juniper-like bracts with salt-secreting glands. Young stems are flexible while older stems become woody. Trunks typically grow to 15 cm in diameter and rarely to 30 cm but often with smaller, multiple stems.

**Flowers:** Saltcedars have attractive, insect-pollinated, four- or five-petaled pink flowers with terminal inflorescences of racemes in dense to loose panicles that grow up to 25 cm long. Basal racemes are sometimes intermingled with foliage twigs. In spring, foliage appears first followed by flowers except in *T. parviflora* where flowers appear first and in more dense clusters of racemes along the stems. Flowers are produced almost throughout the growing season, from mid-spring to a peak in June, then flowers decrease until fall, except in *T. parviflora* which blooms only in the spring.

**Fruits and seeds:** Saltcedars produce capsules with many tiny seeds, each with an apical pappus. The seeds may be windblown or water-borne and germinate rapidly when wet, especially on mud banks after flood waters recede. The seeds do not survive to the next growing season.

## Infestations

**Worst infested states:** Arizona, California, Colorado, Kansas, Nevada, New Mexico, Oklahoma, Texas, and Utah. Also found increasingly in Idaho, Montana, Nebraska, North Dakota, Oregon, South Dakota, and Wyoming.

**Habitat:** Saltcedars are found in riparian areas, including saline or nonsaline river valleys, and around springs, small streams, playa lakes, and reservoir shorelines.

**Impacts:** Saltcedars are among the most destructive exotic, invasive plants of western riparian areas. They rapidly exclude desirable native vegetation (especially following wildfires or floods) in saline soils and in both disturbed and undisturbed areas. They interact synergistically with altered hydrologic cycles below dams, and with droughts, wildfires, high soil salinity, overgrazing, and with some control methods. Saltcedar stands use an average of 4.0 – 5.5 acre-feet of water per year, depending on the depth to water table, salinity of the groundwater, and length of the growing season.

Saltcedars dry up desert springs and small streams, alter stream geomorphology and water quality, and increase soil salinity, all of which seriously degrade wildlife habitat, including that of many sensitive species of birds, fish, and other animals and plants. They interfere with recreational usage of parks and natural areas.

## Comments

The drought in the Southwest since 1992 caused serious water shortages for agriculture and municipalities in several states and Mexico resulting in defaults in water compacts and treaties between states and between the United States and Mexico. Saltcedars' heavy use of water has exacerbated the problem.

The large (to 20 m tall and 1 m trunk diameter), evergreen, cold-intolerant, and less aggressive *Tamarix aphylla* (athel) is not considered a saltcedar.

# *Diorhabda elongata*

C. J. DeLoach and R. I. Carruthers

**Common name:** Saltcedar leaf beetle.

**Type of agent:** Insect: Beetle (Coleoptera: Chrysomelidae).

**Native distribution:** Five forms are found from central Asia to central China, through the Middle East, and to the Mediterranean area and to Senegal.

**Original sources:** Northwestern China (Fukang and Turpan), Greece (Crete), southeastern Kazakhstan (Chilik), Tunisia, and Uzbekistan.

## Biology

**Generations per year:** North of the 38th parallel, two generations of the beetles from Fukang and Chilik are produced. A generation requires about 40 days at 24° C. In field cages, adults emerged from overwintering mid-April to mid-May, first-generation adults appeared in early July, and second-generation adults in late August. Second-generation adults overwinter without (or with little) reproducing. Laboratory and field-cage tests indicate that types from Turpan, Crete, Uzbekistan, and Tunisia can establish in southern areas (south of the 37th parallel) where three or four generations are expected.

**Overwintering stage:** Adult.

**Egg stage:** Eggs are tan, spherical, and laid singly or in masses of up to 25 (average three to 10). They are usually laid from April to early fall on young foliage but sometimes on small stems. Females lay between 100 and 200 eggs. Eggs hatch in five to six days.

**Larval stage:** Larval development requires about 22 days. They have three instars. Larvae are black, the second instar is up to 4 mm long with yellowish mid-lateral spots, and the third instar is up to 9 mm long with a conspicuous mid-lateral yellow stripe. Larvae crawl or fall to the ground and pupate under litter or in loose soil 1 to 2 cm below the surface. They prepare cocoons of litter loosely held together with silk or they prepare cells from soil.

313

D. elongata adults and first and third instar larvae. (Photo credit: USDA-ARS)

**Pupal stage:** Pupae develop in about seven days and are bright yellow.

**Adult stage:** Adult males are about 5.6 mm and females 5.9 mm long with yellowish bodies and two dark brown stripes on each elytron in the Chinese and Kazakhstan beetles and indistinct stripes in the Crete, Tunisia, and Uzbekistan beetles. Adults emerge from overwintering over a two-week period in March or April when saltcedar buds and leaves appear. Adults live 15 to 20 days.

## Effect

**Destructive stages:** Larval and adult (feed on foliage).

**Additional plant species attacked:** Probably all species of deciduous *Tamarix* in the United States.

**Site of attack:** Third instar larvae and sometimes adults may kill more foliage than they eat by scraping the bark on small twigs. Dead foliage is left hanging on the branches.

**Impact on host:** In Kazakhstan and at U.S. release sites in Colorado, Nevada, Utah, and Wyoming defoliated plants suffer severe stem dieback but plants resprout from the base. In field cages in the United States, heavy defoliation for two years killed some large plants. When food becomes scarce, adults will fly en masse a few hundred meters to feed and oviposit on uninfested plants.

**Nontarget effects:** *Diorhabda elongata* is expected to feed and reproduce to a minor extent on the exotic, large, evergreen tree athel (*Tamarix aphylla*), but is not expected to cause important damage to this plant. This tree is of minor beneficial value for windbreaks, as an ornamental shade tree (especially in northern Mexico), provides shade for beehives, and opportunistic nesting habitat for a few bird species. It is not presently a target for biological control, but it is becoming weedy in some areas.

The only moderately closely related native plants (order Tamaricales) in the Western Hemisphere are in the genus *Frankenia* (family Frankeniaceae). Three species are native in the United States and three more in Mexico; one species (*F. johnstonii*) in southern Texas was endangered but has been delisted. Extensive testing of *F. jamesii* and *F. johnstonii* at USDA laboratories in Temple, Texas, and Albany, California, indicated that larvae can feed and develop on them at about a third to half the rate as on *Tamarix*. However, in large outdoor cages at Temple under almost natural conditions, adults rarely were attracted to the *Frankenia* plants and rarely laid eggs on them. The beetles are not expected to damage *Frankenia* in the field and probably cannot maintain a population on it. At Albany, the beetles fed and developed at a somewhat greater rate on *F. salina* than on *F. jamesii* or *F. johnstonii*.

Saltcedar defoliated by *D. elongata*. (Photo credit: E. Coombs, Oregon Department of Agriculture)

## Releases

**First introduced into the United States:** 1999 and 2000, Fukang/Chilik biotypes released into field cages at 10 sites in California, Colorado, Nevada, Texas, Utah, and Wyoming. These beetles were released in the open field at seven of these sites in 2001. Beetles from Crete were released into field cages and in the open field in California, New Mexico, and Texas in 2003. Beetles from China (Turpan), Tunisia, and Uzbekistan were released into field cages only in Texas in 2003 but only the Turpan beetles were released into the open field at one site in Texas.

**Established in:** Research sites in Colorado, Nevada, Oregon, Utah, and Wyoming (Fukang/Chilik biotype).

**Habitat:** Along lakeshores and stream margins, overwintering adults and pupae may drown during floods because these stages occur on the soil surface. Adults of the China and Kazakhstan biotypes require 15 hours or more of daylight to avoid entering overwintering diapause. At shorter day lengths (south of the 38th parallel), these adults enter diapause in midsummer and fail to overwinter. Different types of *Diorhabda* beetles (possibly different subspecies) from China (Turpan), Crete, Tunisia, and Uzbekistan do not require such a long day length and probably can overwinter and establish in areas south of the 38th parallel and to southern Texas. Crete beetles successfully overwintered in outdoor cages in Texas (31° N) during the winter of 2002-2003.

**Availability:** Research is still underway at all sites; beetles may not be removed from research sites. Beetles are expected to be available for distribution in 2004 or 2005.

**Stages to transfer:** Adult, large larval, or egg. (Pupae die when handled but if left undisturbed in cells made of soil, they may be shipped satisfactorily.)

**Redistribution:** Beetles may be redistributed by placing the eggs, larvae, or adults on the foliage of healthy plants. Placement in nylon mesh sleeve bags tied over a branch for a few days will deter predators and adult dispersal until they establish a small population, but may not be necessary.

## Comments

Predation by ants and other insects, spiders, birds, small mammals, and lizards reduced beetle populations at some research sites during the first and second years in the field, but beetle populations overwhelmed predator populations during the third summer.

Preliminary results indicate that the beetles defoliate up to 162 ha of saltcedar during the third year after release and could defoliate a large area within three years if release points are established 1.0 km apart.

## *References*

Baum, B. R. 1978. The Genus *Tamarix*. Israel Academy of Sciences and Humanities, Jerusalem.

DeLoach, C. J., R. I. Carruthers, J. E. Lovitch, T. L. Dudley, and S. D. Smith. 2000. Ecological interactions in the biological control of saltcedar (*Tamarix* spp.) in the United States: toward a new understanding. Pages 819-73 *in* N. R. Spencer, ed. Proc. X Int. Symp. Biol. Contr. Weeds, 4-14 July 1999, Montana State Univ., Bozeman, MT.

DeLoach, C. J., P. A. Lewis, J. C. Herr, R. I. Carruthers, J. L. Tracy, and J. Johnson. 2003. Host specificity of the leaf beetle, *Diorhabda elongata deserticola* (Coleoptera: Chrysomelidae) from Asia, a biological control agent for saltcedars (*Tamarix*: Tamaricaceae) in the western United States. Biol. Control 27: 117-47.

Gaskin, J. F., and B. A. Schall. 2002. Hybrid *Tamarix* widespread in U.S. invasion and undetected in native Asian range. Proc. Nat. Acad. Sci. 99:11256-59.

Kovalev, O.V. 1995. Co-evolution of the tamarisks (Tamaricaceae) and pest arthropods (Insecta; Arachnida: Acarina) with special reference to biological control prospects. Proc. Zool. Inst., Russian Acad. Sci., St. Petersburg, Russia, Moscow, Russia. Pensoft Publishers, Vol. 29.

Lewis, P. A., C. J. DeLoach, J. C. Herr, T. L. Dudley, and R. I. Carruthers. 2003. Assessment of risk to native *Frankenia* shrubs from an Asian leaf beetle, *Diorhabda elongata deserticola* (Coleoptera: Chrysomelidae), introduced for biological control of saltcedars (*Tamarix* spp.) in the western United States. Biol. Control 27: 148-66.

Lewis, P. A., C. J. DeLoach, A. E. Knutson, J. L. Tracy, and T. G. Robbins. 2003. Biology of *Diorhabda elongata deserticola* (Coleoptera: Chrysomelidae), an Asian leaf beetle for biological control of saltcedars (*Tamarix* spp.) in the United States. Biol. Control 27: 101-66.

Robinson, T. W. 1965. Introduction, spread and areal extent of saltcedar (*Tamarix*) in the western states. USDI Geological Survey Professional 491-A.

# Smooth cordgrass

*Spartina alterniflora*

Grass family—Poaceae

*F. S. Grevstad*

**Additional common names:** Spartina, salt marsh cordgrass.

**Native range:** Atlantic and Gulf coasts of North America.

**Entry into the western United States:** This grass was accidentally introduced into Willapa Bay, Washington, in the 1890s, possibly as packing material used in transport of oysters from the Atlantic coast. It was purposefully planted in San Francisco Bay, California, in the 1970s with the intention of stabilizing shorelines.

## Biology

**Life duration/habit:** Smooth cordgrass is a perennial, rhizomatous grass that grows in estuarine intertidal areas. It forms dense, monospecific meadows. Aboveground parts of the plant die back each year and resprout from roots each spring.

**Reproduction:** Reproduction is both vegetative through rhizomes and by seeds that are carried by water.

**Roots:** The fibrous roots of this plant penetrate about 30 cm into the mud and form a dense mat below the surface.

**Stems and leaves:** Smooth cordgrass reaches heights of 1 to 2 m. Stems are round, hollow, and often tinged with red. Leaves are typical of grass blades (long, flat, and taper to a point). At the base of the leaf is a ligule composed of fine hairs.

**Flowers:** Flowering takes place in August and September. The flowers are arranged one per spikelet in compact spikes. The whole inflorescence is 12 to 20 cm long.

**Fruits and seeds:** Seeds are produced throughout the fall. Viability of seeds is low and varies from year to year. This species does not build a seed bank; the seeds are viable for less than one year.

Smooth cordgrass. (Photo credit: F. Grevstad, University of Washington)

Smooth cordgrass infestation in Willapa Bay, Washington. (Photo credit: F. Grevstad, University of Washington)

## Infestations

**Worst infested states:** California and Washington.

**Habitat:** Smooth cordgrass invades sandy or muddy estuarine tidal flats with low to moderate wave energy. On the U.S. West Coast, it primarily occupies tidal flats below the level of native salt marsh, although an occasional patch can be found within the native marsh.

**Impacts:** Invasion by smooth cordgrass results in a dramatic change from open unvegetated mudflat to densely vegetated meadows. It displaces shorebird, wading bird, and waterfowl foraging habitat. It also reduces the available habitat for commercially harvested clams and oysters. Its presence may affect a variety of fish species that use the shallow estuarine waters, although this has not been well studied.

## Comments

The worst infestation of this grass is in Willapa Bay, Washington, where in 2003, over 3,200 ha of smooth cordgrass cover was scattered among 7,200 ha of affected mudflat. The infested area has increased by about 17% per year. Traditional control methods, including herbicides, hand pulling, mowing, and tilling, have not kept pace with the rapid rate of invasion.

In San Francisco Bay, smooth cordgrass hybridizes with the native *S. foliosa* and the resulting hybrids are more vigorous and invasive than the parent species.

Other species of *Spartina* are also invasive on the West Coast. These include *S. anglica* from Europe, *S. patens* from the Atlantic coast, and *S. densiflora* from South America.

In its native range, more than two dozen herbivores have been identified that feed on smooth cordgrass. Several of these are being considered as possible biological control agents for use on the U.S. West Coast. Among the more promising are two stem-boring flies, *Chaetopsis aenea* and *C. apicalis* (Diptera: Ottitidae); a scale insect, *Haliaspis spartinae* (Homoptera: Coccidae); and a plant bug, *Trigonotylus uhleri* (Heteroptera: Miridae). Host specificity testing for these insects has not yet been completed.

318

# Prokelisia marginata

F. S. Grevstad

**Common name:** None widely accepted.

**Type of agent:** Insect: Planthopper (Homoptera: Delphacidae).

**Native distribution:** Atlantic and Gulf coasts of North America.

**Original Source:** San Francisco Bay, California (for Washington introductions).

## Biology

**Generations per year:** Two to three.

**Overwintering stage:** Nymphs.

**Egg stage:** Eggs are inserted into smooth cordgrass leaves with a piercing ovipositor. Eggs are yellow, oblong, about 0.5 mm long, and are usually laid in groups of two to five. Oviposition results in a small brown scar on the leaf surface. Hatching occurs within two to three weeks.

**Nymphal stage:** The nymphs pass through five instars before molting into adults. They use sucking mouthparts to extract sap from the phloem. The nymphs are pale to medium gray, sometimes appearing mottled. They often congregate in dense groups on the leaf surface.

**Adult stage:** Adult *P. marginata* are 3 to 4 mm long, winged, and light gray. Like the nymphs, adults also feed on the sap. Adults of this species are wing dimorphic. Macropterous individuals have two pairs of long wings and can fly long distances. Brachypterous individuals have only one pair of long wings (the hind wings are greatly reduced) and are incapable of flight. The frequency of macroptery increases when nymphs develop under crowded conditions. Each female can produce up to 300 eggs during her lifetime.

## Effect

**Destructive stages:** Nymphal and adult; oviposition causes scars.

**Additional plant species attacked:** English cordgrass (*S. anglica*) and California cordgrass (*S. foliosa*).

**Site of attack:** *P. marginata* will feed anywhere on the leaf, but tends to do so in greater concentrations near the base of the leaf. Occasionally, they feed at the base of inflorescences.

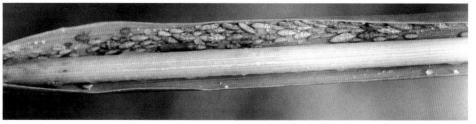

*P. marginata* nymphs. (Photo credit: F. Grevstad, University of Washington)

**Impact on the host:** The full impact of *P. marginata* on invasive smooth cordgrass in the field is not yet known. Plants from Willapa Bay appear to be more vulnerable than plants from San Francisco Bay and the East Coast. In greenhouse trials, *S. alterniflora* plants collected from Willapa Bay suffered 37% mortality and, for those that survived, an 88% reduction in biomass after two years' exposure to *Prokelisia*. Early field results in Willapa Bay show a 50% reduction in aboveground biomass in field cages after one growing season. Reduced seed set has also been demonstrated in the field.

**Nontarget effects:** *P. marginata* has a high degree of host specificity and no nontarget effects on other native or economically important plants are anticipated. It has potential as a biocontrol agent against *S. anglica*, which is an invasive noxious weed in Puget Sound, Washington. The other acceptable host, *S. foliosa*, does not occur north of central California.

## Releases

**First introduced into the western United States:** 2000, Willapa Bay, Washington. *P. marginata* was already present in San Francisco Bay, probably as an accidental introduction. In 2003, releases of *P. marginata* were made into Puget Sound, Washington, against a different cordgrass, *S. anglica*.

**Established in:** California and Washington.

**Habitat:** *P. marginata* is found in upper and middle tidal elevations where smooth cordgrass is present.

*P. marginata* adults. (Photo credit: F. Grevstad, University of Washington)

**Availability:** Availability is limited because of its recent introduction.

**Stages to transfer:** Nymphal and adult.

**Redistribution:** Adults and nymphs can be collected from the plant surface from June to October using a vacuum, aspirator, or sweep net. While *Prokelisia* is well established in San Francisco Bay and nearby estuaries and on the Atlantic and Gulf coasts, it should not be collected for redistribution from these areas because these populations are parasitized (by *Anagrus sophiae* and others). The accidental redistribution of parasitoids with *P. marginata* could prevent successful biocontrol of *Spartina* in other areas.

Damage to smooth cordgrass caused by *P. marginata*. (Photo credit: F. Grevstad, University of Washington)

## Comments
Although this biocontrol project did not import an exotic insect into the United States, but simply transported an indigenous insect between states, the agent underwent complete host specificity testing and a full review by the Technical Advisory Group for Biological Control Agents of Weeds and was approved by USDA-APHIS-PPQ before it was released in Washington.

## *References*

Daehler, C. C., and D. R. Strong. 1995. Impact of high herbivore densities on introduced smooth cordgrass, *Spartina alterniflora*, invading San Francisco Bay, California. Estuaries 18: 409-17.

Daehler, C. C., and D. R. Strong. 1997. Reduced herbivore resistance in introduced smooth cordgrass (*Spartina alterniflora*) after a century of herbivore-free growth. Oecologia 110: 99-108.

Denno, R. F., and E. E. Grissell. 1979. The adaptiveness of wing-dimorphism in the salt marsh-inhabiting planthopper, *Prokelisia marginata* (Homoptera: Delphacidae). Ecology 60: 221-36.

Feist, B. E., and C. A. Simenstad. 2000. Expansion rates and recruitment frequency of exotic smooth cordgrass, *Spartina alterniflora* (Loisel), colonizing unvegetated littoral flats in Willapa Bay, Washington. Estuaries 23: 267-74.

Grevstad, F. S., D. R. Strong, D. Garcia-Rossi, and M. Wecker. 2002. Biological control of *Spartina alterniflora* in Willapa Bay, WA: Agent specificity and early results. Biol. Control 27: 32-42.

Stiller, J. W., and A. L. Denton. 1995. One hundred years of *Spartina alterniflora* (Poaceae) in Willapa Bay, Washington: random amplified polymorphic DNA analysis of an invasive population. Mol. Ecol. 4: 355-63.

Wu, M., S. Hacker, D. Ayres, and D.R. Strong. 1999. Potential of *Prokelisia* spp. as biological control agents of English cordgrass, *Spartina anglica*. Biol. Control 16: 267-73.

# St. Johnswort

*Hypericum perforatum*

St. Johnswort family—Clusiaceae

*G. L. Piper*

**Additional common names:** Goatweed, Klamath weed.

**Native range:** Western Europe, northern Africa, and parts of Asia, including India, China, and Japan.

**Entry into the United States:** St. Johnswort became established in the United States in the late 1700s and was noted in Oregon about 1850, Montana about 1880, California by 1900, Washington since the early 1900s, and Idaho before 1920.

## Biology

**Life duration/habit:** St. Johnswort is a somewhat long-lived herbaceous perennial.

**Reproduction:** By seed and from new shoots that arise from shallow, radiating roots.

**Roots:** The plant has an abbreviated taproot system.

**Stems and leaves:** St. Johnswort produces many erect stems that are 30 to 120 cm tall. The stems bearing the seeds turn rust-colored and woody with numerous branches from the base and at the top. Leaves are oblong or elliptic and contain transparent glands with tiny black dots around the edge.

**Flowers:** Flowers are 1 to 2 cm in diameter, yellow, and develop in clusters on the ends of the branches. Each flower consists of five petals with minute black dots around the edges. The flowering period extends from May through September.

**Fruits and seeds:** Seed capsules are about 5 mm long, rounded at the end and rust-brown. Within each capsule are numerous dark brown (almost black) somewhat cylindrical seeds that are 0.6 to 0.7 mm long. The seeds contain a germination inhibitor. Germination increases during rainy periods because the inhibitor is washed off.

St. Johnswort in bloom. (Photo credit: N. Poritz, bio-control.com)

## Infestations

**Worst infested states:** St. Johnswort is a weed of major importance in several locations in the western United States, especially California, Montana, Oregon, and Washington.

**Habitat:** It invades croplands, grass and rangelands, and open forest areas, and is common along transportation rights-of-way. The plant grows best at low elevations where the annual precipitation is between 38 and 76 cm. It prefers south- and southeast-facing slopes and well-drained gravelly or sandy soils. St. Johnswort does poorly where soils remain continually moist or in heavily shaded sites.

St. Johnswort leaves with oil-producing glands. (Photo credit: S. Dewey, Utah State University)

**Impacts:** Leaves of this plant contain glands that produce an oil (hypericin). When the leaves are consumed by livestock, the animals become sensitized to sunlight, blister on the exposed parts of the body, develop rashes, and lose weight. White-haired animals can lose their hair. If the plant is consumed in large enough amounts, animals can die. It is also a vigorous competitor.

## Comments

St. Johnswort is the only weedy representative of the Clusiaceae family found in California and the northwestern United States. It also has infested Australia, Canada, New Zealand, and South Africa.

Extracts of the plant have long been valued and widely used by folk medicine practitioners to treat depression and a host of other afflictions.

St. Johnswort infestation. (Photo credit: N. Poritz, bio-control.com)

323

# Agrilus hyperici

G. L. Piper

**Common name:** St. Johnswort root borer.

**Type of agent:** Insect: Beetle (Coleoptera: Buprestidae).

**Native distribution:** Southern, central, and eastern Europe.

**Original source:** France.

## Biology

**Generations per year:** One.

**Overwintering stage:** Larval (within the root).

**Egg stage:** Eggs are deposited on stems between soil level and 20 cm above during July and August.

**Larval stage:** The larvae feed within the roots from July through May or June of the following year. The long, white larvae have a flattened wider segment behind the dark brown heads.

**Pupal stage:** Pupation occurs within the damaged root during early May to June. The pupal period lasts nine to 15 days under laboratory conditions. Pupae are creamy white initially.

**Adult stage:** Adults can be found during July or early August and are active during the heat of the day. Adults are about 5 mm long, reddish-bronze, and somewhat flattened and tapered toward the rear.

## Effect

**Destructive stage:** Larval.

**Additional plant species attacked:** *Hypericum concinnum.*

**Site of attack:** Stem and roots.

**Impact on the host:** Larvae feeding within a root may completely consume the tissues. Any stems produced from an infested root crown are stunted and flower production is reduced. Most plants infested by *Agrilus* perish.

**Nontarget effects:** The beetle attacks *H. concinnum* in California, but no long-term impacts have been reported.

*A. hyperici* larva. (Photo credit: E. Coombs, Oregon Department of Agriculture)

## Releases

**First introduced into the United States:** 1950, California.

**Established in:** California, Idaho, Montana, Oregon, and Washington.

**Habitat:** In North America, the insect is found mostly in mountain areas. In Europe, it is only found in the drier regions of the above-noted distribution and is most abundant in the southern part of its range. Larvae are subject to fungal attack in damp sites.

**Availability:** The beetle can be collected in limited numbers.

**Stage to transfer:** Adult.

*A. hyperici* adult. (Photo credit: E. Coombs, Oregon Department of Agriculture)

**Redistribution:** Adults are collected with a sweep net from St. Johnswort plants on hot days during July and August. It is best to release the adults as soon as possible.

## Comments

This beetle also has been established in Australia. It disperses widely and is often not recovered for years after release, but then the population seems to explode. This species will attack plants growing in the shade, plants that are normally untouched by *Chrysolina* species.

# *Aplocera plagiata (=Anaitis plagiata)*

G. L. Piper

**Common name:** None widely accepted (St. Johnswort moth).

**Type of agent:** Insect: Moth (Lepidoptera: Geometridae).

**Native distribution:** Northern Europe from central France into Sweden.

**Original source:** Europe.

## Biology

**Generations per year:** Two.

**Overwintering stage:** Larval.

**Egg stage:** Eggs are laid on the foliage. A female can lay up to 300 eggs in her lifetime. Eggs are pearly white and oval-shaped. They hatch in five to seven days.

**Larval stage:** The first-generation larvae are present during July, the second-generation from mid-August through September. Larvae develop rapidly except in the fall when cool temperatures slow feeding and eventually cause them to seek protection in the soil. Larvae are commonly referred to as inchworms and are weakly striped reddish-brown. Mature larvae are about 2.5 cm long.

*A. plagiata* larva on St. Johnswort. (Photo credit: N. Poritz, bio-control.com)

**Pupal stage:** Pupation occurs amongst the soil litter or, more commonly, within the soil. The pupal period lasts 15 to 17 days. The pupae are slender and a light greenish-golden brown.

**Adult stage:** Generally, the first-generation adults are few in number in the northern areas, such as southern Canada and northern Idaho, while second-generation adult densities are larger. Harsh winter weather may affect larval survival; in high elevations and northern locations, winter weather arrives before most second-generation larvae have matured sufficiently to survive. The adults are gray, triangular-shaped moths with characteristic dark wing bands and a wingspan of 23 to 25 mm.

## Effect

**Destructive stage:** Larval.

**Additional plant species attacked:** None.

**Site of attack:** Leaves and flowers.

*A. plagiata* adult. (Photo credit: E. Coombs, Oregon Department of Agriculture)

**Impact on the host:** This moth can be very effective in favorable areas. Large populations of larvae can defoliate plants, thus inhibiting flower and seed formation.

**Nontarget effects:** No instances of nontarget plant feeding have been reported.

## Releases

**First introduced into the United States:** 1989, Montana. The moth moved unassisted into northern Washington and Idaho from Canada.

**Established in:** Idaho, Montana, Oregon, and Washington.

**Habitat:** Dry areas such as rocky ground, open sandy places, and limestone regions are favored while areas receiving high rainfall are not.

**Availability:** Populations are still somewhat limited. The moth can readily be collected in northern Idaho, eastern Oregon, and eastern Washington.

**Stages to transfer:** Larval and adult.

**Redistribution:** Larvae can be collected from the plants with a sweep net and kept in cool storage for several days before redistribution. If larvae are ready for winter, they can be kept cool for several months and reared in the laboratory to obtain adults for releases.

## Comments

This species can be reared continuously in the laboratory with proper rearing conditions. In nature, larvae feed mainly at night, while in the uniform temperatures of the laboratory they will also feed during the day.

Thousands of *Aplocera plagiata* were introduced into Australia between 1936 and 1938, but the species did not establish. It was believed that ants, wasps, and other insects preyed upon the moth larvae. However, introductions into Canada in 1967 were successful after a disease was eliminated from the colony and pathogen-free breeding stock was released. This moth has also been described as *Anaitis plagiata*.

The effectiveness of this agent is quite variable. It appears to need warm dry areas with a summer season long enough to complete both generations. Nevertheless, in some of the older Canadian establishments, St. Johnswort populations have been greatly reduced.

# *Chrysolina hyperici*

G. L. Piper

**Common name:** None widely accepted (Klamathweed beetle).

**Type of agent:** Insect: Beetle (Coleoptera: Chrysomelidae).

**Native distribution:** Northern and central Europe and western Asia.

**Original source:** England via Australia.

## Biology

**Generations per year:** One.

**Overwintering stages:** Egg, larval, and sometimes adult.

**Egg stage:** In its native homeland and in California and Oregon the eggs are laid on the leaves of St. Johnswort in the fall, while in the interior northwestern United States the eggs are laid in the spring. Each female produces several hundred eggs which are deposited singly or in clusters and hatch in six to seven days. Eggs are elongated and orange.

**Larval stage:** Upon hatching, the larvae migrate to the leaf buds and immature leaves. Larvae can completely defoliate a plant before they reach maturity, which forces them to move to other plants. When mature, larvae burrow into the soil where they create cells and pupate. Larvae are somewhat humpbacked or C-shaped and plump. They are orange at first, becoming a dirty grayish-pink with age. Larvae of this species closely resemble those of *Chrysolina quadrigemina*.

*C. hyperici* adult. (Photo credit: E. Coombs, Oregon Department of Agriculture)

**Pupal stage:** Pupation occurs in the soil during late spring. The pupal period lasts about 12 days.

**Adult stage:** Adult beetles emerge in the spring, feed for several weeks, and then enter the soil to rest during the summer. Fall rains activate the adults to mate and lay eggs in their endemic areas and some areas of California. If fall rains do not occur (as is often the case in the interior Pacific Northwest), spring rains will induce mating and egg laying. Adults are generally shiny metallic green, black, bronze, or blue, about 5 mm long, and robust. This species is only slightly smaller than *C. quadrigemina* and it is difficult to differentiate the two species.

## Effect

**Destructive stages:** Larval and adult.

**Additional plant species attacked:** None.

**Site of attack:** Leaves and flowers.

**Impact on the host:** Heavy larval feeding in the fall reduces the foliage and lowers the root reserves which makes it difficult for the plant to survive the winter. However, if feeding occurs in the spring, the plant can sometimes outgrow the feeding injury inflicted by the insect, especially in areas with summer rains.

**Nontarget effects:** No instances of nontarget plant feeding have been reported, although it may develop on several *Hypericum* species attacked by *C. quadrigemina*.

## Releases

**First introduced into the United States:** 1945, California.

**Established in:** The beetle is found in many eastern, midwestern, and western states.

**Habitat:** The beetle prefers more moist conditions than *C. quadrigemina* and avoids shaded or barren, rocky locations. It tolerates cold winter weather better than does *C. quadrigemina*.

**Availability:** The beetle is readily available throughout the United States.

**Stage to transfer:** Adult.

**Redistribution:** Handpick or collect with a sweep net. The beetles can be kept for several weeks in cool storage and several days in transit without ill effects.

## Comments

This species is more moisture-loving than *C. quadrigemina*. Fall moisture patterns may be responsible for the variation in the effectiveness of this agent, as well as with *C. quadrigemina*. In its native locations, there are strong fall rainstorms as there sometimes are in western Oregon and California. These storms may stimulate the beetles to mate and lay eggs. In the interior Pacific Northwest, however, rains sometimes do not come until the spring, and eggs are often not laid until then. This is believed to be the reason that this agent has been effective in California and western Oregon and less effective in Idaho and Washington.

Another theory suggests that continuous feeding in the fall and winter stresses the plants so that they cannot survive the heat stress of a hot summer. The same climatic factors inhibiting the effectiveness of *C. hyperici* also appear to affect *C. quadrigemina* populations.

## *Chrysolina quadrigemina*

G. L. Piper

**Common name:** Klamath weed beetle.

**Type of agent:** Insect: Beetle (Coleoptera: Chrysomelidae).

**Native distribution:** From North Africa to Denmark, but does not reach the Swedish mainland.

**Original source:** Australia (via France).

### Biology

**Generations per year:** One.

**Overwintering stages:** Egg, larval, and sometimes adult.

**Egg stage:** Eggs are usually deposited on leaves of St. Johnswort in the fall, but sometimes in the spring. They are laid singly or in clusters of two to four on the underside of the leaves. Eggs that are deposited on foliage during the late fall or early winter can survive and hatch the following spring. Each female may lay several hundred eggs during her lifetime. Eggs are oval and orange to reddish.

**Larval stage:** Eggs hatch about three weeks after they are laid. The larvae migrate to the leaf buds and immature leaves. Larvae can completely defoliate a plant before they reach maturity, which forces them to move to other plants. Larvae are somewhat humpbacked or C-shaped and plump. They are orange at first and become a dirty grayish-pink with age. Larvae of this species closely resemble those of *Chrysolina hyperici*.

**Pupal stage:** Mature larvae burrow into the soil in February and March and pupate in oval cells. They emerge from late April until June. The orange pupae are oval; the wing pads, legs, head, and antennae are readily apparent.

**Adult stage:** Adult emergence and behavior is very similar to that of *C. hyperici*. Adult beetles emerge in the spring, feed for several weeks, and then enter the soil to rest. Fall rains

*C. quadrigemina* adults. (Photo credit: N. Poritz, bio-control.com)

activate the adults to mate and lay eggs. The oval beetles are shiny, metallic black, blue, green or bronze, and 5 to 7 mm long.

## Effect

**Destructive stages:** Larval and adult.

**Additional plant species attacked:** *H. calycinum* and *H. concinnum.*

**Site of attack:** Both the adults and larvae attack the leaves.

**Impact on the host:** Larval feeding in the fall and spring reduces the foliage and lowers root reserves making it difficult for the plants to survive the a harsh winter or dry summer conditions.

**Nontarget effects:** Adults will feed and oviposit on *H. calycinum*, an introduced ornamental, and on *H. concinnum*, a native species. No population-level impacts have been reported on the native species.

## Releases

**First introduced into the United States:** 1946, California.

**Established in:** Occurs in many eastern, midwestern, and western states.

**Habitat:** It is found in mountainous, open, sunny and warm areas. In its native homeland it thrives in a Mediterranean climate with dry summers and mild, moist

Northern California St. Johnswort infestation in 1948, two years after release of *C. quadrigemina*. (Photo credit: USDA-ARS)

Same vista in 1949, three years after release of *C. quadrigemina*. (Photo credit: USDA-ARS)

winters. It apparently does not do well in shaded, barren, and excessively rocky locations.

**Availability:** The beetle is readily available for collection.

**Stage to transfer:** Adult.

**Redistribution:** Collect the adults with a sweep net or handpick from infested stems. The adults readily keep for several weeks in cool storage and several days in transit.

## Comments

The introduction of *C. quadrigemina* into California in 1946 caused such an impact upon the St. Johnswort population that the weed was removed from the noxious weed list there. The subsequent redistribution of this agent throughout much of northwestern Canada and the United States raised false hopes that this agent could control all populations of St. Johnswort. However, at many locations where the insect is established, many of the St. Johnswort populations are still increasing in size and density, while at others there is little change. The same climatic factors inhibiting the effectiveness of *C. hyperici* also appear to affect *C. quadrigemina* populations.

331

Another *Chrysolina* beetle, *C. varians*, was released in limited quantities California and Idaho in the 1950s. This insect did not establish populations in the United States and is no longer pursued as a biocontrol agent.

# *Zeuxidiplosis giardi*

G. L. Piper

**Common name:** None widely accepted (St. Johnswort gall midge).

**Type of agent:** Insect: Fly, gall midge (Diptera: Cecidomyiidae).

**Native distribution:** Central and southern Europe, including England, France, Germany, Italy, and Portugal.

**Original source:** France.

## Biology

**Generations per year:** Between five and seven generations, depending on climatic conditions.

**Overwintering stages:** Larval and pupal.

**Egg stage:** Eggs are laid on leaves or stems. A female will produce about 170 pale red elongate eggs. Eggs hatch in about 12 days.

**Larval stage:** Newly emerged larvae enter the leaf buds. Their feeding causes the leaves to grow together creating a hollow chamber or bivalved gall in which the 2 mm-long, red-orange larvae develop. Several larvae are often found within a gall.

**Pupal stage:** Pupation occurs inside the gall. Depending on the time of year, the pupal period may last from six to 20 days. Pupae are yellow-red but become darker red as they mature.

**Adult stage:** Adults survive for only a few days. The adults are very small (3.0 mm), delicate, gray flies similar in appearance to fungus gnats.

## Effect

**Destructive stage:** Larval.

**Additional plant species attacked:** None.

**Site of attack:** Larvae attack the leaf buds which results in the formation of galls that provide the larvae with a protective environment as well as nutrition.

**Impact on the host:** Heavily infested plants exhibit a

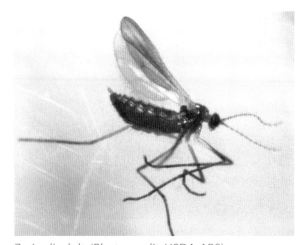

*Z. giardi* adult. (Photo credit: USDA-ARS)

Galls caused by *Z. giardi.* (Photo credit: USDA-ARS)

marked loss of vigor and a drastic reduction of foliage and root development. Such plants may die during the dry season because of their inability to obtain moisture. This agent reduces overall growth and seedling survival of St. Johnswort. It has not been successful in many areas where released, but in Hawaii it is credited with the reduction of St. Johnswort.

**Nontarget effects:** The midge is capable of forming galls on *H. concinnum*, but the damage to the plant is insignificant.

## Releases
**First introduced into the United States:** 1951, California.

**Established in:** California and Hawaii.

**Habitat:** The midge seems to prefer damp locations with moderate to high relative humidity and high elevations. It also apparently does not like dry summers or continuously windy areas. It does not persist in areas heavily grazed by livestock.

**Availability:** Limited collections are possible in California; the midge is readily available in Hawaii.

**Stage to transfer:** Larval (in galled plants).

**Redistribution:** Pot plants of St. Johnswort from the area that will receive the insect and infest these prior to transplantation. Because the insect is in the plant galls and subject to continuous colonization in the laboratory, no storage is needed. Shipments of infested plants with roots wrapped in damp paper towels is most effective. Shipment should not exceed three days.

## Comments
There appears to be a resting period for the midge during the summer and the literature is not clear whether this is a true dormancy or a heat-induced response. The effect that this species has on the plant is not fully understood. It has been suggested that when larvae feed, as many gall-forming insects do, they inject a chemical substance into the sap of the plant. This, of course, causes the plant to form

galls. In this case, however, more damage is produced than can be attributed to gall formation alone because the roots are impacted and are reduced in size and vigor. A sufficient number of larvae feeding within the galls along a stem may cause death of the stem and root system. Quite often the number of galls per plant is fewer than 17.

In Hawaii, where both *Chrysolina quadrigemina* and *Z. giardi* have been established for the control of St. Johnswort, most of the plants are now less than 30 cm tall and occur in tolerable numbers.

## References

Campbell, C. L., J. P. McCaffrey, and H. W. Homan. 1987. Collection and redistribution of biological control agents of St. Johnswort. Univ. Idaho Coll. Agric. Current Info. Series No. 798.

Harris, P. 1967. Suitability of *Anaitis plagiata* (Geometridae) for biocontrol of *Hypericum perforatum* in dry grassland of British Columbia. Can. Entomol. 99: 1304-10.

Harris, P., and M. Maw. 1984. *Hypericum perforatum* L., St. John's-wort (Hypericaceae). Pages 171-77 *in* J. S. Kelleher and M. A. Hulme, eds. Biological Control Programmes against Insects and Weeds in Canada 1969-1980. CAB, Farnham Royal, Slough, England.

Harris, P., D. Peschken, and J. Milroy. 1969. The status of biological control of the weed *Hypericum perforatum* in British Columbia. Can. Entomol. 101: 1-15.

Holloway, J. K. 1957. Weed control by insects. Sci. Am. 197: 56-62.

Huffaker, C. B., and C. E. Kennett. 1959. A ten-year study of vegetational changes associated with biological control of Klamath weed. J. Range Manage. 12: 69-82.

Johansson, S. 1962. Insects associated with *Hypericum* L. 2. Lepidoptera, Diptera, Hymenoptera, Homoptera, and general remarks. Opusc. Entomol. 27: 175-92.

McCaffrey, J. P., C. L. Campbell, and L. A. Andres. 1995. 74. St. Johnswort *Hypericum perforatum* L. (Hypericaceae). Pages 281-85 *in* J. R. Nechols, L. A. Andres, J. W. Beardsley, R. D. Goeden, and C. G. Jackson, eds. Biological Control in the Western United States: Accomplishments and Benefits of Regional Research Project W-84, 1964-1989. Univ. Calif. Div. Agric. Nat. Res. Pub. 3361. Oakland, CA.

Piper, G. L. 1999. St. Johnswort. Pages 372-81 *in* R. L. Sheley and J. K. Petroff, eds. Biology and Management of Noxious Rangeland Weeds. Oregon State Univ. Press, Corvallis.

Shepherd, R. C. H. 1984/85. The present status of St. John's-wort (*Hypericum perforatum* L.) and its biological control agents in Victoria, Australia. Agric. Ecosyst. Environ. 12: 141-49.

Tisdale, E. W. 1976. Vegetational responses following biological control of *Hypericum perforatum* in Idaho. Northwest Sci. 50: 61-75.

# Tansy ragwort

*Senecio jacobaea*

Sunflower family—Asteraceae

*E. M. Coombs, P. B. McEvoy, and G. P. Markin*

**Additional common names:** Ragwort, tansy, stinking Willy.

**Native range:** Europe and Asia.

**Entry into the United States:** Tansy ragwort was introduced in pproximately 1900.

## Biology

**Life duration/habit:** Tansy ragwort is à biennial or short-lived perennial herb growing 0.3 to 2.0 m tall.

**Reproduction:** This plant reproduces by seed and regrowth from the root crown.

**Roots:** Small to medium taproot.

**Stems and leaves:** Stems are solitary or several and bear dark green, deeply divided, 5- to 15-cm-long leaves. The undersurface of the leaf is somewhat hairy during the early growth stages.

**Flowers:** Bright yellow flowers (usually with 13 ray flowers) are borne in terminal clusters from July to September.

**Fruits and seeds:** Approximately 150,000 tufted seeds per plant are produced.

## Infestations

**Worst infested states:** California, Idaho, Montana, Oregon, and Washington.

**Habitat:** Tansy ragwort can be found on rangeland and pastures, in forest clearings, and along roadsides.

**Impacts:** The toxic properties of ragwort have caused heavy losses of cattle and horses along the U.S. West Coast. Ragwort has forced desirable vegetation out of many pastures. Livestock do not often feed on ragwort exclusively, but consume small quantities as they graze on grass. Tansy ragwort contains pyrrolizidine alkaloids

Tansy ragwort. (Photo credit: N. Poritz, bio-control.com)

335

that are converted to pyrroles in the liver, the accumulation of which causes liver failure. Ingestion of 3 to 7% of the body weight in ragwort can cause death. Horses and cattle are most prone to poisoning; sheep and goats are less susceptible and have been used as cultural biocontrol agents. In the 1970s, it was estimated that Oregon lost more than $5 million a year in livestock due to ragwort poisoning and crop contamination. Annual losses of 5 to 10% of cattle herds were not uncommon in coastal counties during the 1970s, forcing some dairies out of business.

## Comments

Weed control districts in Washington are increasing their use of biological control to combat the weed. These efforts have been highly successful in northern California and western Oregon. The ragwort flea beetle (*Longitarsus jacobaeae*) and the cinnabar moth (*Tyria jacobaeae*) have achieved spectacular regional control of ragwort and have saved millions of dollars by reducing cattle losses. The cost-benefit ratio of biologically controlling tansy ragwort in Oregon was conservatively estimated at 1:15. Cases of poisoning in cattle have been rare in Oregon since 1992. Discoveries in the 1990s of large infestations in western Montana have been cause for alarm. All three biocontrol agents were released in Montana and are showing some success.

## *Botanophila seneciella*
## *(=Hylemyia seneciella, Pegohylemyia seneciella)*

E. M. Coombs, P. B. McEvoy, and G. P. Markin

**Common name:** Ragwort seed head fly.

**Type of agent:** Insect: Fly (Diptera: Anthomyiidae).

**Native distribution:** Eurasia.

**Original source:** France.

## Biology

**Generations per year:** One.

**Overwintering stage:** Pupal.

**Egg stage:** Females lay eggs on the young flower buds in the late spring and early summer.

**Larval stage:** Larvae emerge in April to May (or June and July at high elevations), burrow into the flower bud and mine the receptacle. They feed in the seed head and often destroy all of the developing seeds. Infested flowers often exude frothy spittle. The larvae mature in late

*B. seneciella* larva. (Photo credit: E. Coombs, Oregon Department of Agriculture)

summer and exit the flower to pupate in the soil. The larvae are creamy white and 4 to 6 mm long.

**Pupal stage:** Pupation begins in late summer (August to September) and lasts through the winter. The puparium resembles that of the common house fly and is dark brown and about 5 mm long.

**Adult stage:** The adults emerge in the spring and look similar to house flies. They are most easily observed when they walk upon the flower buds of bolting ragwort. Including their wings, they are 5 to 7 mm long, and they have reddish eyes.

## Effect

**Destructive stage:** Larval.

**Additional plant species attacked:** None.

**Site of attack:** Flower receptacle.

**Impact on the host:** Infestation rates of 30 to 40% have been documented. The fly reduces seed production and may help to reduce spread.

**Nontarget effects:** No nontarget effects have been reported.

## Releases

**First introduced into the United States:** 1966, California.

**Established in:** California, Idaho, Montana, Oregon, and Washington.

**Habitat:** The fly prefers infestations in meadows and forest clearings. However, plants in the open are often stripped by the cinnabar moth, which leaves few seed heads other than those on plants growing in shaded, moist areas.

**Availability:** This insect is easily collected as adults in the late spring or as larvae-infested seed heads in early summer.

Flower (upper left) with characteristic spittle following infestation by *B. seneciella* larva. (Photo credit: E. Coombs, Oregon Department of Agriculture)

*B. seneciella* adult. (Photo credit: P. McEvoy, Oregon State University)

**Stages to transfer:** Larval and, to a lesser extent, adult.

**Redistribution:** Transplant infested plants, collect larvae for pupation, or sweep net the adults. The easiest method of redistributing the seed fly is to transplant infested plants. Another method involves harvesting infested plants in late summer. Plant stems are kept in a bucket of water while the flowers are positioned over a bed of fine sand. The larvae exit the flowers and fall into the sand to pupate. The sand is kept in a cooler and placed in cages or in the field in the spring. Adult flies can easily be captured on bolting plants in the early mornings. Flies can be stored a week if kept cool and dry and provided sugar water.

## Comments

Because the flies are so well distributed on their own, no further releases are being made in the Pacific Northwest. The seed fly has been found at nearly every ragwort infestation. It has been able to spread rapidly and find infestations more than 100 km away from the core area.

   This insect is also known under the genera of *Hylemyia* and *Pegohylemyia*.

   The fly is not an effective biocontrol agent by itself, but it is the only biocontrol agent that is well established on infestations east of the Cascade Mountains in the Pacific Northwest.

# Longitarsus jacobaeae

E. M. Coombs, P. B. McEvoy, and G. P. Markin

**Common name:** Tansy ragwort flea beetle.

**Type of agent:** Insect: Beetle, flea beetle (Coleoptera: Chrysomelidae).

**Native distribution:** Eurasia.

**Original source:** Italy and Switzerland.

## Biology

**Generations per year:** One.

**Overwintering stages:** Larval and often adult.

**Egg stage:** The Italian biotype lays eggs during October and November, whereas the Swiss biotype produces eggs in the spring. Eggs are laid on the plant near the base of the rosette or on the ground. Eggs are small, white, round, and less than 1 mm in diameter.

**Larval stage:** The larvae hatch after a couple of weeks and feed on and in the roots. The larvae are white, slender, and about 2 to 4 mm long.

**Pupal stage:** Pupation occurs in the soil and usually starts in the spring (but in the fall for the Swiss biotype). The white pupae are 2 to 4 mm long.

**Adult stage:** Adults emerge briefly in the spring and then rest during the summer. The flea beetles become active following rain storms and begin mating in the fall. The biology of the Swiss strain differs from the original Italian strain in that the adults emerge from pupae in midsummer, immediately begin to feed, mate, and lay eggs that remain dormant during the summer and fall and hatch the following spring. The adults are 2 to 4 mm long. The females are about 1 mm longer than males, especially when about to lay eggs. Beetles are a light golden color and have enlarged hind legs for hopping.

## Effect

**Destructive stages:** Larval and adult.

**Additional plant species attacked:** None.

*L. jacobaeae* adult. (Photo credit: N. Poritz, bio-control.com)

**Site of attack:** Larvae mine the roots of the rosettes which may cause plant mortality in the spring when infested plants begin to bolt. The adults feed on the leaves and cause a typical shot-holed appearance. Heavy adult feeding on rosettes during the late fall and winter can kill plants.

**Impact on the host:** This flea beetle is a highly successful biological control agent. It is able to maintain colonies in low host densities. In Oregon, 90% control of flowering plants was achieved within six years of release. The insect has been so successful that flowering plants at many sites have not been observed for more than six years. At most sites where small resurgences of the host have occurred, the beetle population controlled the outbreak within a year or two after being reported to authorities.

**Nontarget effects:** No nontarget effects have been reported. Effects of the ragwort flea beetle should not be confused with those of a native flea beetle (*Longitarsus ganglbaueri*) that feeds on native plants, including *Packera pseudaurea* (=*Senecio pseudaureus*).

"Shotholing" damage by adult *L. jacobaeae*. (Photo credit: E. Coombs, Oregon Department of Agriculture)

**339**

Tansy ragwort infestation in Oregon in 1978 before release of *L. jacobaeae.* (Photo credit: Oregon Department of Agriculture)

Same vista nine years after release of *L. jacobaeae.* (Photo credit: Oregon Department of Agriculture)

## Releases

**Introduced into the United States:** 1969, California.

**Established in:** California, Oregon, Montana, and Washington.

**Habitat:** This insect is found in sunny pastures that are not prone to flooding. It does not do well in heavily shaded areas and at elevations over 800 m. The beetle can survive in colder climates where snow prevents the ground from deeply freezing.

**Availability:** The insect is easily collected in October and November. The Swiss biotype can be collected in midsummer.

**Stage to transfer:** Adult (best transferred in groups of 100 to 500).

**Redistribution:** An insect-collecting vacuum is used to collect adults from large infested rosettes in the fall after first rains. Summer collections can be done by sweep netting bolted plants. Adults that are kept cool and dry and provided with fresh plant material can survive for a couple of weeks. When they are kept in large numbers, it is important to have sufficient tissue paper for these active insects to crawl on or they rapidly wear themselves out by disturbing each other.

## Comments

The success of the ragwort flea beetle rivals that of the Klamath weed beetle in California. Striking before-and-after photographs and scientific studies have documented outstanding control of tansy ragwort. The flea beetle has greatly reduced the occurrence of livestock deaths, remote infestations, and flowering plants in many areas. This has led to a very high expectation of control of weeds by biological agents. The public now has less tolerance of small tansy ragwort infestations in areas where it once was controlled.

Because of the failure of this insect to establish in areas east of the Cascade Mountains in Oregon and Washington, a new strain of this beetle was identified in Switzerland, tested, and approved for release in North America in 2002. The first release of this new strain was made in Montana in 2002.

## *Tyria jacobaeae*

E. M. Coombs, P. B. McEvoy, and G. P. Markin

**Common name:** Cinnabar moth.

**Type of agent:** Insect: Moth, tiger moth (Lepidoptera: Arctiidae).

**Native distribution:** Eurasia.

**Original source:** France.

### Biology

**Generations per year:** One.

**Overwintering stage:** Pupal.

**Egg stage:** Eggs are laid in late spring to midsummer in clusters on the undersides of basal leaves of rosettes. Eggs are 1 mm wide, round, yellow, turning black with age.

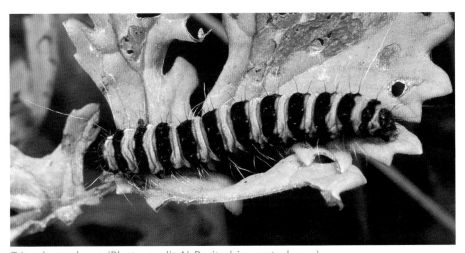

*T. jacobaeae* larva. (Photo credit: N. Poritz, bio-control.com)

341

**Larval stage:** The larvae hatch several weeks after the eggs are deposited, from mid-spring through summer. At first they feed under the leaf near the eggs and then make their way to the top of bolting plants. Larvae consume the foliage and destroy the flowers, often leaving bare stalks. After stripping a plant, they crawl to other nearby plants. The larvae are banded with orange and black and often occur in groups of 10 to 30. A full-grown larva is about 2.5 cm long after five larval growth stages.

**Pupal stage:** The larvae generally pupate in debris, under bark, or in the soil. The pupal period extends from summer to the following spring. The pupae are dark reddish-brown and are about 20 to 25 mm long.

**Adult stage:** The adults emerge in mid- to late spring and mate, and the females lay eggs within two weeks. Adults have been observed in late fall in colder areas such as along the Pacific coast and in the Cascade Mountains. They often fly in the day, especially when they are disturbed in the egg-laying areas. The striking red and black adults are 15 to 22 mm long with a 27 to 35 mm wingspan. The forewing is mostly black with a couple of red spots and a narrow red line that parallels the leading edge. The hind or underwing is bright crimson.

Plant being defoliated by *T. jacobaeae* larvae. (Photo credit: E. Coombs, Oregon Department of Agriculture)

## Effect

**Destructive stage:** Larval.

**Additional plant species attacked:** Two native species, *Senecio triangularis* and *Packera pseudaurea* (=*S. pseudaureus*), the introduced weed common groundsel (*S. vulgaris*), and the ornamental dusty miller (*S. bicolor*) are occasional hosts.

**Site of attack:** Leaves, terminal buds, and flowers.

**Impact on the host:** The moth has been effective in reducing stand densities and seed production, but works best in conjunction with the flea beetle *Longitarsus jacobaeae*.

**Nontarget effects:** In some high-elevation sites (more than 1,000 m), the moth occasionally attacks native species of *Senecio*, such as *S. triangularis* and *Packera pseudaurea*, but no population effects have been documented as defoliation occurs

Adult *T. jacobaeae* moth. (Photo credit: E. Coombs, Oregon Department of Agriculture)

after seeding. The cinnabar moth should not be released into new areas with native *Senecio* and *Packera* species on which it has not been tested for host specificity.

## Releases

**First introduced into the United States:** 1959, California.

**Established in:** California, Montana, Oregon, and Washington.

**Habitat:** The cinnabar moth does best in sunny, warm areas where tansy ragwort densities are greater than 4 plants per m² at elevations less than 1,000 m, and west of the Cascade Mountains. It does not do as well in shady areas under trees, steep canyons, stands with a density less than 0.5 plants per m², and near the Pacific coast.

**Availability:** The moth is available for mass release if sources can be found. Often the tansy ragwort infestations are controlled within a couple of years, so it is difficult to find large numbers of cinnabar moths in the same area year after year.

**Stages to transfer:** Adult and larval.

**Redistribution:** Shake plants so nearly grown larvae fall into a polystyrene cooler or a paper bag. Fresh plant material should be provided and care taken to prevent larvae from escaping during shipment. Larvae may be kept a day or two in large numbers. If adequate food is present and the larvae are kept cool and dry, they may be stored for up to a week. Several adults can be kept in a covered paper cup. Tissue paper must be provided for hiding places and to reduce damage to the wings. Adults that are kept cool and dry may survive a week in storage. Adults can be moved, but it takes more effort and they are difficult to keep for long periods of time. From 500 to 1,000 larvae and/or 20 to 100 adults (1:1 sex ratio) should be released at a site.

Older populations of the moth along the U.S. West Coast were sometimes found to be infested by a virus suspected of causing periodic crashes of the moth population. Eggs moved to a new area should be surface-sterilized by soaking in a weak bleach solution (5 parts bleach to 95 parts water) for one minute, then rinsing them in running water for at least 10 minutes. Leaves with treated egg masses then can be attached to tansy plants and allowed to hatch in the field or the larvae allowed to hatch in the laboratory, then released.

343

## Comments

Cinnabar moths are very active in the larval stage and require a lot of care and food. Do not release larvae in areas where aphids and ants are abundant because the ants will kill larvae. The larvae and adults are able to sequester the toxic alkaloids from the host plant. This makes them toxic to birds and other animals, hence their warning colors of black and red.

The cinnabar moth, in conjunction with the ragwort flea beetle, has proven very effective in controlling tansy ragwort in northern California, Oregon, and Washington west of the Cascade Mountains. Early efforts to establish this moth east of the Cascades always failed. However, a release of the moth in 1997 in northwestern Montana (1,200 to 1,650 m altitude) in an area that had burned three years earlier readily established and helped reduce the tansy ragwort densities.

## *References*

Coombs, E. M., P. B. McEvoy, and C. E. Turner. 1999. Tansy ragwort. Pages 389-400 *in* R. L. Sheley and J. K. Petroff, eds. Biology and Management of Noxious Rangeland Weeds. Oregon State Univ. Press, Corvallis.

Coombs, E. M., H. Radtke, D. L. Isaacson, and S. P. Snyder. 1996. Economic and regional benefits from the biological control of tansy ragwort, *Senecio jacobaea* (Asteraceae), in Oregon. Pages 489-94 *in* V. C. Moran and J. H. Hoffmann, eds. Proc. IX Int. Symp. Biol. Contr. Weeds, 19-26 January 1996, Stellenbosch, South Africa. Univ. Cape Town.

Dempster, J. P. 1982. The ecology of the cinnabar moth, *Tyria jacobaeae* L. (Lepidoptera, Arctiidae). Advan. Ecol. Res. 12: 1-36.

Diehl, J., and P. B. McEvoy. 1990. Impact of the cinnabar moth (*Tyria jacobaeae*) on *Senecio triangularis*, a non-target native plant in Oregon. Pages 119-26 in E. S. Delfosse, ed. Proc. VII Int. Symp. Biol. Contr. Weeds, 6-11 March 1988, Rome, Italy. Ist. Sper. Patol. Veg. (MAF), Rome.

Frick, K. E. 1969. Tansy ragwort control aided by the establishment of seedfly from Paris. Calif. Agric. 23 (12): 10-11.

Frick, K. E. 1971. *Longitarsus jacobaeae* (Coleoptera: Chrysomelidae), a flea beetle for the biological control of tansy ragwort. II. Life history of a Swiss biotype. Ann. Entomol. Soc. Am. 64: 834-40.

Frick, K. E., and G. R. Johnson. 1973. *Longitarsus jacobaeae* (Coleoptera: Chrysomelidae), a flea beetle for the biological control of tansy ragwort. 4. Life history and adult aestivation of an Italian biotype. Ann. Entomol. Soc. Am. 66: 358-67.

McEvoy, P., C. Cox, and E. Coombs. 1991. Successful biological control of ragwort, *Senecio jacobaea*, by introduced insects. Ecol. Appl. 1: 430-42.

McEvoy, P. B., and E. M. Coombs. 1999. Biological control of plant invaders: Regional patterns, field experiments, and structured population models. Ecol. Appl. 9: 387-401.

Turner, C. E., and P. B. McEvoy. 1995. 71. Tansy ragwort, *Senecio jacobaea* L. (Asteraceae). Pages 264-69 *in* J. R. Nechols, L. A. Andres, J. W. Beardsley, R. D. Goeden, and C. G. Jackson, eds. Biological Control in the U.S. Western Region: Accomplishments and Benefits of Regional Research Project W-84, 1964-1989. Univ. Calif. Div. Agric. Nat. Res. Pub. No. 3361. Oakland, CA.

# Thistles

## G. L. Piper and E. M. Coombs

Thistles, by their physical nature, are unpleasant plants to handle because of their sharp spines. Although horses will occasionally nibble the flowers of young thistle plants and cattle will reluctantly feed on thistles during extreme drought conditions, livestock generally avoid areas where thistles grow. Therefore, their presence occupies space and denies access to forage plants for livestock and some forms of wildlife.

The introduced thistle species presently being studied for biological control include bull thistle (*Cirsium vulgare*), Canada thistle (*Cirsium arvense*), Italian thistle (*Carduus pycnocephalus*), milk thistle (*Silybum marianum*), musk thistle (*Carduus nutans, C. macrocephalus,* and *C. thoermeri* = *C. nutans* ssp. *leiophyllus*), plumeless thistle (*Carduus acanthoides*), Scotch thistle (*Onopordum acanthium*), and slenderflower thistle (*Carduus tenuiflorus*). Yellow starthistle (*Centaurea solstitialis*) and Russian thistle (*Salsola tragus*) are not true thistles and are listed in other sections of this book.

## Bull thistle                    *Cirsium vulgare*

Sunflower family—Asteraceae

### E. M. Coombs and G. L. Piper

**Additional common names:** Common thistle, Fuller's thistle, spear thistle.

**Native range:** Eurasia.

**Entry into the United States:** Bull thistle was introduced multiple times as a contaminant of crop seeds.

### Biology

**Life duration/habit:** Bull thistle is a biennial.

**Reproduction:** Reproduction is by seeds.

**Roots:** The plant has a short, fleshy taproot.

**Stems and leaves:** The stem grows 0.7 to 1.5 m tall and supports many spreading branches. The leaves are prickly and hairy above and cottony below, green or

Bull thistle. (Photo credit: E. Coombs, Oregon Department of Agriculture)

brownish, spiny, and irregularly winged. Some have referred to the appearance of the upper leaf surface as "frog skin."

**Flowers:** Flowers are deep purple to pinkish-purple, 3.8 to 5 cm wide, and clustered at the ends of branches. The bracts are narrow and spine-tipped and extend away from the flower head. Flowering generally occurs from July through September, variations depending on elevation.

**Fruits and seeds:** Seeds are capped with a circle of plume-like white hairs (pappus) and can be wind-blown for long distances.

### Infestations

**Worst infested states:** Bull thistle can be found throughout the United States.

**Habitat:** It is common along roadsides, in logged areas, vacant fields, and pastures. Heavy infestations can exclude livestock from areas. Bull thistle is distributed widely in north and south temperate zones. It occurs at elevations as high as 2,800 m.

**Impacts:** Bull thistle can be troublesome in overgrazed pastures. It is often a transient species, appearing in disturbed and degraded land and in recent clearcuts where it becomes a dominant species for several years.

### Comments

This plant is an important source of pollen and nectar for bees, especially in the fall when other sources are dwindling.

Bull thistle is attacked by several biocontrol insects that were originally intended to target other thistles, as well as by native insects, particularly the larvae of the painted lady butterfly (*Vanessa cardui*) and the artichoke plume moth (*Platyptilia carduidactyla*), but the damage levels seldom control the plants. The accidentally introduced natural enemy, the weevil *Larinus planus,* occasionally uses bull thistle as a host.

Bull thistle infestation. (Photo credit: E. Coombs, Oregon Department of Agriculture)

# Canada thistle

*Cirsium arvense*

Sunflower family—Asteraceae

*G. L. Piper and E. M. Coombs*

**Additional common names:** California thistle, creeping thistle, field thistle.

**Native range:** This species is native to North Africa, Europe, and Asia, and ranges from Scandinavia through Siberia, into China and Japan, and south into Afghanistan.

**Entry into the United States:** Canada thistle was introduced to North America in the early 1700s.

## Biology

**Life duration/habit:** The plant is a colony-forming, aggressive perennial.

**Reproduction:** Canada thistle reproduces by seed, but it mostly spreads by lateral roots that send up new shoots each year. Cutting the roots with cultivation tools only produces more plants unless cultivation is repeated frequently.

**Roots:** The Canada thistle plant consists of several well-separated shoots connected by a deep and extensive lateral root system.

**Stems and leaves:** Stems can grow 1.2 to 1.5 m tall and are ridged and branched. The aboveground portion of the shoot dies during the winter, but the underground part generally survives to produce new shoots the following season. New shoots are also developed from lateral root buds. The leaves are spiny with serrated and ruffled edges.

**Flowers:** Flowers are unisexual, i.e., plants contain either all male or all female flowers. (This condition is unique among thistles in North America.) The flower head is urn-shaped and the bracts are spineless. The color of the flowers may vary from plant to plant — either purple, pink, or white. Flowering occurs June through August.

**Fruits and seeds:** Seeds are 2.5 to 4 mm long and a straw to light brown color.

Canada thistle flowers. (Photo credit: E. Coombs, Oregon Department of Agriculture)

347

## Infestations

**Worst infested states:** Canada thistle grows throughout the northern half of the United States.

**Habitat:** This weed grows in cultivated fields, gardens, flower beds, pastures, rangelands, forests, and along roadsides, ditches, and river banks. Canada thistle thrives under diverse environmental conditions. It tolerates temperatures of 0 to 32° C and precipitation of 40 to 75 cm per year. The plant requires good light intensity for optimal growth.

**Impacts:** It was reported in 1952 to infest more acreage than any other noxious weed in the states of Idaho, Montana, Oregon, and Washington. In heavy concentrations it effectively prevents grazing, thus reducing a site's potential to support livestock. It is also highly competitive with crops.

## Comments

This species is extremely difficult to control. There can also be great diversity in the morphology, phenology, and response to herbicides of the plants, even within the same site.

Canada thistle is attacked by several accidentally introduced insects—the seed head weevil *Larinus planus*, the beetle *Cassida rubiginosa*, the native painted lady butterfly (*Vanessa cardui*), and the crown root weevil *Baris subsimilis*. A rust fungus (*Puccinia carduorum*) is often found in some stands and kills some plants. The seed head fly *Terellia ruficauda* (=*Orellia ruficauda*) may become abundant in some areas but causes little damage.

The seed head weevils *L. planus* and *Rhinocyllus conicus* have become widespread on Canada thistle. *L. planus* was accidentally introduced into the United States at an unknown time. This weevil resembles *R. conicus*, but is larger and has a longer, narrower snout. It is not recorded as a pest of any economically important plant. The weevil was tested by a Canadian researcher and found to accept mainly foliage of *Cirsium* and *Carduus* spp. and

Canada thistle infestation. (Photo credit: N. Poritz, bio-control.com)

related genera such as *Arctium, Onopordum,* and *Silybum.* Tests also indicated that the favored plant was Canada thistle, although others were acceptable if Canada thistle was not available.

The leaf-feeding beetle *Altica carduorum,* another thistle agent, was released in 1963 in Canada and 1966 in the United States, but never established. In 1994, Canadian scientists sought approval for release. The host specificity of the agent was not sufficient to warrant release even though it had been approved in 1963 and 1966. Because of the more stringent current standards, the TAG and the USDA held firm. Out of consideration of U.S. interests, it was never approved for release in Canada.

# Italian thistle

*Carduus pycnocephalus*

Sunflower family— Asteraceae

## E. M. Coombs

**Additional common names:** Italian plumeless thistle, Plymouth thistle, slender thistle.

**Native range:** The Mediterranean area.

**Entry into the United States:** Unknown.

### Biology

**Life duration/habit:** This thistle is a winter annual that sometimes acts as a biennial.

**Reproduction:** Italian thistle reproduces by seeds.

**Roots:** The plant possesses a stout taproot.

**Stems and leaves:** Plants can grow from 0.3 to 1.3 m tall. The stems and upper surface of the leaves are spiny. The leaves may also have a woolly appearance and have light-colored veins. The plant may have many stems that branch from the lower part of the plant and give rise to elongated stems with clusters of two to five flower

Italian thistle. (Photo credit: E. Coombs, Oregon Department of Agriculture)

heads. The stems are semi-winged with small leaves, but not as winged as slenderflower thistle and its more robust relatives.

**Flowers:** The flowers are usually purple to lavender. Flowering occurs from late April to early June.

**Fruits and seeds:** Seeds are 4 to 5 mm long, 1 to 1.5 mm wide, and light brown with a parachute-like pappus that is 15 to 20 mm wide.

## Infestations

**Worst infested states:** California, Idaho, Oregon, and Texas.

**Habitat:** This thistle is generally associated with disturbed areas, livestock pastures, and vacant lots. Irrigated pastures that are not heavily grazed or burned annually do not have severe infestations. Drought favors an increase in Italian thistle density.

**Impacts:** This species, like other *Carduus* thistles, interferes with livestock grazing. Infestations can become so severe that livestock are unable to graze large areas of their pastures. It is often spread by sheep and as a contaminant in hay and soil.

## Comments

Italian thistle is generally less robust in appearance than slenderflower thistle and appears to require soils that are less fertile and moister. *Rhinocyllus conicus* has had a significant impact on controlling the thistle, especially in unburned areas. Also, the stem-boring fly *Cheilosia corydon* attacks plants with stem diameters >10 mm.

# Milk thistle                                  *Silybum marianum*

## Sunflower family—Asteraceae

### E. M. Coombs

**Additional common names:** Blessed milk thistle, holy thistle, lady's thistle, spotted thistle, variegated thistle.

**Native range:** Southwestern Europe and the Mediterranean area.

**Entry into the United States:** Unknown.

## Biology

**Life duration/habit:** The plant is a winter annual that sometimes acts as a biennial.

**Reproduction:** Milk thistle reproduces by seeds.

**Roots:** The plant possesses a deeply seated taproot.

Milk thistle flower head. (Photo credit: E. Coombs, Oregon Department of Agriculture)

**Stems and leaves:** Plants can grow up to 2 m tall and leaves sometimes reach more than 0.3 m long. The broad, lobed, white-marbled leaves clasp the stem. The marbling follows the veins and appears as if someone had poured milk on the leaves.

**Flowers:** The flowers are red to purple, 4 to 8 cm wide, and have stiff, elongated bracts with smaller spines. Flowering occurs from late April to August.

**Fruits and seeds:** The large black to brown mottled seeds are about 12 mm long and are tipped with a pappus of white, barbed hairs.

## Infestations

**Worst infested states:** Western California, Oregon, and Washington. It is also reported in southwestern states through Texas and Oklahoma, and some northeastern states.

**Habitat:** This thistle is generally associated with disturbed areas,

Milk thistle plant. (Photo credit: E. Coombs, Oregon Department of Agriculture)

livestock pastures, and vacant lots. It often grows in fertile soils around dairies and in abandoned corrals. In hotter areas, it grows best in shaded locations.

**Impacts:** Milk thistle interferes with livestocks' access to forage.

## Comments

Milk thistle is often associated with Italian and slenderflower thistles, especially in the more fertile areas of pastures. The large receptacle provides enough room that *Rhinocyllus conicus* larvae do not always make their way into the developing seed tissues. Otherwise, there are no other introduced natural enemies for this species.

Milk thistle contains the compound silymarin which can have a dramatic regenerative effect on the liver. It is used in the treatment of hepatitis, cirrhosis, and in death cap mushroom (*Amanita phalloides*) poisoning and other forms of liver poisoning.

# Musk thistle

*Carduus nutans* (group)

Sunflower family—Asteraceae

## J. L. Littlefield and W. L. Bruckart

**Additional common name:** Nodding thistle.

**Native range:** Southern Europe and western Asia.

**Entry into the United States:** The earliest collection record is from Harrisburg, Pennsylvania, in 1853. Introductions on the eastern seaboard are reported to have been made by European ships dumping seed-bearing soil that had been used as ballast. Multiple introductions of this weed probably have occurred over the years.

## Biology

**Life duration/habit:** Musk thistle plants are usually biennials, but can act as either winter annuals or annuals. The majority of seeds can germinate six to eight weeks after they have fallen to the ground, but some can remain dormant in the soil for five to seven years. Plants may reach heights in excess of 2 m given adequate moisture and fertile soil.

**Reproduction:** Reproduction of musk thistle is by seeds.

**Roots:** The plant grows numerous small roots in the fall and extends a large, fleshy taproot in the spring.

**Stems and leaves:** The stem is erect, branched, and has spiny leaves extending down the side giving it the appearance of being winged. Leaves are deep to light green.

**Flowers:** The flowers of musk thistle are solitary, bent slightly, and reddish-purple. The flower head is flat on the backside where it attaches to the stem. A key characteristic is the large, wide, brown bracts (when dry) that resemble those found on pine cones.

**Fruits and seeds:** The straw-colored, oblong seeds are produced from June to July until the plant dies from dry conditions or freezing. An average plant may produce up to 3,000 seeds, while a larger plant may produce more than 10,000 seeds. Seeds are often attached to pappus and are borne on the wind. Unattached seeds will fall to

Musk thistle. (Photo credit: N. Poritz, bio-control.com)

Musk thistle infestation. (Photo credit: N.Poritz, bio-control.com)

the ground close to the parent plant, or are distributed by birds, small animals, and running water.

## Infestations

**Worst infested states:** In 1981 musk thistle was reported to have spread to 12% or more of the counties of the United States, infesting more than 730,000 ha of land in 40 states. It is particularly troublesome along the eastern seaboard and in the Great Plains.

**Habitat:** Musk thistle can be found on all types of land except deserts, dense forests, high mountains, coastal areas, and newly cultivated lands. The plant grows under a wide range of environmental conditions. It occurs at elevations up to 3,000 m and does well with as little as 25 cm of annual rainfall.

**Impacts:** Dense populations of musk thistle prevent animals from utilizing infested areas. It reduces available forage. One musk thistle plant per 1 m$^2$ can reduce forage production by 16%.

## Comments

Musk thistle is the common name for several species or subspecies of plants belonging to the large-headed *Carduus nutans* group, also known as *C. thoermeri*. Included in this group are *C. nutans*, *C. macrocephala*, and *C. thoermeri=C. nutans* ssp. *leiophyllus*. In many areas of the eastern United States and Great Plains, the severity of musk thistle infestations has been reduced by the action of biological agents, e.g., *Rhinocyllus conicus* often in conjunction with *Trichosirocalus horridus*. Unfortunately, *R. conicus* will also utilize native *Cirsium* thistles, leading to undesirable impacts on these nontarget species. Because of concerns, the redistribution of *R. conicus* across state lines is no longer permitted by USDA-APHIS.

# Plumeless thistle

*Carduus acanthoides*

## Sunflower family— Asteraceae

### G. L. Piper

**Additional common names:** Bristly thistle, spiny plumeless thistle.

**Native range:** Europe and Asia.

**Entry into the United States:** Plumeless thistle was introduced to U.S. East Coast states in the 1870s in ship ballast.

## Biology

**Life duration/habit:** The plant is an annual, winter annual, or biennial.

**Reproduction:** Plumeless thistle reproduces by seeds.

**Roots:** The plant has a stout, fleshy taproot.

**Stems and leaves:** The thistle can grow from 0.3 to 1.2 m tall. Stems are much branched toward the top of the plant and are covered with 6.2 to 18.5 mm spiny wings extending up to or almost to the flower heads. Stem leaves are alternately arranged and blend into the stem.

**Flowers:** Flower heads occur individually or in clusters of two to five at the ends of stem branches. The 1- to 3-cm-diameter heads contain purplish-pink or sometimes white to cream-colored flowers between May and July.

**Fruits and seeds:** The 2- to 3-mm-long, pappused seeds are straw-colored to light brown with dark brown longitudinal striations.

Plumeless thistle. (Photo credit: N. Poritz, bio-control.com)

Plumeless thistle infestation. (Photo credit: N. Poritz, bio-control.com)

## Infestations

**Worst infested states:** Plumeless thistle is abundant in several mid-Atlantic and upper midwestern states and also has been recorded in Colorado, Idaho, Oregon, Washington, and Wyoming.

**Habitat:** This thistle is normally found in disturbed areas, rangelands, livestock pastures, stream valleys, and along transportation rights-of-way. Like musk thistle, plumeless thistle tolerates a wide range of environmental conditions.

**Impacts:** Plumeless thistle, like other *Carduus* thistles, interferes with livestocks' access to forage. Extremely dense infestations are not uncommon.

## Comments

Musk and plumeless thistle often can be found growing close to one another. Both *Rhinocyllus conicus* and *Trichosirocalus horridus* attack plumeless thistle.

# Slenderflower thistle                              *Carduus tenuiflorus*

## Sunflower family—Asteraceae

### E. M. Coombs

**Additional common name:** Seaside thistle.

**Native range:** Southern Europe and the Mediterranean area.

**Entry into the United States:** Unknown.

## Biology

**Life duration/habit:** This thistle is a winter annual that sometimes acts as a biennial.

**Reproduction:** Slenderflower thistle reproduces by seeds.

**Roots:** The plant has a shallow, branched, slender taproot.

355

**Stems and leaves:** Plants can grow from 0.3 to 2 m tall. The stems and upper surface of the leaves are spiny; the leaves may have a woolly appearance and light-colored veins. It may have many stems branching from the lower part of the plant with clusters of 5 to 20 flowers. The stems are winged with small leaves.

**Flowers:** The flowers are usually purple to lavender, and bloom occurs from April to early June. Slenderflower thistle usually has more than five heads per cluster and the bracts are not hairy.

Slenderflower thistle. (Photo credit: E. Coombs, Oregon Department of Agriculture)

**Fruits and seeds:** Two types of seeds are produced. Numerous centrally positioned seeds are sticky and plumose and are dispersed rapidly after maturation, whereas peripheral seeds are nonsticky and plumeless and tend to remain in the heads longer.

## Infestations

**Worst infested states:** California, Oregon, Texas, and Washington.

**Habitat:** This thistle is generally associated with disturbed areas, livestock pastures, and vacant lots. It prefers high-fertility soils. Infested pastures that are annually burned in western Oregon often become dominated by slenderflower thistle.

**Impacts:** Slenderflower thistle, like other *Carduus* thistles, interferes with livestocks' access to forage. Infestations can become so severe that livestock are unable to graze

Slenderflower thistle infestation. (Photo credit: E. Coombs, Oregon Department of Agriculture)

large areas of their pastures. It is often spread by sheep and as a contaminant in hay and soil.

## Comments

Slenderflower thistle is more robust in appearance than Italian thistle, and apparently requires fertile, moist soils. *Rhinocyllus conicus* has helped control the thistle, especially in unburned areas. The crown/stem-boring fly *Cheilosia corydon*, a biocontrol agent intended for musk thistle, has adapted well to this thistle when stem diameters are greater than 10 mm. Also, the accidentally introduced seed head weevil *Larinus planus* has been found occasionally attacking the plant.

# *Ceutorhynchus litura (=Hadroplontus litura)*

G. L. Piper and E. M. Coombs

**Common name:** Canada thistle stem weevil.

**Type of agent:** Insect: Beetle, weevil (Coleoptera: Curculionidae).

**Native distribution:** Europe.

**Original source:** Germany.

## Biology

**Generations per year:** One.

**Overwintering stage:** Adult (generally in the soil near the plant, or on the plant next to the soil).

**Egg stage:** Eggs are deposited in March and April until mid-May, generally until young plants are about 5 cm tall, after which time the plant becomes undesirable for egg-laying. Each female lays an average of 120 eggs, generally on the smaller, younger shoots.

**Larval stage:** The newly hatched larvae mine the tissues of the leaf toward the main vein. Older larvae mine the stem, root crown, and upper root. There are three larval growth stages. Larvae are whitish, grub-like, and pointed at the anterior end.

**Pupal stage:** Pupation occurs in the soil in an oval cocoon constructed from small soil particles. The pupal period lasts two to three weeks.

*C. litura* larvae in a stem. (Photo credit: N. Poritz, bio-control.com)

357

**Adult stage:** Adults can be found in the field from August until the following May, June, and in some areas July. Adults are 3- to 4-mm-long, long-snouted, black weevils with whitish hairs and a very pronounced white thunderbird or T-shaped marking on the back.

*C. litura* adult. (Photo credit: N. Poritz, bio-control.com)

## Effect

**Destructive stages:** Limited feeding by *C. litura* adults in the early spring has little effect on the plant. Larval feeding during the spring and early summer also does not cause sufficient damage to be outwardly apparent. Secondary damage to the plant, however, is caused by other organisms such as opportunistic pathogens that enter the stem in the winter through the exit holes made by *C. litura* larvae.

**Additional plant species attacked:** None.

**Site of attack:** Adults feed on leaf and stem tissue and make concavities in the midrib of the leaf in which females deposit eggs. The greatest damage to the plant is caused by the larvae feeding within the stem and crown of the plant and then chewing exit or escape holes below the soil surface.

**Impact on the host:** Impact on Canada thistle is mostly indirect. Departing larvae create emergence holes below the soil surface which provides access for small insects, other arthropods, nematodes, and pathogens. The plant's appearance may change slightly with dark staining showing in the lower stem fibers. Rotting of the underground shoot during the winter months reduces shoot abundance the following spring.

**Nontarget effects:** None have been reported. The host range of the beetle was found to be narrow, restricted to plants of the tribe Cardueae.

## Releases

**First introduced into the United States:** 1972, Montana.

**Established in:** Idaho, Montana, Nebraska, North Dakota, Oregon, South Dakota, Utah, Virginia, Washington, and Wyoming.

**Habitat:** Favorable conditions include disturbed areas where Canada thistle is dense and where the plant is not stressed by grazing, dry conditions, flooding, mowing, or herbicides. In Europe, it is more commonly encountered in cultivated lands than in grasslands or forests.

**Availability:** This weevil is readily available in many states early in the spring.

**Stage to transfer:** Adult.

**Redistribution:** Collect adults from the early spring shoots either with fingers, forceps, or an aspirator. Place the beetles into paperboard containers provisioned with Canada thistle leaves and store them in a cool place. The insects can be shipped or transported for up to a week if kept cool and provided with moderately fresh leaves. It is best to release the adults over young plant material similar in size to the plants from which the weevils were collected.

## Comments

A *Nosema* pathogen was detected in some *C. litura* in the Gallatin Valley of Montana in 1991 through 1994. The effect of this pathogen on the weevil has not been fully ascertained.

A study in Montana established that: 1) *C. litura* can spread up to 9 km in 10 years; 2) *C. litura* in a 10-year period increased to infest more than 80% of Canada thistle stems; 3) the infestation level of Canada thistle stems is not influenced by the presence or absence of surrounding vegetation; 4) the underground parts of shoots attacked by *C. litura* generally do not survive the winter; 5) underground parts of some unattacked plants also may die if they are connected by lateral roots to attacked plants; and 6) the roots of plants with at least one shoot attacked by *C. litura* produce less than two shoots the following year compared to more than nine shoots produced by plants that have not been attacked. It is concluded that *C. litura* is an effective agent because it reduces the overwintering survival of Canada thistle. Thistle stands are often able to maintain themselves by reinfestation from nonattacked plants, but if the effects of the weevil are augmented by another biocontrol agent, thistle populations can decrease.

## *Cheilosia corydon (=Cheilosia grossa)*

G. L. Piper and E. M Coombs

**Common name:** Thistle stem hover fly.

**Type of agent:** Insect: Fly, flower fly (Diptera: Syrphidae).

**Native distribution:** Throughout Europe, including Bosnia, Bulgaria, and Turkey.

**Original source:** Italy.

### Biology

**Generations per year:** One.

**Overwintering stage:** Pupal.

**Egg stage:** In Italy, eggs are laid on young, hairy leaves and young shoots, usually at the center of the plant, between mid-March and early April. Eggs are deposited singly or in groups of three to four and generally hatch in six to seven days. Eggs measure 1.2 by 0.48 mm and are elongate and whitish, becoming darker with age.

**Larval stage:** There are three larval growth stages. Newly hatched larvae mine directly into tender, young shoots. The first growth stage lasts about 11 days. As the shoot

359

*C. corydon* larva in a thistle stem. (Photo credit: E. Coombs, Oregon Department of Agriculture)

Damage to slenderflower thistle by larvae of *C. corydon*. (Photo crdit: E. Coombs, Oregon Department of Agriculture)

grows, the second and third growth stage larvae enter the stem and mine up and down. The second stage lasts nearly seven days and the third feeds until November when it pupates. The plants in Italy grow old and deteriorate during the summer awaiting the fall rains. During this time, larvae in the crown remain dormant.

**Pupal stage:** Pupation occurs within the damaged root or in the soil.

**Adult stage:** This is a fuzzy, yellow- and orange-haired, black fly that superficially resembles a bee. It is 13 to 15 mm long, including its wings.

## Effect

**Destructive stage:** Larval.

**Plant species attacked:** Musk thistle, Italian thistle, slenderflower thistle, and plumeless thistle. *Cirsium crassicaule* is a marginally suitable laboratory host. In Oregon, recent evidence indicates that this fly may occasionally attack large rosettes of bull thistle.

**Site of attack:** Leaves, stems, and crown of the plant. In slenderflower and Italian thistles, larvae attack the early buds (usually causing that portion of the plant to shrivel), then tunnel down the stem to the root crowns.

**Impact on the host:** Feeding by the larvae interrupts the plant's moisture and nutrient transport system, impairs inflorescence development, and eventually lowers seed production. Root lesions allow the introduction of soil microorganisms. When more than one larva infests the root, the plant often dies.

**Nontarget effects:** The host range of this fly was found to be narrow, restricted to plants of the tribe Cardueae. *C. crassicaule*, a threatened species in the United States, was marginally suitable as a host in laboratory trials. However, in open-field egg-laying tests using replicated and randomized plantings, no egg-laying on any native *Cirsium* species occurred.

*C. corydon* adult on musk thistle. (Photo credit: E. Coombs, Oregon Department of Agriculture)

## Releases

**First introduced into the United States:** 1990, Maryland.

**Established in:** The fly is established in Oregon on slenderflower and Italian thistle.

**Habitat:** In Europe this fly is common in a range of climates similar to those in the United States in which *Carduus* spp. are found. Adults of *C. corydon* may need early-flowering plants from which to obtain nectar. However, because this species emerges very early in the spring, the absence of the proper flowering plants may limit the areas where it may be effective.

**Availability:** The fly is available in Douglas County, Oregon, as of 2003.

**Stages to transfer:** Adult and pupal.

**Redistribution:** Use a sweep net to collect adults in early spring (March through April) or dig roots in the late summer or fall, collect the pupae, and keep them at 4 to 8°C until spring when the thistle plants begin growth. Adults can be kept in cool storage for only a few days.

## Comments

In Italy, larvae infest the flower buds, stems, and root crowns of musk, slenderflower, and Italian thistles. In Oregon, this fly shows a distinct preference for the larger slenderflower thistle (infestation up to 50%) and rarely attacks Italian thistle stems that are less than 10 mm in diameter. Once established, the fly spreads rapidly, more than 20 km in five years.

Early information in the literature concerning *C. corydon* was given for *C. grossa*.

# Psylliodes chalcomera

G. L. Piper and J. R. Nechols

**Common name:** None widely accepted.

**Type of agent:** Insect: Beetle, flea beetle (Coleoptera: Chrysomelidae).

**Native distribution:** Central Europe and Asia.

**Original source:** Italy.

## Biology

**Generations per year:** One.

**Overwintering stage:** Sexually mature, egg-laying adult.

**Egg stage:** In Italy, eggs are laid from January to June at the base of the plants or on the soil at the base of the plants. Females lay an average of 296 eggs each. Eggs are yellow, oval, and measure 0.67 by 0.37 mm.

**Larval stage:** Larvae feed in flower and leaf buds, and mature in mid-May. Larvae are slender, elongate, and whitish except for their brown head capsules.

**Pupal stage:** Pupation occurs in the soil near the host plant.

**Adult stage:** Adults emerge from May to early June in Italy. Dormancy is synchronized with the hot, dry summers when the plants dry. In November, adults mate and lay eggs on the leaves of the rosettes of next year's flowering plants. Adults are shiny, dark, metallic blue-green, and
about 1.5 mm wide by 3 mm long.

## Effect

**Destructive stage:** Larval.

**Plant species attacked:** Musk, Italian, plumeless, and Illyrian (*Onopordum illyricum*) thistles.

**Site of attack:** Leaves, flowers, and leaf buds.

**Impact on the host:** Larvae destroy the growing tips of buds and stems and severely damage the vascular system of attacked plants. An average of 30 larvae per plant are usually found in the field.

**Nontarget effects:** No nontarget effects have been reported.

Late bud damage caused by *P. chalcomera*. (Photo credit: J. Nechols, Kansas State University)

## Releases

**First introduced into the United States:** 1997, Kansas, Maryland, and Texas.

**Established in:** Establishment of this insect is unknown.

**Habitat:** The flea beetle is found in both cold and hot areas of Italy.

**Availability:** This insect is unavailable.

**Stage to transfer:** Adult.

**Redistribution:** Collect the adults with a sweep net or a mechanical aspirator such as a power vacuum. Sort and store at cool temperatures with food. Release the flea beetles as soon as possible.

*P. chalcomera* adult. (Photo credit: USDA-ARS)

## Comments

This flea beetle has been approved for field release, but only limited releases have been made because of the inability to collect large numbers at overseas locations.

# *Puccinia carduorum*

J. L. Littlefield, W. L. Bruckart, D. M. Woods, and A. B. A. M. Baudoin

**Common name:** Musk thistle rust.

**Type of agent:** Rust fungus (Uredinales: Pucciniaceae).

**Native distribution:** Throughout Eurasia and North Africa.

**Original source:** Turkey.

## Biology

**Life cycle of the fungus:** *P. carduorum* has five stages in its life cycle. It makes teliospores, which are resistant to cold and freezing weather, to overwinter. In the spring, teliospores germinate and produce new spores (basidiospores) that infect musk thistle plants. From this infection, the pathogen makes rust-colored aeciospores and then urediniospores, hence the name for the disease. Urediniospores are powdery and easily blown to healthy plants. When the weather is optimal—at temperatures between 18 to 21° C and a dew period of eight to 12 hours—new urediniospores can be made within two weeks. Because urediniospores are wind-blown, they can be dispersed very long distances. When the plants begin to die after seed production, the fungus makes teliospores in preparation for bad weather, thus completing the life cycle.

**Urediniospore description:** This is the most characteristic spore of *P. carduorum*. Single spores are golden to brown, round or slightly flattened, about 25 microns in diameter, with a thick wall that is thicker at the point of attachment (hilum), nearly always with

363

*P. carduorum* pustules on musk thistle leaf (right). (Photo credit: USDA-ARS)

three germpores on the "equator" of the spore (halfway between the hilum and the top of the spore), and the spore surface covered with short spines (echinulations). A mass of spores, as would be seen in a pustule on an infected leaf, is dark to reddish-brown and powdery in appearance.

## Effect

**Plant species attacked:** Musk thistle.

**Site of attack:** Leaves, stems, and bract leaves throughout the season.

**Impact on the host:** High levels of infection reduce seed set and seed quality. Also, damage by *P. carduorum* and insect biocontrol agents (particularly *Rhinocyllus conicus*) are additive, based on field data from a Virginia study.

**Nontarget effects:** The pathogen completes its life cycle only on musk thistle. Other strains of *P. carduorum* attack other species of *Carduus* thistles.

## Releases

**First introduced into the United States:** 1987, Virginia.

**Established in:** California, Delaware, Georgia, Indiana, Kentucky, Maryland, Ohio, South Carolina, Tennessee, Virginia, and Wyoming, and probably throughout the distribution of musk thistle in North America.

**Habitat:** *P. carduorum* generally is associated with *Carduus* thistles wherever they grow in Eurasia.

**Availability:** Permission for general use in the United States was still pending as of 2003.

**Stage to transfer:** Urediniospores.

**Redistribution:** Either vacuum-harvest urediniospores, or collect infected leaves and dry them before shipment. Urediniospores on dry leaves were viable after at least two months. Urediniospores can be frozen for several years at very cold temperatures (-80°C). Plants can be field-inoculated by spraying urediniospores suspended in water with a wetting agent (e.g., Tween-20® at 0.25% v/v), or spores can be dusted onto healthy leaves. Good results have been achieved with rust fungi by inoculating in the evening when dew or good moisture is expected. To insure high humidity, inoculated plants (which are wet) can be covered in a plastic tent overnight. Once *P. carduorum* infects a few plants in a stand, it should spread rapidly with favorable weather conditions. The fungus is likely to perform better as stand density increases, but it has colonized musk thistle stands that are several kilometers apart.

## Comments

*Puccinia carduorum* was permitted for a field study in Virginia in 1987. In three years of field tests, there were no infections of native *Cirsium* thistles and only one very small pustule on one of 32 artichoke plants, thus supporting data from containment greenhouse and Swiss field studies that this strain of *P. carduorum* infects only musk thistle. It was very aggressive on musk thistle, and increased on musk thistle most rapidly in the spring when plants were bolting; fall inoculations of musk thistle resulted in very few infections. Stands of musk thistle were monitored over two years in Virginia and, during the second season (1989), infected musk thistle plants were found at the farthest site, about 11 km away. During a more comprehensive survey between 1992 and 1994, infected plants were located in nine mid-Atlantic states. No infected musk thistles were found in 1992 surveys from Arkansas, Iowa, Missouri, or North Dakota. The rust was detected later in Missouri (1994) and in California (1998).

Data from the field study not only support host specificity of the musk thistle rust, but also indicate that high levels of infection reduce seed production and quality. Damage by the rust fungus enhances effects from *R. conicus*, and it does not interfere with any of the insect agents already present on musk thistle.

## *Rhinocyllus conicus*

G. L. Piper and E. M Coombs

**Common name:** None widely accepted (Thistle seed head weevil).

**Type of agent:** Insect: Beetle, weevil (Coleoptera: Curculionidae).

**Native distribution:** Europe, western Asia, and North Africa between latitudes 30 and 50° N.

**Original source:** Europe.

## Biology

**Generations per year:** One.

**Overwintering stage:** Adult. It has been suggested in the literature that *R. conicus* overwinters under rocks, logs, etc., but they are very difficult to find there. They generally are found in sheltered locations such as caves, the hollows of trees, or occasionally the attics of homes.

**Egg stage:** Each female produces from 100 to 150 eggs and generally deposits them on bud bracts (the modified leaves on the back of the flower). The bracts of musk thistle are preferred locations, but when the bracts are saturated with eggs, the musk thistle stems receive the next largest number of eggs. If other host thistle species are in the area, the insect will then move to them. Depending on the host and the region, eggs are laid from April through July. Eggs are covered with chewed plant material that becomes tan with age and appears as "warts" on the buds and stems, thus protecting the eggs from predators. Eggs hatch in six to eight days.

*R. conicus* larvae in a musk thistle seed head. (Photo credit: N. Poritz, bio-control.com)

**Larval stage:** The larvae attack the seed heads or stems from early April to August. Here they develop for 25 to 40 days, feeding on the receptacle and maturing seed tissue. Each larva feeds within the chamber or cell it forms. The feeding stimulates the plant to concentrate nutrients and tissue in the affected area. The mature larvae eventually coat the inner cell walls with feces and masticated plant material to produce hard, protective chambers for the pupal stage. Larvae that tunnel into stems of musk thistle do not construct cells. Larvae are C-shaped, creamy white, and have amber-brown head capsules.

**Pupal stage:** Pupation occurs within the plant tissue in which the larvae developed. The pupal stage lasts from eight to 14 days. Pupae are whitish to creamy white and darken as they mature.

**Adult stage:** Adults remain within the pupal cells for several weeks. During this time, they turn from a reddish-tan to almost black. When weevils emerge from the plant, their body hair is a patchy mixture of black and yellow, which gives the appearance that the weevils are covered with pollen. Weevils chew their way out of the seed heads through the face of the receptacle (that part of the flower that produces seeds), whereas those in the stems exit through several small openings chewed near the attachment of the seed head. Adults are present for only a short period after emerging from the plants. They can occasionally be seen flying about on warm fall days. This weevil has a short snout, and although the size is variable, most weevils are no more than 5 to 6 mm long.

## Effect

**Destructive stages:** Larval in the seed head and, to a lesser extent, adult when they chew holes in the leaves and notch the stems.

*R. conicus* adult. (Photo credit: N. Poritz, bio-control.com)

**Plant species attacked:** Host thistles include species belonging to the genera *Carduus, Cirsium, Onopordum,* and *Silybum,* many of which are exotic plants that have been introduced into North America. Among these are plumeless thistle, welted thistle (*Carduus crispus*), musk thistle, Italian thistle, slenderflower thistle, Canada thistle, bull thistle, and milk thistle. The weevil also infests numerous native *Cirsium* species, including several threatened and endangered species.

**Site of attack:** Seed head and sometimes the stem. Adults may slightly defoliate plants.

**Impact on the host:** Because the weevil attacks the seed-producing tissue and because musk thistle reproduces totally by seed, it is effective by itself in areas where the plant and insect life cycles are synchronized. In those plant species that it attacks and especially those that reproduce by other means (Canada thistle), it only affects the seed production potential.

**Nontarget effects:** Since *R. conicus* has been introduced from several European locations and established at various locations in the United States, scientists have been able to observe how strains vary. It appears that *R. conicus* collected in Europe from Italian thistle will also utilize slenderflower thistle, whereas the strain collected from European musk thistle will not. The milk thistle strain was collected in Rome and released directly on milk thistle in California. More than 25% of the native *Cirsium* species in the United States are attacked by *R. conicus* and the list is expected to grow as monitoring projects continue. However, USDA-APHIS revoked all permits for the interstate shipment of *R. conicus* in 2000 and release of the weevil may be prohibited in areas or states with sensitive, threatened, or endangered *Cirsium* species.

## Releases

**First introduced into the United States:** 1969, Montana and Virginia.

**Established in:** The weevil is widely established throughout the country.

**Habitat:** Meadows and areas with adequate moisture and moderate temperatures are best for the weevil, while areas of extreme heat where the plant is under moisture stress greatly limit this insect's population. Areas where summer arrives quickly do not allow the weevil to utilize secondary and later seed heads.

367

**Availability:** Large numbers can be collected for redistribution during May and June from almost any established location.

**Stage to transfer:** Adult.

**Redistribution:** This insect is not recommended for redistribution because of its nontarget impacts on native thistles.

## Comments

This was the first species released for the biocontrol of musk thistle in the United States. Experiments have shown that mortality of *R. conicus* is not significantly increased when the herbicide 2,4-D is sprayed on bolting and bolted musk thistle plants with developing larvae in the seed heads.

# *Trichosirocalus horridus (=Ceuthorhynchidius horridus, T. mortadelo)*

G. L. Piper and E. M Coombs

**Common name:** None widely accepted (Musk thistle crown weevil).

**Type of agent:** Insect: Beetle, weevil (Coleoptera: Curculionidae).

**Native distribution:** Western, central, and southern Europe.

**Original source:** Italy.

## Biology

**Generations per year:** One.

**Overwintering stages:** Adult and larval.

**Egg stage:** Eggs are white, opaque, and measure about 0.54 by 0.33 mm at first. They are inserted into the tiny punctures made by the female on the underside of leaves along the midrib and primary veins. Eggs are deposited singly or in clusters of two to four, although as many as 14 have been reported. The incubation period requires about 13 days. As the eggs mature, they become more yellowish and soon several tiny brown dots appear which indicate the position of the head capsule. A female may lay up to 2,000 eggs.

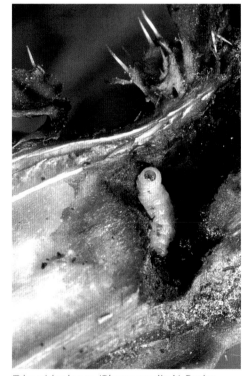

*T. horridus* larva. (Photo credit: N. Poritz, bio-control.com)

**Larval stage:** There are three larval growth stages. Larvae are creamy white with dark brown or black head capsules. The larvae move down the petiole toward the growth point soon after hatching. As they mature, they feed within the growth point of the thistle at the root-stem junction. Damaged tissue blackens after 15 to 20 days. The larval period ranges between 40 to 164 days, depending upon temperature. The mature larvae exit the plants and enter the soil near the roots, where they create cells made from silk and soil particles in which pupation occurs.

*T. horridus* adult. (Photo credit: N. Poritz, bio-control.com)

**Pupal stage:** Duration of the pupal period is about 18 days. Pupae average 4.3 mm long.

**Adult stage:** Adults emerge from late June into July, feed briefly on foliage, then aestivate for the remainder of the summer. They become active again from late September until the onset of colder temperatures. Adults become active again in the spring and can be found until the emergence of the next generation. They are small (3.5 to 4.5 mm long), stout, brown weevils with strong tubercles on the prothorax.

## Effect

**Destructive stage:** Larval (feeds on the growing tip of the thistle rosette). Adults may slightly defoliate plants.

**Plant species attacked:** Thistles of the subtribe Carduinae, including musk thistle, plumeless thistle, Italian thistle, Canada thistle, and bull thistle are acceptable hosts.

**Site of attack:** Adults are usually found on the young rosette leaves and begin feeding immediately after emergence, producing circular, hollowed, feeding punctures in leaf midribs. Each weevil produces about 10 punctures per week, mainly on the underside of the leaves. Larvae attack the rosette's growing tips and developing stems but will not attack small rosettes or seedlings.

**Impact on the host:** Weak plants often succumb to moderate populations of this natural enemy. There are several reports of areas where the plant population has been greatly reduced. Feeding by *T. horridus* larvae on musk thistle rosettes disrupts the growing tips, causing the plants to produce multiple shoots instead of the main one. Because there is more demand on the root system with multiple shoots, the seed heads may be smaller than normal and produce fewer seeds. Reductions in plant density have been observed in Idaho, Kansas, Montana, Oregon, Virginia, and Wyoming several years after the release of *T. horridus*.

**Nontarget effects:** None have been reported where this weevil is established.

369

## Releases

**First introduced into the United States:** 1974, Virginia.

**Established in:** Colorado, Idaho, Kansas, Maryland, Missouri, Montana, Oregon, Virginia, Washington, and Wyoming.

**Habitat:** The most favorable sites for this weevil are on the fringe of open infestations of musk thistle where the plant height is shorter. High elevation locations (>3,000 m) and marshy areas are probably not suitable.

**Availability:** The weevil is generally available.

**Stage to transfer:** Adult.

**Redistribution:** Use a sweep net to collect the adults from bolted plants in July, or collect the adults by hand in the early spring before the musk thistle has started to bolt. Adults can be shipped in cardboard containers containing leaves of the host plant. Keep the insects cool and release them as soon as possible.

## Comments

This was the second insect species introduced into the United States for the biological control of musk thistle. Recent studies on *T. horridus* indicate that it is a species complex and the weevils introduced into the United States are likely *T. mortadelo*.

Populations of *T. horridus* develop slowly, requiring three to five years to build up suitable numbers. Once well established, weevils may disperse well over 8 km to new musk thistle sites. This species is seldom effective by itself; it requires another agent such as *Rhinocyllus conicus* to decrease plant populations. Release in California is prohibited because of concerns of attack on artichokes.

# *Urophora cardui*

G. L. Piper and E. M Coombs

**Common name:** Canada thistle stem gall fly.

**Type of agent:** Insect: Fly (Diptera: Tephritidae).

**Native distribution:** Sweden south to the Mediterranean, and from France east to near the Crimea.

**Original source:** Central Europe.

## Biology

**Generations per year:** One.

**Overwintering stage:** Mature larval.

**Egg stage:** From one to 30 eggs are laid in the vegetative shoots at any time during the plant's active growing season.

**Larval stage:** Newly hatched larvae spend the first growth stage within the egg. Second-stage larvae tunnel into the stems and cause galls to form. Larvae grow slowly in the second stage while the gall is growing. As the gall matures, the larvae molt to the third growth stage and quickly attain 98% of their total body weight. Larvae are white and barrel-shaped with dark anal plates. Multiple larvae (three to 10) can be found within large galls.

**Pupal stage:** Pupation occurs in early spring. The length of the pupal period is 24 to 35 days at 18 to 20° C. Puparia are dark reddish-brown and are located within the gall.

**Adult stage:** The adult flies emerge from deteriorating galls in late spring to early summer. Adult flies are 5 to 8 mm long and have very distinct, black, W-shaped markings on the wings.

## Effect

**Destructive stage:** Larval.

**Plant species attacked:** Canada thistle. Although several other *Cirsium* species are listed in the literature, none of these has been noted recently or confirmed with testing.

**Site of attack:** Adults deposit eggs in the stems of the plant. Developing larvae within the stem cause the plant to form a gall which looks like a small green crabapple in the middle of the stem or on one of the side branches. Gall size varies considerably; they are generally marble- to walnut-sized depending on the number of larvae within the gall. After the plant dies or freezes in late fall or early winter, the gall becomes a brown to grey woody structure that is very hard except when wet.

*U. cardui* gall cut open to show larva. (Photo credit: N. Poritz, bio-control.com)

*U. cardui* adult female. (Photo credit: N. Poritz, bio-control.com)

**Impact on the host:** Most of the effect of *Urophora cardui* is in the formation of a metabolic sink preventing the plant's nutrients from being allocated to other areas (roots, flowers). This reduces the plant's vigor, making it less able to compete and resist attacks by pathogens or other insects. Stems above the galls are often retarded and may not produce flowers. However, by itself, this insect does not kill the plant.

**Nontarget effects:** None have been reported.

### Releases

**First introduced into the United States:** 1980, Oregon.

**Established in:** California, Maryland, Montana, Nevada, Oregon, Virginia, Washington, and Wyoming.

**Habitat:** This fly does best in moist, disturbed areas with scattered Canada thistle plants. Semi-shaded areas seem to be slightly preferred over those in full sun. Fields subject to flooding, grazing, mowing, or chemical treatments are not conducive to fly survival.

Damage caused to thistles by *U. cardui*. (Photo credit: N. Poritz, bio-control.com)

**Availability:** In some locations where it has been established for many years, the fly is quite prevalent. It is most abundant in western Oregon and Washington.

**Stages to transfer:** Adult or pupae in galls.

**Redistribution:** Galls are collected in the fall, winter, or early spring by snipping them from the previous year's plants. These are stored in paper sacks or cardboard boxes in the refrigerator at 4 to 8° C. If the refrigerator has ventilation and/or a drying effect, it may become necessary to mist the contents of the containers with water every two to four weeks.

It is best to place the galls in cages or sleeve boxes to confine emerging flies. Galls can be misted up to five times a day to assist in emergence of the adults. A sponge soaked in sugar-water will provide nourishment to newly emerged adults. This technique allows adults to be counted and sorted from other insects and parasitic wasps that may emerge from the galls. A sweet, fruity smell is noticeable when adults emerge in the cage.

In local areas, 50 to 100 galls can be placed in infested areas in early spring.

Adults can be collected with a sweep net, though this is very time-consuming when they are few in number.

Release the flies onto Canada thistle plants in the morning or evening.

## Comments

Cold-adapted strains of this insect have been developed in recent years; therefore fly galls should be collected from habitats similar to the intended release site. Flies do best in high humidity/rainfall areas or along watercourses. Substantial bird and rodent predation of the galls occurs during the fall and winter months at some locations.

## *Urophora solstitialis*

J. L. Littlefield

**Common name:** Musk thistle seed head fly.

**Type of agent:** Insect: Fly (Diptera: Tephritidae).

**Native distribution:** Found throughout Europe, from the United Kingdom extending into Eurasia and Central Asia to the east.

**Original sources:** Italy (Canadian releases from Austria).

### Biology

**Generations per year:** One, although two in some locations.

**Overwintering stage:** Mature larvae in galls within seed heads.

**Egg stage:** Females may produce up to 100 eggs. Eggs are small and elongated and are laid within the developing floret. Several eggs may be laid per flower bud.

**Larval stage:** Larvae hatch within several days and mine the floret to the receptacle. Once in the receptacle, a lignified gall forms. Galls may coalesce and adjoining seeds may attach to the gall. Larvae are white with a distinct brown spiracular plate. There are three larval instars.

*U. solstitialis* adult. (Photo credit: E. Coombs, Oregon Department of Agriculture)

373

**Pupal stage:** Pupation occurs within the gall. Early maturing larvae may pupate in early summer and emerge as a second generation. Most larvae overwinter and pupate the following spring.

**Adult stage:** First-generation adults are present during the spring and the second-generation adults appear as early as June in some locations. Adults live for several weeks and lay eggs as long as there are suitable flower buds. Adults are about 3 to 5 mm long; females have a long ovipositor. Adults have a black abdomen, a brownish thorax, and yellowish head and legs. Flies have two V-shaped bands on the edge of the wings.

## Effect

**Destructive stage:** Larval (galls).

**Plant species attacked:** In Europe, flies are associated primarily with *Carduus* thistles: *C. nutans*, *C. acanthoides*, *C. crispus*, *C. defloratus*, *C. nigrescens*, and *C. personata*.

**Site of attack:** Flower buds.

**Impact on the host:** Galls reduce seed production and act as a metabolic sink reducing available nutrients that would otherwise go to seed production.

**Nontarget effects:** No nontarget effects have been reported.

## Releases

**First introduced into the United States:** 1993, Maryland.

**Established in:** Establishment has not been reported.

**Habitat:** Specific site requirements are not known.

**Availability:** This fly is not available as of 2003.

**Stages to transfer:** Adults and mature larvae (within seed heads).

**Redistribution:** Infested seed heads can be collected in late fall. The infested heads may be distinguished from uninfested flower heads by the presence of hard galls. The heads can be stored through the winter if they are kept cool and dry. From these, adults can be reared in sleeve cages in the spring. Another method for release is to place infested flower heads near thistle infestations. Adults can also be collected with a sweep net during May and June.

## Comments

The fly has been released in Maryland, Montana, and Oregon.

Both *Rhinocyllus conicus* and *Urophora solstitialis* can develop within the same flower head, although on musk thistle *R. conicus* may outcompete the gall fly. *U. solstitialis* may be an effective agent in combination with *Trichosirocalus horridus* in areas with low *R. conicus* populations or where thistles have an extended period of flower production. In Canada, the fly has been effectively established on plumeless thistle.

Burdock (*Arctium lappa*) and bull thistle have been reported as being possible "aberrant" hosts, although there is some confusion as to the identity of the *Urophora* sp. attacking these plant species. In laboratory tests, a few galls were also observed on burdock, *Centaurea cyanus*, and *Cirsium heterophyllum*.

# Urophora stylata

G. L. Piper and E. M Coombs

**Common name:** Bull thistle seed head gall fly.

**Type of agent:** Insect: Fly (Diptera: Tephritidae).

**Native distribution:** Western Europe.

**Original sources:** Germany and Switzerland, via Canada.

## Biology

**Generations per year:** One.

**Overwintering stage:** Third-stage larval.

**Egg stage:** Eggs are laid on top of developing flower buds and hatch after one week.

**Larval stage:** The larvae burrow through the flower ovaries into the seed-producing tissues where the plant is induced to produce the gall tissue on which the larvae feed. Each larva resides in a separate chamber in the gall. Galls may contain from five to 20 or more larvae. The larvae are cream-colored with dark anal plates. They are thick-bodied and 3 to 5 mm long.

**Pupal stage:** Pupation occurs inside the gall. The puparia are dark tan-brown.

**Adult stage:** Adults emerge in late May through early July and may live up to two months. Males set up territories where they characteristically display their patterned wings. Females are distinctive with their telescoping abdomen. Both are found on bolting bull thistle plants with flower buds forming. The adults are about the size of a housefly; the males are 4 to 5 mm long and the females are up to 7 mm long. The body is a light gray, the scutellum (located on the back behind the head) is often light brown to yellowish. The wings are clear with a dark grayish-brown "IV" marking, the "V" being at the apex of the wing.

## Effect

**Destructive stage:** Larval.

**Plant species attacked:** Bull thistle.

**Site of attack:** The larvae feed within the seed-producing tissues of the developing seed heads.

**Impact on the host:** From 60 to 90% of the seed heads have been attacked in some areas, which has reduced seed production by up to 60%.

**Nontarget effects:** No nontarget effects have been reported.

*U. stylata* adult. (Photo credit: E. Coombs, Oregon Department of Agriculture)

375

## Releases

**First introduced into the United States:** 1983, Washington.

**Established in:** Colorado, Maryland, Oregon, and Washington.

**Habitat:** Favorable sites appear to be open meadows with large plants, while heavily grazed areas, areas that flood, canyons with high winds, and dense stands of plants are not desirable habitats.

**Availability:** The galls are available in Oregon and Washington. It generally takes three years for a nursery site to become collectible.

**Stages to transfer:** Adult and larval (within the seed heads).

Infested (left) vs. non-infested (right) seed heads. (Photo credit: E. Coombs, Oregon Department of Agriculture)

**Redistribution:** Infested seed heads can be collected in late fall by cutting the heads with clippers or they can be pulled off by hand. The infested heads are easy to distinguish from uninfested flower heads by their fluffy appearance, and a gentle squeeze with gloved hands will reveal the hard, golf ball-sized galls. The heads can be stored through the winter if they are kept cool and dry. From these, adults can be reared in sleeve cages in the spring. Another method for release is to place 25 to 50 infested galls on the ground near large rosettes. Adults can also be collected with a sweep net between June and August.

## Comments

Since bull thistle is a transient species, it is difficult to maintain fly populations more than several years at one location. The flies spread rapidly from the release site within a few years. Although no parasitoids have yet been found, it is best to rear adults from galls when moving flies into new areas.

# References

Alonso-Zarazaga, M. A., and M. Sanchez-Ruiz. 2002. Revision of the *Trichosirocalus horridus* (Panzer) species complex, with a description of two new species infesting thistles (Coleoptera: Curculionidae, Ceutorhynchinae). Aust. J. Entomol. 41: 199-208.

Andres, L. A., and N. E. Rees. 1995. 67. Musk thistle, *Carduus nutans* L. (Asteraceae). Pages 248-51 *in* J. R. Nechols, L. A. Andres, J. W. Beardsley, R. D. Goeden, and C. G. Jackson, eds. Biological Control in the Western United States: Accomplishments and Benefits of Regional Research Project W-84, 1964-1989. Univ. Calif. Div. Agric. Nat. Res. Pub. 3361. Oakland, CA.

Baudoin, A. B. A. M., R. G. Abad, L. T. Kok, and W. L. Bruckart. 1993. Field evaluation of *Puccinia carduorum* for biological control of musk thistle. Biol. Control 3: 53-60.

Beck, K. G. 1999. Biennial thistles. Pages 145-61 *in* R. L. Sheley and J. K. Petroff, eds. Biology and Management of Noxious Rangeland Weeds. Oregon State Univ. Press, Corvallis.

Boldt, P. E., and G. Campobasso. 1978. Phytophagous insects on *Carduus macrocephalus* in Italy. Environ. Entomol. 7: 904-09.

Briese, D. T. 1989. A new biological control programme against thistles of the genus *Onopordum* in Australia. Pages 155-163 *in* E. S. Delfosse, ed. Proc. VII Int. Symp. Biol. Contr. Weeds, 6-11 March 1988, Rome, Italy. 1st. Sper. Patol. Veg. (MAF), Rome.

Gassmann, A., and L. T. Kok. 2002. Musk Thistle. Pages 229-45 *in* R. Van Driesche, S. Lyon, B. Blossey, M. Hoddle, and R. Reardon, eds. Biological Control of Invasive Plants in the Eastern United States. USDA Forest Serv. Pub. FHTET-2002-04. Morgantown, WV.

Gassmann, A., and L. T. Kok. 2002. Slenderflower Thistle. Pages 251-53 *in* R. Van Driesche, S. Lyon, B. Blossey, M. Hoddle, and R. Reardon, eds. Biological Control of Invasive Plants in the Eastern United States. USDA Forest Serv. Pub. FHTET-2002-04. Morgantown, WV.

Goeden, R. D. 1995. 65. Italian thistle, *Carduus pycnocephalus* L. (Asteraceae). Pages 242-44 *in* J. R. Nechols, L. A. Andres, J. W. Beardsley, R. D. Goeden, and C. G. Jackson, eds. Biological Control in the Western United States: Accomplishments and Benefits of Regional Research Project W-84, 1964-1989. Univ. Calif. Div. Agric. Nat. Res. Pub. 3361. Oakland, CA.

Goeden, R. D. 1995. 66. Milk thistle, *Silybum marianum* (L.) Gaertner (Asteraceae). Pages 245-47 *in* J. R. Nechols, L. A. Andres, J. W. Beardsley, R. D. Goeden, and C. G. Jackson, eds. Biological Control in the Western United States: Accomplishments and Benefits of Regional Research Project W-84, 1964-1989. Univ. Calif. Div. Agric. Nat. Res. Pub. 3361. Oakland, CA.

Kok, L. T., and A. Gassmann. 2002. Bull Thistle. Pages 247-50 *in* R. Van Driesche, S. Lyon, B. Blossey, M. Hoddle, and R. Reardon, eds. Biological Control of Invasive Plants in the Eastern United States. USDA Forest Serv. Pub. FHTET-2002-04. Morgantown, WV.

Kok, L. T., and A. Gassmann. 2002. Plumeless Thistle. Pages 255-61 *in* R. Van Driesche, S. Lyon, B. Blossey, M. Hoddle, and R. Reardon, eds. Biological Control of Invasive Plants in the Eastern United States. USDA Forest Serv. Pub. FHTET-2002-04. Morgantown, WV.

Kok, L. T., and J. T. Trumble. 1979. Establishment of *Ceuthorhynchidius horridus* (Coleoptera: Curculionidae), an imported thistle-feeding weevil in Virginia. Environ. Entomol. 8: 221-23.

Kok, L. T., R. G. Abad, and A. B. A. M. Baudoin. 1996. Effects of *Puccinia carduorum* on musk thistle herbivores. Biol. Control 6: 123-29.

Littlefield, J. L., W. L. Bruckart, D. G. Luster, P. W. Pratt, and V. L. Scogin. 1998. First report of musk thistle rust (*Puccinia carduorum*) in Oklahoma. Plant Dis. 82: 832.

Louda, S. M. 2000. Negative ecological effects of the musk thistle biological control agent, *Rhinocyllus conicus*. Pages 215-43 *in* P. A. Follett and J. J. Duan, eds. Nontarget Effects of Biological Control. Kluwer Academic Publishers, Boston, MA.

Louda, S. M., D. M. Kendall, J. Conner, and D. Simberloff. 1997. Ecological effects of an insect introduced for the biological control of weeds. Science 277: 1088-90.

McClay, A. S. 2002. Canada Thistle. Pages 217-28 *in* R. Van Driesche, S. Lyon, B. Blossey, M. Hoddle, and R. Reardon, eds. Biological Control of Invasive Plants in the Eastern United States. USDA Forest Serv. Pub. FHTET-2002-04. Morgantown, WV.

Möller-Joop, H., and D. Schroeder. 1986. *Urophora solstitialis* (L.) (Diptera: Tephritidae), a candidate for the biological control of plumeless thistle (*Carduus acanthoides* L.) in Canada. CIBC Report, Delémont, Switzerland.

Morishita, D. W. 1999. Canada thistle. Pages 162-74 *in* R. L. Sheley and J. K. Petroff, eds. Biology and Management of Noxious Rangeland Weeds. Oregon State Univ. Press, Corvallis.

Peschken, D. P., and R. W. Beecher. 1973. *Ceutorhynchus litura* (Coleoptera: Curculionidae): Biology and first releases for biological control of the weed Canada thistle (*Cirsium arvense*) in Ontario, Canada. Can. Entomol. 105: 1489-94.

Peschken, D. P., and P. Harris. 1975. Host specificity and biology of *Urophora cardui* (Diptera: Tephritidae). A biocontrol agent for Canada thistle (*Cirsium arvense*). Can. Entomol. 107: 1101-10.

Peschken, D. P., and A. T. S. Wilkinson. 1981. Biological control of Canada thistle (*Cirsium arvense*): Releases and effectiveness of *Ceutorhynchus litura* (Coleoptera: Curculionidae) in Canada. Can. Entomol. 113: 777-85.

Piper, G. L., and L. A. Andres. 1995. 63. Canada thistle, *Cirsium arvense* (L.) Scop. (Asteraceae). Pages 233-36 *in* J. R. Nechols, L. A. Andres, J. W. Beardsley, R. D. Goeden, and C. G. Jackson, eds. Biological Control in the Western United States: Accomplishments and Benefits of Regional Research Project W-84, 1964-1989. Univ. Calif. Div. Agric. Nat. Res. Pub. 3361. Oakland, CA.

Rees, N. E. 1990. Establishment, dispersal, and influence of *Ceutorhynchus litura* on Canada thistle (*Cirsium arvense*) in the Gallatin Valley of Montana. Weed Sci. 38: 198-200.

Rees, N. E. 1991. Biological control of thistles. Pages 264-73 *in* L. F. James, J. O. Evans, M. H. Ralphs, and R. D. Child, eds. Noxious Range Weeds. Westview Press, Boulder, CO.

Rizza, A., G. Campobasso, P. H. Dunn, and M. Stazi. 1988. *Cheilosia corydon* (Diptera: Syrphidae), a candidate for the biological control of musk thistle in North America. Ann. Entomol. Soc. Am. 81: 225-32.

Rosenthal, S. S., and G. L. Piper. 1995. 62. Bull thistle, *Cirsium vulgare* (Savi) Ten. (Asteraceae). Pages 231-32 *in* J. R. Nechols, L. A. Andres, J. W. Beardsley, R. D. Goeden, and C. G. Jackson, eds. Biological Control in the Western United States: Accomplishments and Benefits of Regional Research Project W-84, 1964-1989. Univ. Calif. Div. Agric. Nat. Res. Pub. 3361. Oakland, CA.

Turner, C. E., and J. C. Herr. 1996. Impact of *Rhinocyllus conicus* on a non-target, rare, native thistle (*Cirsium fontinale*) in California. Page 103 *in* V. C. Moran and J. H. Hoffman, eds. Proc. IX Int. Symp. Bio. Contr. Weeds, 19-26 January 1996, Stellenbosch, South Africa. Univ. Cape Town.

Turner, C. E., R. W. Pemberton, and S. S. Rosenthal. 1987. Host utilization of native *Cirsium* thistles (Asteraceae) by the introduced weevil *Rhinocyllus conicus* (Coleoptera: Curculionidae) in California. Environ. Entomol. 16: 111-15.

Woodburn, T. L. 1993. Host specificity testing, release and establishment of *Urophora solstitialis* (L.) (Diptera: Tephritidae), a potential biological control agent of *Carduus nutans* in Australia. Biocontrol Sci. Technol. 3: 419-26.

# Toadflax

## R.M. Nowierski

Dalmatian toadflax (*Linaria genistifolia* ssp. *dalmatica*) and yellow toadflax (*L. vulgaris*) are perennial forbs of European origin that have become naturalized in North America. They were introduced as ornamentals, but have since become serious weeds in certain soil and climatic zones. Dalmatian toadflax is a problem in uncultivated, summer-dry, coarse soils (rocky, gravelly, and sandy) in California, Colorado, Idaho, Montana, Oregon, Washington, and Wyoming, while yellow toadflax is more problematic in relatively summer-moist, coarse soils in the northwestern and northcentral United States. Consistent control has not been achieved for Dalmatian toadflax with herbicides or other alternative methods. For yellow toadflax, there currently is no effective means for control when it occurs in orchards, alfalfa, hay, and in strawberry fields, though it can be controlled in fields that are summer fallowed.

*Brachypterolus pulicarius* and *Gymnetron antirrhini* are two natural enemies of toadflax that were accidentally introduced into North America, apparently as they accompanied introductions of yellow and/or Dalmatian toadflax for ornamental purposes. Six other natural enemy species have gone through host specificity testing and have been approved by USDA-APHIS-PPQ for introduction including: *Calophasia lunula, Mecinus janthinus*, the Dalmatian toadflax-adapted strain of *G. antirrhini*, *G. linariae, Eteobalea intermediella*, and *E. serratella*. Permits can be obtained for all six of these insect species, as well as for *B. pulicarius*. Two additional natural enemies of toadflax are undergoing host specificity testing (as of 2003) by CABI Bioscience. They are the stem-galling weevils *G. hispidum* and *G. thapsicola*.

# Dalmatian toadflax

*Linaria genistifolia* ssp. *dalmatica*

## Snapdragon family—Scrophulariaceae

*R. M. Nowierski*

**Additional common name:** Broad-leaved toadflax.

**Native range:** Mediterranean regions of Europe and western Asia.

**Entry into the United States:** The plant was intentionally brought to North America as an ornamental in 1874.

## Biology

**Life duration/habit:** Dalmatian toadflax is a broad-leaved perennial herb adapted to cool, semiarid climates and coarse-textured soils. It is found most often on sparsely vegetated soils and degraded rangelands.

**Reproduction:** Reproduction occurs both by seed and secondary crown points along the lateral root system which may also develop floral stems.

**Roots:** The plant has a deeply penetrating taproot and lateral roots.

**Stems and leaves:** Stems of mature plants often grow from 0.3 to 0.9 m tall. Leaves are broad, alternate, and heart-shaped, with the upper leaves being conspicuously broad-based. Light green leaves clasp the stem and appear waxy in the spring.

**Flowers:** Flowers appear from midsummer until early fall and are borne in axils of the upper leaves. Flowers are two-lipped, 2 to 4 cm long, and have a long spur. The flowers are yellow with an orange, bearded throat.

**Fruits and seeds:** Dalmatian toadflax produces egg-shaped to nearly round fruit. Large plants are capable of producing half a million seeds that may remain viable in the soil for up to 10 years. Seeds are tiny, angular, approximately 1 mm in diameter, and have irregular wings.

## Infestations

**Worst infested states:** California, Colorado, Idaho, Montana, Oregon, Washington, and Wyoming.

Dalmation toadflax. (Photo credit: N. Poritz, bio-control.com)

Infestation of Dalmatian toadflax. (Photo credit: E. Coombs, Oregon Department of Agriculture)

**Habitat:** The plant grows along roadsides, in pastures, idle land, and rangeland, particularly in sites with coarse-textured soils.

**Impacts:** Although Dalmatian toadflax seedlings are poor competitors for soil moisture with established perennials and fast-maturing winter annuals, established plants may be extremely competitive and substantially affect the yearly composition of annual vegetation, the production of other perennial herbs, and the recruitment of toadflax seedlings. The survival of new seedlings, combined with the yearly extension of lateral roots and production of floral stems from these roots, allows Dalmatian toadflax to persist in areas of low plant competition.

Dalmatian toadflax and other *Linaria* species reportedly are toxic to livestock because they contain a glucoside antirrhinoside, a quinoline alkaloid, and peganine. Because the plant is not readily used as forage, the productivity of infested rangelands is reduced.

### Comments

The extensive, well-developed root system, waxy leaves, and typical association with coarse-textured soils have made chemical control of this weed very difficult and often inconsistent.

# Yellow toadflax

*Linaria vulgaris*

Snapdragon family—Scrophulariaceae

*R. M. Nowierski*

**Additional common names:** Butter-and-eggs, common toadflax.

**Native range:** Eurasia.

**Entry into the United States:** Yellow toadflax was brought to North America in the mid-1800s as an ornamental and is now widespread across the United States.

Yellow toadflax flowers. (Photo credit: Montana State University)

Infestation of yellow toadflax. (Photo credit: E. Coombs, Oregon Department of Agriculture)

## Biology

**Life duration/habit:** This aggressive, creeping perennial weed frequents roadsides and pastures and is an increasing problem in cultivated crops such as strawberry and mint. Unlike Dalmatian toadflax, which prefers dry, coarse-textured soils, yellow toadflax is more commonly associated with moist, coarse soils.

**Reproduction:** The plant reproduces by seed and creeping rhizomes.

**Roots:** Like Dalmatian toadflax, yellow toadflax has an extensive, well-developed root system with vegetative root buds.

**Stems and leaves:** Stems are generally 0.3 to 0.6 m tall. Leaves are about 5.5 cm long, narrow, and pointed at both ends. They are pale green, numerous, and individually connected to the central stem.

**Flowers:** The flowers are quite similar to those of Dalmatian toadflax. Flowers are yellow, 2.5 cm long with a bearded, orange throat. They possess a spur-like appendage.

**Fruits and seeds:** Yellow toadflax, like Dalmatian toadflax, produces egg-shaped to nearly round fruit. Each fruit is about 5 mm in diameter and contains two compartments and many seeds. Seeds are flattened, approximately 1 mm in diameter, and have papery margins. More than 5,500 seeds have been recorded from a single stem of yellow toadflax. Seed production ranges from 800 to 35,000 seeds per plant.

## Infestations

**Worst infested states:** Idaho, Montana, Oregon, and Washington.

**Habitat:** Roadsides, edges of fields, waste areas, clearcuts, pastures, and cultivated fields.

**Impacts:** The leaves are reputed to have a disagreeable odor and are seldom eaten by livestock.

## Comments

Yellow toadflax is presently found throughout the continental United States, Canada, and Mexico.


# *Brachypterolus pulicarius*

R. M. Nowierski

**Common name:** Toadflax flower-feeding beetle.

**Type of agent:** Insect: Beetle (Coleoptera: Nitidulidae).

**Native distribution:** Europe.

**Original source:** Europe.

## Biology

**Generations per year:** This beetle typically has one generation per year, although two generations per year have been reported in Germany.

**Overwintering stage:** Pupal.

**Egg stage:** Eggs are deposited in the toadflax flowers.

**Larval stage:** Larvae reportedly feed on pollen, anthers, ovaries, and maturing seeds of toadflax which reduces pollination success and seed production.

**Pupal stage:** Pupation occurs in the soil beneath the host plant.

**Adult stage:** The adults emerge in late May. They are small, black, oval beetles about 2 mm long.

Adult *B. pulicarius*. (Photo credit: R. Richard, USDA-APHIS)

## Effect

**Destructive stages:** Larval and adult.

**Plant species attacked:** Yellow toadflax and Dalmatian toadflax.

**Site of attack:** Adults attack the young, succulent shoot tips and some of the reproductive parts. Larvae feed on pollen, anthers, ovaries, and maturing seeds.

**Impact on the host:** Adult feeding on the young shoots causes increased branching of the plants. Larval feeding on the reproductive parts of the plants and the inhibition of early-season flowering by the adults and larvae reduce seed production.

**Nontarget effects:** No nontarget impacts have been reported.

## Releases

**First introduced into the United States:** The beetle was accidentally introduced into New York in 1919. Collections of the Dalmatian toadflax-adapted strain of this beetle have been made in Canada a number of times since 1992 and releases of Canadian material have been made at numerous sites in Montana and sites in Arizona, Idaho, Nevada, and Wyoming.

**Established in:** Colorado, Idaho, Montana, New York, Oregon, Washington, and Wyoming.

**Habitat:** This beetle appears to be well established at most major yellow toadflax infestations in North America. It is found less frequently on Dalmatian toadflax.

**Availability:** These beetles usually can be collected wherever yellow toadflax is found; the beetle is most likely already present at most yellow toadflax sites.

**Stage to transfer:** Adult.

**Redistribution:** Collect with a sweep net or aspirator, sort, and transport.

## Comments

Although the impact of this beetle on Dalmatian toadflax is still unclear, previous studies have shown that *B. pulicarius* can reduce seed production of yellow toadflax by 80 to 90%. It is hoped that the Dalmatian toadflax-adapted strain of this beetle will similarly reduce seed production.

# *Calophasia lunula*

R. M. Nowierski

**Common name:** Toadflax moth.

**Type of agent:** Insect: Moth (Lepidoptera: Noctuidae).

**Native distribution:** Eurasia.

**Original source:** Europe.

## Biology

**Generations per year:** One to three.

**Overwintering stage:** Pupal.

**Egg stage:** The eggs are laid singly on foliage and flowers. They are off-white to pale yellow, strongly ribbed all around from the base to the top, and are about 1 mm in diameter. Fertilized eggs typically turn reddish-brown with age. Hatching occurs within seven to 11 days.

**Larval stage:** Larvae progress through five molts in approximately one month and increase in size to about 4 cm long when they reach the fifth larval growth stage. Newly hatched caterpillars are gray-black and about 5 mm long. Second through fifth growth stage larvae are brightly colored with white spots and black and yellow parallel stripes.

**Pupal stage:** Pupation occurs within a cocoon constructed from chewed leaves, litter, or soil debris. The cocoon is found on the soil surface or attached to the lower stem. The pupa within the cocoon is golden to reddish-brown.

**Adult stage:** The adult stage lasts from a few days to nearly three weeks in captivity. A typical adult female lays from 30 to 80 eggs. Adult *C. lunula* are nondescript gray moths that measure 11 to 13.4 mm long. They feed on the nectar of Dalmatian toadflax, yellow toadflax, and other plant species.

*C. lunula* larva. (Photo credit: N. Poritz, bio-control.com)

## Effect

**Destructive stage:** Larval.

**Plant species attacked:** Dalmatian toadflax and yellow toadflax.

**Site of attack:** Larvae prefer to feed on new vegetative shoots with tender leaves, but will also consume terminal portions of stems, flowers, and older foliage.

**Impact on the host:** Defoliation of toadflax plants can be quite spectacular in localized areas once this biocontrol agent becomes abundant. However, the level of damage is inconsistent from one year to the next. Defoliation possibly reduces root reserves and the general vigor of the toadflax in the year following attack. Seed production is also decreased when buds and flowers are eaten.

**Nontarget effects:** No nontarget impacts have been reported.

### Releases

**First introduced into the United States:** Multiple releases of this insect were made against Dalmatian toadflax in Idaho, Oregon, Montana, Washington, and Wyoming from the 1960s through the 1980s. Releases were made in Arizona, Colorado, Idaho, Nevada, and Wyoming in the 1990s.

**Established in:** Idaho, Montana, and Washington.

**Habitat:** Dry, xeric sites with coarse-textured soils are preferred. The caterpillars of this moth are typically found in relatively low numbers scattered across relatively isolated toadflax plants.

*C. lunula* adult. (Photo credit: Montana State University)

**Availability:** Availability of the moth is limited in Idaho and Montana, but the insect is readily obtainable in Washington.

**Stage to transfer:** Larval.

**Redistribution:** Larvae collected from stems can be stored in paperboard containers provisioned with cut stems and leaves of toadflax until the larvae can be transported to other locations for release.

### Comments

As of 2003, the moth is established at a few sites in Idaho and western Montana, and is widely distributed throughout northeastern Washington. The lack of establishment of the moth at higher elevations and results of cold-tolerance studies suggest that this insect may be adversely affected by cold temperatures. Hence, warmer release sites may be more conducive to *Calophasia* establishment and survival. Releases of *Calophasia* should not be made at toadflax sites that have high ant activity.

## *Eteobalea intermediella*

R. M. Nowierski

**Common name:** Toadflax root-boring moth.

**Type of agent:** Insect: Moth (Lepidoptera: Cosmopterygidae).

**Native distribution:** Mediterranean region extending into central Europe.

**Original sources:** Croatia, Italy, and Slovenia.

### Biology

**Generations per year:** Two or more.

**Overwintering stage:** Larval (in the root).

**Egg stage:** Eggs are deposited in the leaf axils or at the base of stems and are laid from June to August in Italy. Eggs are reticulate with a sticky substance on the egg covering and are approximately 0.3 by 0.5 mm in size. Initially the eggs are white and change to a cream color as development proceeds.

**Larval stage:** Larval mining occurs mainly in the root crown area. Larvae are cream-colored with brown head capsules.

**Pupal stage:** Pupation occurs within the root.

**Adult stage:** It appears that mating and egg-laying may occur during twilight periods. The adult stage lasts about two weeks in the field and up to four weeks under laboratory conditions. The adults are small, 8-mm-long, black moths with white speckles.

## Effect
**Destructive stage:** Larval.

**Plant species attacked:** Dalmatian toadflax and yellow toadflax.

**Site of attack:** Roots.

**Impact on the host:** Larval feeding substantially damages the root system. Tillers of attacked plants break easily under dry conditions. From this point, no regrowth occurs though under ideal conditions the plant can produce new shoots from below the damaged areas.

**Nontarget effects:** No nontarget impacts have been reported.

## Releases
**First introduced into the United States:** 1996, Montana.

**Established in:** Efforts to establish this insect in the United States have been unsuccessful as of 2003.

**Habitat:** The optimum habitat for the moth is undetermined.

**Availability:** The moth is unavailable.

**Stages to transfer:** Possibly pupal or adult.

**Redistribution:** Redistribution probably requires techniques and conditions similar to other lepidopteran larvae and adults for collection, packaging, and release.

*E. intermediella* adult. (Photo credit: Montana State University)

387

## Comments

Studies by the International Institute of Biological Control (now the Commonwealth Agricultural Bureau International [CABI Bioscience]) on *Eteobalea intermediella* and *E. serratella* showed no obvious differences between the two species in the lifespan of the adults, sex ratio, and the number of eggs deposited. The main difference between the two insects is that *E. intermediella* has at least two generations per year while *E. serratella* has only one. Furthermore, the two moth species differ in the types of eggs produced and in their egg-laying behaviors. The egg covering of *E. intermediella* is reticulate, relatively thin, and covered with a sticky substance, while that of *E. serratella* is striate, relatively thick, and has a dry surface. Although both species may lay eggs at the base of stems, *E. serratella* may also lay eggs on or in the soil. CABI Bioscience is conducting studies to develop better rearing techniques for *E. intermediella* and *E. serratella* to improve their chances for establishment in North America.

## *Eteobalea serratella*

R. M. Nowierski

**Common name:** Yellow toadflax root-boring moth.

**Type of agent:** Insect: Moth (Lepidoptera: Cosmopterygidae).

**Native distribution:** Northern Eurasia.

**Original sources:** Croatia, Italy, and Slovenia.

### Biology

**Generations per year:** One.

**Overwintering stage:** Larval (in the root).

**Egg stage:** Eggs are generally deposited at the base of stems, on the soil surface, or in the soil. Eggs are generally striate and dry. Hatching occurs in nine to 10 days at 25° C.

**Larval stage:** Larval mining occurs mainly in the root crown area. Larvae are cream-colored with brown head capsules.

**Pupal stage:** Pupation occurs within the root.

**Adult stage:** It appears that mating and egg-laying may occur during twilight periods. Adults are small black moths with white speckles.

### Effect

**Destructive stage:** Larval.

**Plant species attacked:** Yellow toadflax.

**Site of attack:** Roots.

**Impact on the host:** Larval feeding substantially damages the root system. Tillers of attacked plants break very easily under dry conditions. From this point, no regrowth occurs, though under ideal conditions the plant can produce new shoots below the damaged sections.

**Nontarget effects:** No nontarget impacts have been reported.

## Releases
**First introduced into the United States:** 1996, Montana.

**Established in:** Attempts to establish this insect in the United States and Canada have been unsuccessful as of 2003.

**Habitat:** The optimum habitat for this insect is undetermined.

**Availability:** This insect is unavailable.

**Stages to transfer:** Possibly pupal or adult.

**Redistribution:** Redistribution probably requires techniques and conditions similar to other lepidopteran larvae and adults for collection, packaging, and release.

## Comments
Studies by the International Institute of Biological Control (now the Commonwealth Agricultural Bureau International [CABI Bioscience]) on *Eteobalea intermediella* and *E. serratella* showed no obvious differences between the two species in the lifespan of the adults, sex ratio, and the number of eggs deposited. The main difference between the two insects is that *E. intermediella* has at least two generations per year while *E. serratella* has only one. Furthermore, the two moth species differ in the types of eggs produced and in their egg-laying behaviors. The egg covering of *E. intermediella* is reticulate, relatively thin, and covered with a sticky substance, while that of *E. serratella* is striate, relatively thick, and has a dry surface. Although both species may lay eggs at the base of stems, *E. serratella* may also lay eggs on or in the soil. CABI Bioscience is conducting studies to develop better rearing techniques for *E. serratella* and *E. intermediella* to improve their chances for establishment in North America.

# *Gymnetron antirrhini*

R. M. Nowierski

**Common name:** Toadflax seed capsule weevil.

**Type of agent:** Insect: Beetle, weevil (Coleoptera: Curculionidae).

**Native distribution:** Eurasia.

**Original source:** Unknown.

## Biology
**Generations per year:** One.

**Overwintering stage:** The adults overwinter beneath debris or in old toadflax seed capsules where they developed the previous autumn.

389

*G. antirrhini* larvae in seed capsules. (Photo credit: E. Coombs, Oregon Department of Agriculture)

**Egg stage:** Eggs are laid in the ovaries of toadflax plants during flowering. Seeds surrounding the weevil eggs become swollen and have a pale, watery appearance.

**Larval stage:** The weevil has three larval growth stages. Larvae initially feed on the swollen seeds, but the last larval stage also may consume normal seeds.

**Pupal stage:** The mature larvae construct oval cells within the seed capsules, where pupation occurs.

**Adult stage:** Adults are small, 2.5-mm-long, gray weevils that feed on the young shoots of toadflax in May and early June.

## Effect

**Destructive stages:** Larval, and adult to a lesser extent.

**Plant species attacked:** Yellow toadflax. A strain of *G. antirrhini* adapted to Dalmatian toadflax was approved for release in the United States and first released in Montana in 1996.

**Site of attack:** Immature seeds inside the seed capsules are destroyed.

**Impact on the host:** Substantial damage to shoots and flowers of yellow toadflax may occur from weevil attack. These weevils can reduce seed production in yellow toadflax by 85 to 90% (from observations of weevil impact in Washington). However, in Canada seed reductions of 20 to 25% are more typical.

**Nontarget effects:** No nontarget impacts have been reported.

*G. antirrhini* adults. (Photo credit: E. Coombs, Oregon Department of Agriculture)

## Releases

**First introduced into the United States:** The weevil was first recorded in Massachusetts in 1909. The Dalmatian toadflax-adapted strain of this weevil was first released in Montana in 1996.

**Established in:** Idaho, Montana, Oregon, Washington, and Wyoming (yellow toadflax strain). The weevil is well established where yellow toadflax occurs in the northwestern and northeastern United States.

**Habitat:** The optimum habitat for the weevil is unknown.

**Availability:** These weevils are generally collectible wherever significant infestations of yellow toadflax occur.

**Stage to transfer:** Adult.

**Redistribution:** The weevil is widespread and usually common at most major yellow toadflax sites already, so redistribution may be unnecessary.

## Comments

A small release of the Dalmatian toadflax-adapted strain of this weevil occurred in Wyoming in 1998. As of 2003, it is not known whether this weevil strain established on Dalmatian toadflax in either Montana or Wyoming.

# *Gymnetron linariae*

R. M. Nowierski

**Common name:** Toadflax root-galling weevil.

**Type of agent:** Insect: Beetle, weevil (Coleoptera: Curculionidae).

**Native distribution:** Eurasia.

**Original source:** Upper Rhine Valley in Germany.

## Biology

**Generations per year:** One.

**Overwintering stage:** Adult.

**Egg stage:** Eggs are laid singly in pockets chewed in the root tissue by the adult female and then are covered with excrement. Eggs are pale yellow with a smooth surface, pear-like in shape, and measure about 0.39 mm long and 0.22 mm at their widest. The majority of eggs are laid in the root crown.

**Larval stage:** Larvae develop in galls on the roots and rhizomes of a few *Linaria* species. Larval development and pupation are completed in approximately eight to 10 weeks.

*G. linariae* adult. (Photo credit: CABI Bioscience, Switzerland Center)

391

**Pupal stage:** Pupation occurs within the galls.

**Adult stage:** Adults appear in spring and begin laying eggs after feeding approximately three weeks on toadflax shoots.

## Effect

**Destructive stages:** Larval and adult.

**Plant species attacked:** Dalmatian toadflax and yellow toadflax.

**Site of attack:** Adults attack the shoots, while larvae develop in galls formed on the roots and rhizomes of the two toadflax species.

**Impact on the host:** Impact of adult feeding is not known. It is anticipated that gall formation may affect growth and reproduction through the diversion of nutrients and biomass to gall production.

**Nontarget effects:** No nontarget impacts have been reported.

## Releases

**First introduced into the United States:** 1996, Montana.

**Established in:** Wyoming and possibly Colorado.

**Preferred habitat:** According to European records, *Gymnetron linariae* is common in grassland habitats.

**Availability:** This insect is unavailable as of 2003.

**Stage to transfer:** Adult.

**Redistribution:** Adults can be collected with a sweep net while feeding on shoots, then sorted and shipped in cartons with shoots and packing material.

## Comments

The only definitive record of this weevil establishing in the United States is at a single Dalmatian toadflax site in Wyoming. Adult weevils were first released inside a cage from which all the old vegetation had been removed. The release site was open and sunny.

# *Mecinus janthinus*

R. M. Nowierski

**Common name:** Toadflax stem weevil.

**Type of agent:** Insect: Beetle, weevil (Coleoptera: Curculionidae).

**Native distribution:** Central and southern Europe and Turkmenistan, Uzbekistan, and Kazakhstan.

**Original sources:** Croatia, Italy, and Slovenia.

## Biology

**Generations per year:** One.

**Overwintering stage:** Adult (in stems inside pupal cell).

**Egg stage:** Eggs are deposited inside cavities chewed in the shoots by the females and covered with what is assumed to be chewed plant material. Eggs are laid from June to mid-July. They are oval and 0.6 by 0.65 mm. The incubation period is six to seven days.

**Larval stage:** Larvae develop successfully only in shoots with a diameter of more than 0.9 mm, although the weevils may lay eggs in smaller shoots. Larval development takes between 23 and 34 days. The larvae are C-shaped and white with pale brown head capsules.

**Pupal stage:** Pupation occurs within the larval mine. The pupae are 3 to 4.5 mm long and white at first but gradually turning black.

**Adult stage:** Adults emerge from overwintering in the stems in May, feed, copulate, and lay eggs from the end of May to mid-July. The adults are small bluish-black weevils that appear somewhat elongated. They are 3.6 to 4 mm long.

## Effect

**Destructive stages:** Larval and adult.

**Plant species attacked:** Dalmatian toadflax and yellow toadflax.

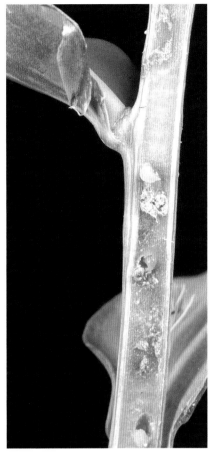

*M. janthinus* larvae in a toadflax stem. (Photo credit: N. Poritz, bio-control.com)

**Site of attack:** Adults feed on leaves and stems; larvae mine the stems.

**Impact on the host:** Adult feeding on the leaves and stems of toadflax apparently has a limited effect under field conditions in Europe. However, mining of the stems by the larvae causes premature wilting of shoots and suppresses flower formation, particularly under conditions of high weevil density and cases of multiple attack. Effects of the weevil on the plant reportedly are enhanced under drought stress.

**Nontarget effects:** No nontarget impacts have been reported.

## Releases

**First introduced into the United States:** 1996, Montana. Adventive movement of the weevil from British Columbia into Washington has occurred.

**Established in:** Colorado, Idaho, Montana, Oregon, Washington, and Wyoming.

**Habitat:** According to Canadian researchers, the weevil prefers hot, dry, forested areas or grasslands with large-stemmed Dalmatian toadflax plants.

393

**Availability:** The availability of this weevil in the United States is somewhat limited as of 2003, but it should become more widely available in the future.

**Stage to transfer:** Adult.

**Redistribution:** Adults can be collected using a sweep net while they are actively feeding and ovipositing. They may be shipped in paperboard containers with fresh toadflax shoots.

## Comments

This weevil has great potential for damaging Dalmatian toadflax, and has significantly reduced plant densities at a number of sites in British Columbia, Canada, and in Washington. The Canadians report that overwintering survival has been greatest at more southern sites in British Columbia or sites that maintain significant snow

*M. janthinus* adult. (Photo credit: N. Poritz, bio-control.com)

cover over the course of the winter. Several sites in the northwestern United States are also beginning to show significant depression of toadflax at *Mecinus* release sites.

## References

DeClerck-Floate, R., and V. Miller. 2002. Overwintering mortality of and host attack by the stem-boring weevil, *Mecinus janthinus* Germar, on Dalmatian toadflax (*Linaria dalmatica* (L.) Mill.) in western Canada. Biol. Control 24: 65-74.

Grieshop, M. J., and R. M. Nowierski. 2002. Selected factors affecting seedling recruitment of Dalmatian toadflax. J. Range Manage. 55: 612-19.

Grubb, R. T., R. M. Nowierski, and R. L. Sheley. 2002. Effects of *Brachypterolus pulicarius* (L.) (Coleoptera: Nitidulidae) on growth and seed production of Dalmatian toadflax, *Linaria genistifolia* ssp. *dalmatica* (L.) Maire and Petitmengin (Scrophulariaceae). Biol. Control 23: 107-14.

Harris, P. 1963. Host specificity of *Calophasia lunula* (Hufn.) (Lepidoptera: Noctuidae). Can. Entomol. 95: 101-5.

Harris, P. 1984. *Linaria vulgaris* Miller, yellow toadflax and *L. dalmatica* (L.) Mill., broad-leaved toadflax (Scrophulariaceae). Pages 179-82 *in* J. S. Kelleher and M. A. Hulme, eds. Biological Control Programmes Against Insects and Weeds in Canada 1969-1980. CAB, Farnham Royal, Slough, England.

Jeanneret, P., and D. Schroeder. 1992. Biology and host specificity of *Mecinus janthinus* Germar (Col.: Curculionidae), a candidate for the biological control of yellow and Dalmatian toadflax, *Linaria vulgaris* (L.) Mill. and *Linaria dalmatica* (L.) Mill. (Scrophulariaceae) in North America. Biocontrol Sci. Technol. 2: 25-34.

Lajeunesse, S. 1999. Dalmatian and yellow toadflax. Pages 202-16 *in* R. L. Sheley and J.K. Petroff, eds. Biology and Management of Noxious Rangeland Weeds. Oregon State Univ. Press, Corvallis.

Lajeunesse, S. E., P. K. Fay, J. R. Lacey, R. M. Nowierski, D. Zamora, and D. Cooksey. 1993. Dalmatian toadflax – a weed of pasture and rangeland: Identification, biology, and management. Montana Agric. Ext. Bull. 115.

McClay, A. S., and R. B. Hughes. 1995. Effects of temperature on developmental rate, distribution, and establishment of *Calophasia lunula* (Lepidoptera: Noctuidae), a biocontrol agent for toadflax (*Linaria* spp.). Biol. Control 5: 368-77.

Nowierski, R. M. 1995. 82. Dalmatian toadflax, *Linaria genistifolia* ssp. *dalmatica* (L.) Maire and Petitmengin (Scrophulariaceae). Pages 312-17 *in* J. R. Nechols, L. A. Andres, J. W. Beardsley, R. D. Goeden, and C. G. Jackson, eds. Biological Control in the Western United States: Accomplishments and Benefits of Regional Research Project W-84 (1964-1989). Univ. Calif. Div. Agric. Nat. Res. Pub. 3361. Oakland, CA.

Robocker, W. C. 1974. Life history, ecology, and control of Dalmatian toadflax. Wash. Agric. Expt. Sta. Tech. Bull. 79.

Vujnovic, K., and R. W. Wein. 1997. The biology of Canadian weeds. 106. *Linaria dalmatica* (L.) Mill. Can. J. Plant Sci. 77: 483-91.

# Tropical soda apple

*Solanum viarum*

Nightshade family—Solanaceae

*J. P. Cuda, N. C. Coile, D. Gandolfo, J. C. Medal, and J. J. Mullahey*

**Additional common names:** Yu-a, tutia de vibora (Argentina); joa bravo, joa amarelo pequeno (Brazil); sodom apple.

**Native range:** Argentina, Brazil, Paraguay, and Uruguay.

**Entry into the United States:** Initial introduction of tropical soda apple into North America probably occurred from seed adhering to shoes or it escaped from cultivation. The earliest record is from Florida in 1988.

## Biology

**Life duration/habit:** Tropical soda apple is an herbaceous, prickly, perennial plant that persists in areas with mild winters. Aboveground plant parts are sensitive to cold temperatures and will die back during the winter months in infested areas in the southeastern United States that experience freezing temperatures.

**Reproduction:** The plant reproduces mainly by seed but also is capable of regenerating vegetatively from its extensive root system.

**Roots:** Root buds regenerate new shoots. The root system can be extensive with feeder roots 0.6 to 2.5 cm in diameter located a few centimeters below ground and extending 0.9 to 1.8 m from the crown of the plant.

**Stems and leaves:** The stems of tropical soda apple are sturdy and have scattered, small, hooked prickles. The simple, lobed, alternate leaves are up to 20 cm long and 15 cm wide. They are covered with fine, soft hairs that give a velvety sheen to the leaves. Rigid, yellowish prickles up to 20 mm long are scattered along the midvein and the secondary veins on both surfaces of leaf blades and are more concentrated on the petiole.

Tropical soda apple plant. (Photo credit: J.Mullahey, University of Florida - IFAS)

**Flowers:** Flowers and fruits are produced primarily from September through May in the United States with few fruits developing during the summer months. Flowers are in small terminal clusters. The calyx is five-lobed with tiny prickles on the surface, corollas are white with five recurved petals, and cream-colored anthers and yellow stamens surround a single pistil.

**Fruits and seeds:** A single plant produces about 150 fruits per year. Each mature fruit contains about 400 seeds, and one plant can produce an average of 45,000 seeds with a germination rate of 70%. In one growing season, a single plant can yield enough viable seed to produce 28,000 to 35,000 new plants. Fruits are glabrous, globular, and about 2 to 3 cm in diameter. Immature fruits are pale green with dark green veining and resemble tiny striped watermelons. Mature fruits are dull, medium yellow, with a leathery skin. The pulp band is narrow, pale green, and mucilaginous. Seeds are numerous, light reddish-brown, moderately flattened, about 2.5 mm in diameter, and are surrounded by a mucilaginous layer that contains the glycoalkaloid solasodine.

## Infestations

**Worst infested states:** Alabama, Florida, Georgia, Louisiana, and Mississippi. It also has been reported from North and South Carolina, Pennsylvania, Tennessee, and Puerto Rico.

**Habitat:** Tropical soda apple appears to be restricted to semidisturbed sites, and is mainly a weed of pastures and rangeland. In Florida, the highest incidence has occurred in improved pastures, such as Bahia grass (*Paspalum notatum*). It also invades hammocks, ditch banks, citrus groves, sugarcane, vegetable, and watermelon fields, and roadsides. In south Florida, tropical soda apple typically is found in nearly level, somewhat poorly drained, sandy soil. Cattle and wildlife such as raccoons, deer, and feral pigs feed on the fruits and are vectors for spreading the seed through defecation. Tropical soda apple populations have been observed to increase rapidly following extended periods of dry weather followed by several years of normal rainfall. Seeds can survive in dry soil moisture conditions for two or more years. Moving water, seed-contaminated hay, grass seed, sod, and machinery also contribute to spreading the plant.

Tropical soda apple infestation in a south Florida cattle pasture. (Photo credit: J. Mullahey, University of Florida - IFAS)

397

**Impacts:** Because tropical soda apple typically invades improved pastures, livestock carrying capacity is severely affected. Infestations in Florida have been estimated at 405,000 ha. The foliage and stems are unpalatable to livestock, and dense stands of this prickly shrub deny cattle access to shaded areas, which results in heat stress. Pasture production declines and stocking rates are drastically reduced if tropical soda apple is left uncontrolled. Bahia grass does not tolerate shade well and its productivity declines when forced to compete with tropical soda apple. In 1993, production losses to Florida cattle ranchers attributed to tropical soda apple infestations were estimated at $11 million annually, or about 1% of total Florida beef sales. Economic losses from heat stress alone approached $2 million.

In addition to causing economic problems, tropical soda apple reduces the biological diversity in natural areas by displacing native plants and disrupting the ecological integrity of the natural system. Wooded areas (oak hammocks, cabbage palm hammocks, cypress heads) comprise about 10% of the total land infested by tropical soda apple in Florida. When the plant is allowed to cover large areas, wildlife habitat also declines. Prickles on the plants create a physical barrier to animals, preventing them from passing through the infested area. Tropical soda apple also interferes with restoration efforts in Florida by invading tracts of land that are reclaimed following phosphate-mining operations.

## Comments

Tropical soda apple was placed on the Florida Noxious Weed List in 1994 and the Federal Noxious Weed List in 1995, and is listed as one of the most invasive species in Florida (Category I) by the Florida Exotic Pest Plant Council.

Tropical soda apple contains the glycoalkaloid solasodine that is found in the mucilaginous layer surrounding the seeds. Solasodine, a nitrogen analogue of diosgenin, is used in the production of steroid hormones. These steroids have been useful in the treatment of cancer, Addison's disease, rheumatic arthritis, and in the production of contraceptives. Maximum content of solasodine in tropical soda apple fruits occurs when the fruits change color from green to yellow. Although the plant was intensively cultivated as a source of solasodine in Mexico and India, propagation for this purpose has significantly declined or ceased altogether. Solasodine is poisonous to humans. Symptoms of poisoning appear after consumption of 10 fruits, and a lethal dose would require approximately 200 fruits.

Tropical soda apple also serves as a reservoir for various diseases and insect pests of solanaceous crop plants. At least six plant viruses (cucumber mosaic virus, potato leaf-roll virus, potato virus Y, tobacco etch virus, tomato mosaic virus, and tomato mottle virus) and the potato early blight fungus *Alternaria solani* use tropical soda apple as a host and are vectored during the growing season to cultivated crops. In addition, the following major crop pests use tropical soda apple as an alternate host: the tobacco hornworm (*Manduca sexta*), the tomato hornworm (*M. quinquemaculata*), the Colorado potato beetle (*Leptinotarsa decemlineata*), the tobacco budworm (*Helicoverpa virescens*), the tomato pinworm (*Keiferia lycopersicella*), the green peach aphid (*Myzus persicae*), the sweetpotato whitefly (*Bemisia tabaci*), the silverleaf whitefly (*B. argentifolii*), the soybean looper (*Pseudoplusia includens*), and the southern green stink bug (*Nezara viridula*).

Two promising candidates for biological control of this species are the flower bud weevil *Anthonomus tenebrosus* (Coleoptera: Curculionidae) and another leaf beetle, *Platyphora* sp. (Coleoptera: Chrysomelidae). *A. tenebrosus* feeds and usually lays a single egg inside a flower bud of tropical soda apple. Feeding damage by the developing larva causes the flower bud to abort, inhibiting fruit production. *Platyphora* sp. is a nocturnal leaf-feeder capable of defoliating tropical soda apple. Females exhibit an unusual form of reproduction: instead of laying eggs, they deposit second instar larvae directly on the leaves of tropical soda apple. Biological and host range studies with these two insects are being conducted by the University of Florida in Gainesville and United States Department of Agriculture Agricultural Research Service (USDA-ARS) in Argentina.

## *Gratiana boliviana*

J. C. Medal, J. P. Cuda, and D. Gandolfo

**Common name:** Tropical soda apple leaf beetle.

**Type of agent:** Insect: Beetle (Coleoptera: Chrysomelidae).

**Native distribution:** South America (Argentina, Brazil).

**Original source:** Argentina.

### Biology

**Generations per year:** Unknown, but multiple generations are expected because the life cycle of the beetle can be completed in 26 to 31 days.

**Overwintering stage:** Probably adult, if it can survive winter temperatures. Reproductive diapause may be induced by plant senescence in northern parts of tropical soda apple's distribution in the southeastern United States.

**Egg stage:** Eggs are cylindrical, 1.3 mm long by 0.6 mm wide, and are laid individually. They are white initially but turn light green during incubation. Each egg is deposited with the long axis parallel to the leaf surface and is encased in two translucent brown membranes. The egg stage lasts five to seven days at 25° C.

**Larval stage:** Newly hatched larvae are cream-colored. There are five instars and the larvae usually feed on the underside of younger leaves. The larval stage is completed in 15 to 22 days. Like most tortoise beetles, larvae have an anal fork that arises from the last abdominal segment; they carry the cast skins and feces on this posterior process to deter potential predators. The anal fork is cream-colored in the first four instars, but turns dark brown to black in the fifth instar.

**Pupal stage:** After mature larvae stop feeding, they attach themselves with a secretion from the last abdominal segment to the underside of the leaves near the insertion of the petiole to pupate. Pupae are green, dorsoventrally flattened, and the first thoracic segment is expanded laterally. They are 6.0 mm long by 3.9 mm wide and flex their bodies when disturbed. The pupal stage usually lasts six to seven days.

*G. boliviana* larvae and feeding damage on a tropical soda apple leaf. (Photo credit: J. Lotz, DPI)

**Adult stage:** Adults of *G. boliviana* are elliptical and light green. The expanded areas of the pronotum and elytra have small, rounded, depressed areas without pigmentation in the center and are turquoise along the margins. There is a well-defined yellow band above the expanded areas of the elytra, and the rest of the elytra is light green with an irregularly defined yellowish area. Sexes can be distinguished by examining the underside of the abdomen three to five days after adult emergence. The paired internal sex organs are visible through the translucent exoskeleton; in males, the sex organs are round and orange-colored whereas in females they are white. The preoviposition period is approximately 12 days. Females live about 12 weeks and deposit approximately 300 eggs on tropical soda apple leaves and petioles, peaking at six eggs/female/day between the third and seventh week.

## Effect

**Destructive stages:** Larval and adult.

**Additional plant species attacked:** Sticky nightshade (*Solanum sisymbriifolium*), another invasive weed also introduced from South America.

**Site of attack:** Leaves.

**Impact on the host:** Larvae usually feed in the upper third of the plant. Feeding by larvae and adults damages the leaves of tropical soda apple, and at high densities the insect can completely defoliate the plant. Loss of photosynthetic tissue inhibits the vigor and growth of tropical soda apple and may reduce the competitive advantage of this invasive weed over more desirable native and improved pasture vegetation.

*G. boliviana* adult. (Photo credit: J. Lotz, DPI)

**Nontarget effects:** No nontarget effects have been reported.

## Releases

**First introduced into the United States:** 2003, Florida.

**Established in:** Florida.

**Habitats:** The insect prefers plants growing in areas receiving full sunlight.

**Availability:** This beetle was not available for redistribution as of 2003.

**Stages to transfer:** Adult and larval.

**Redistribution:** Although adults and larvae may be collected with a heavy-duty sweep net, the extensive prickles on tropical soda apple plants are a deterrent to sweeping the plants. Alternative approaches are to aspirate the adults into vials from infested plants and release the adults at other sites, or inoculate potted tropical soda apple plants with the insect and transport plants infested with all life stages of the insect to new release sites.

This is Florida Agricultural Experiment Station Journal Series Number N-02422.

## *References*

Coile, N. C. 1993. Tropical soda apple, *Solanum viarum* Dunal: The plant from hell. Florida Bot. Circ. No. 27. Florida Dept. Agric. Consumer Services, Div. Plant Industry, Gainesville, FL.

Cuda, J. P., D. Gandolfo, J. C. Medal, R. Charudattan, and J. J Mullahey. 2002. Tropical soda apple, wetland nightshade, and turkey berry. Pages 293-309 *in* R. Van Driesche, S. Lyon, B. Blossey, M. Hoddle, and R. Reardon, eds. Biological Control of Invasive Plants in the Eastern United States. USDA Forest Serv. Pub. FHTET-2002-04. Morgantown, WV.

Gandolfo, D., D. Sudbrink, and J. Medal. 2000. Biology and host specificity of the tortoise beetle *Gratiana boliviana*, a candidate for biocontrol of tropical soda apple (*Solanum viarum*). Page 679 *in* N. R. Spencer, ed. Proc. X Int. Symp. Biol. Contr. Weeds, 4-14 July 1999, Montana State Univ., Bozeman, MT.

Habeck, D. H., J. C. Medal, and J. P. Cuda. 1996. Biological control of tropical soda apple. Pages 73-78 *in* J. J. Mullahey, ed. Proc. of the Tropical Soda Apple Symp., 9-10 January 1996. Inst. Food Agric. Sci., Univ. Florida, Gainesville, FL.

McGovern, R. J., J. E. Polston, and J. J. Mullahey. 1996. Tropical soda apple (*Solanum viarum* Dunal); Host of tomato, pepper, and tobacco viruses in Florida. Pages 31-34 *in* J. J. Mullahey, ed. Proc. Tropical Soda Apple Symp., 9-10 January 1996. Inst. Food Agric. Sci., Univ. Florida, Gainesville, FL.

Medal, J. C., J. P. Cuda, and D. Gandolfo. 2002. Classical biological control of tropical soda apple in the USA. University of Florida-IFAS Ext. Circ. ENY-824. http://edis.ifas.ufl.edu/IN457.

Medal, J. C., N. C. Coile, D. Gandolfo, and J.P. Cuda. 2002. Status of biological control of tropical soda apple, *Solanum viarum*, in Florida. Florida Dept. Agric. Consumer Serv., Div. Plant Industry. Bot. Circ. No. 36.

Medal, J. C., D. Sudbrink, D. Gandolfo, D. Ohashi, and J.P. Cuda. 2002. *Gratiana boliviana*, a potential biocontrol agent of *Solanum viarum*: Quarantine host-specificity testing in Florida and field surveys in South America. BioControl 47: 445-61.

Mullahey, J. J., and J. Cornell. 1994. Biology of tropical soda apple (*Solanum viarum*), an introduced weed in Florida. Weed Technol. 8: 465-69.

Mullahey, J. J., D. G. Shilling, P. Mislevy, and R. A. Akanda. 1998. Invasion of tropical soda apple (*Solanum viarum*) into the U.S.: Lessons learned. Weed Technol. 12: 733-36.

# Waterhyacinth

*Eichhornia crassipes*

Pickerelweed family—Pontederiaceae

*T. D. Center*

**Additional common names:** Water hyacinth, water-hyacinth.

**Native range:** Tropical South America, primarily the Amazon Basin.

**Entry into the United States:** Ostensibly first introduced during 1884 at the International Cotton Exposition in New Orleans, Louisiana, where plants were distributed as souvenirs. Some evidence suggests an earlier introduction.

## Biology

**Life duration/habit:** Waterhyacinth is an erect, free-floating, stoloniferous, perennial herb of variable size up to 1.5 m tall, often forming dense monocultures (mats) extensively covering the surface of water bodies. Seedlings and stranded plants root in moist soil. Plants perennate through constant stem elongation with continuous leaf production and turnover.

**Roots:** Numerous, plumose, adventitious roots of variable length grow from the lower portion of the shoot axis forming a diffuse root system. Root length is dependent upon nutrient concentrations, with longer roots produced in response to low nutrients.

**Stems and leaves:** The shoot consists of a short stem (rhizome) with compressed internodes. The buoyant leaves vary in size and structure. Leaf production is sequential from the apical bud and leaves remain functional for four to six weeks. They are displayed in a whorl forming a rosette about the upper portion of the stem. Each rosette bears six to eight (to 12) leaves that vary in length from a few centimeters to over a meter. The aerenchymatous leaf petiole may be inflated and bulbous or long and tapered. The lamina ranges from kidney- or heart-shaped to lanceolate. The erect, elongated petiole form occurs in dense stands. The shorter, inflated petiole form,

Waterhyacinth infestation. (Photo credit: U.S. Army Corps of Engineers)

which produces a stable platform for growth, is associated with newer infestations and open conditions.

**Reproduction:** Waterhyacinth reproduces from seeds and vegetative growth from stolons. Seed capsules contain fewer than 50 seeds each. Each inflorescence can produce more than 3,000 seeds and a single rosette can produce several inflorescences each year.

**Flowers:** Tubular blue flowers are 4 to 6 cm long and six-lobed with the adaxial lobe somewhat larger and bearing a large yellow spot. The spike-like inflorescence consists of 10 or more flowers. The 14-day flowering cycle concludes when the flower stalk bends below the water and releases the seeds.

Waterhyacinth plants. (Photo credit: U.S. Army Corps of Engineers)

**Fruits and seeds:** The fruit is a many-seeded capsule and the ovary is three-locular. The small, long-lived seeds sink and remain viable in sediments for 15 to 20 years. Seeds germinate on moist sediments or in warm (28 to 36° C), shallow water and flowering can occur 10 to 15 weeks thereafter. After emergence of the cotyledon and penetration of the substrate by the radicle, a series of four or five grass-like leaves are produced. Subsequent leaves become wider and thicker with more aerenchyma. When sufficiently buoyant, the seedling abscises from the rootstock and floats to the surface.

Waterhyacinth flowers. (Photo credit: U.S. Army Corps of Engineers)

403

## Infestations

**Worst infested states:** California, Florida, Hawaii, Louisiana, Texas, and Puerto Rico.

**Habitat:** Waterhyacinth grows best in still or slowly moving water bodies (lakes, ponds, canals, and rivers) with neutral pH, high macronutrients, warm temperatures (28 to 30 °C), and high light intensities. It tolerates pH levels from 4.0 to 10.0, but not more than 20 to 25% seawater. The plants survive frost if the rhizomes do not freeze, even though emergent portions may succumb. Prolonged cold kills the plants, but reinfestation occurs from seed. Growth is inhibited above 33° C. Plants rooted on moist sediments can resist desiccation for several months. Mats made up of small plants consist of as many as 200 rosettes/m²; large plants, 60 to 80 rosettes/m².

**Impacts:** Dense waterhyacinth mats create impenetrable barriers that obstruct navigation. Floating mats block drainage, cause flooding, and prevent subsidence of floodwaters. Large rafts that accumulate where channels narrow have been known to collapse bridges. Waterhyacinth hinders irrigation by impeding water flow, clogging irrigation pumps, and interfering with weirs. Infestations can render multimillion-dollar flood control and water supply projects useless.

Infestations block access to recreational areas, decrease waterfront property values, and harm the economies of communities dependent upon water-related activities. Shifting mats can prevent boats from reaching shore, exposing the marooned occupants to environmental hazards. Waterhyacinth infestations intensify mosquito problems by hindering insecticide application, interfering with predators, increasing habitat for attached species, and impeding runoff and water circulation.

Dense mats reduce light to submerged plants, thus depriving aquatic communities of oxygen. The resultant lack of phytoplankton depresses invertebrate communities, ultimately affecting fisheries. Drifting mats scour vegetation and destroy native plants. Waterhyacinth also competes with other plants, often displacing wildlife forage and habitat. Higher sediment loading from siltation and increased detrital production occurs under mats. Herbicidal treatment or mechanical harvesting of waterhyacinth often damages nearby desirable vegetation.

## Comments

Waterhyacinth was introduced during the 19th century as an ornamental plant for the beauty of its flower and its unique floating habit. Rampant growth quickly overran garden pools, however. Discarded surplus plants, presumably dumped into natural water bodies, soon multiplied, causing serious navigation and water management problems.

Emergent portions are easily killed with herbicides, but the persistent seed bank is impervious. Untreated plants, viable fragments, and seeds enable rapid reinfestation.

# *Neochetina bruchi*

T. D. Center

**Common name:** Waterhyacinth weevil.

**Type of agent:** Insect: Beetle, weevil (Coleoptera: Curculionidae).

**Native distribution:** South America.

**Original source:** Argentina.

## Biology

**Generations per year:** Continuously brooded, multiple overlapping generations.

**Overwintering stages:** All stages, if plant material remains present.

**Egg stage:** Eggs are whitish, ovoid, about 0.75 mm long, often deposited in groups and often embedded in bulbous petioles near the water line.

**Larval stage:** Larvae are white or cream-colored with a yellow-orange head. They have no legs or prolegs, only enlarged pedal lobes, each bearing an apical seta. Larvae of *N. bruchi* can be distinguished from *N. eichhorniae* by a small, nipple-like protuberance on each pedal lobe that bears the apical seta. Neonates are about 2 mm long and cylindrical. The fully grown third-stage larva is more grub-like, C-shaped, and 8 to 9 mm long. The posterior end of the abdomen is blunt with a pair of spur-like spiracles projecting upward on the last segment.

**Pupal stage:** Pupae are white and enclosed in cocoons formed among the lateral rootlets. The cocoons are attached to main roots below the water surface. These appear as small balls or nodules about 5 mm in diameter.

**Adult stage:** Adults are long-lived (one year) and occur year-round. They can usually be distinguished from *N. eichhorniae* by the color and pattern of the elytral scales, as well as by genital features. The ground color is usually tan with a lighter crescent-shaped band on the forewings. Short, shiny tubercles on either wing cover near the mid-line are shorter and further back than on *N. eichhorniae.*

## Effect

**Destructive stages:** Adult feeding leaves characteristic rectangular scars, about 2 to 3 mm wide and of variable length. Larvae tunnel downward in the leaf petioles to the crown, often destroying meristematic tissues. Adult and larval feeding reduce the plant's regeneration capacity.

**Additional plant species attacked:** None.

**Site of attack:** Leaves and lateral buds.

*N. bruchi* adult. (Photo credit: A. Cofrancesco, U.S. Army Corps of Engineers)

405

*N. bruchi* larvae. (Photo credit: U.S. Army Corps of Engineers)

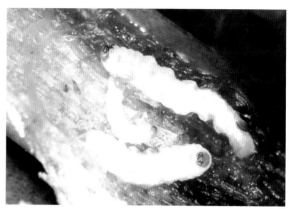

Damage to waterhyacinth caused by adult *N. bruchi.* (Photo credit: L. Bezark, California Department of Food and Agriculture)

**Impact on the host:** Adult feeding reduces photosynthetic area, sometimes girdling the distal end of the petiole and severing the leaf blade. Severe infestations result in the production of curled leaves nearly covered with feeding scars. This desiccation may cause collapse of the mats. Moderate to severe infestations lead to shorter plants with smaller leaves, fewer offsets and flowers, lower tissue nutrient content, and reduced vigor.

**Nontarget effects:** Nontarget effects have not been recorded specifically for *N. bruchi*, but are likely to be overlooked in mixed populations with *N. eichhorniae*. Nontarget effects for the two species are therefore likely to be similar (see *N. eichhorniae*, below).

## Releases

**First introduced into the United States:** 1974, Florida.

**Established in:** This weevil is widely established throughout the range of waterhyacinth. Its status in California and Hawaii is uncertain.

**Habitat:** This insect's habitat is semiaquatic. Climatic limits are unknown, but populations are likely to endure throughout the range of persistent waterhyacinth populations (i.e., where water bodies are likely to remain unfrozen through the winter).

**Availability:** Adults can be collected year-round.

**Stage to transfer:** Adult.

**Redistribution:** Populations do not recover rapidly during weed resurgence following herbicide use or severe winter dieback. Local redistribution may then become necessary. Augmentative serial releases may be useful for increasing impact. Weevils can be hand-collected from plants by pulling back older leaves and searching in the leaf axils under the membranous stipules. Adults often aggregate in the partially open youngest leaf during the day, so special attention should be paid to rosettes that possess a loosely furled central leaf. An alternative method involves submerging the plants and collecting the adults as they rise to the surface. Extraction with Berlese funnels is sometimes useful, but the high moisture content of the plant tissue necessitates modification of the funnel design to vent the humidity and prevent overflow of the container. Sweep netting at night can also be effective, but this requires traversing the mat with "water shoes" or in an airboat. Light trapping can be of value if done during periods of flight activity.

## Comments

*Neochetina bruchi* has been released in approximately 27 countries but its effects are not discernible from the earlier-released *N. eichhorniae*, which usually predominates. Some data suggest that *N. bruchi* prefers plants with higher tissue nitrogen concentrations. Waterhyacinth has been reduced to one-third of its former abundance in the Gulf Coast states. This reduction has sometimes resulted from direct plant mortality, but more often may be attributed to subtle effects such as growth inhibition following diebacks (winter or herbicidal) or suppression of flower and seed production.

As with *N. eichhorniae*, adults undergo reversible generation and degeneration of indirect flight muscles, related to alternating dispersive and reproductive phases. Plant quality may influence the propensity to switch.

## *Neochetina eichhorniae*

T. D. Center

**Common name:** Waterhyacinth weevil.

**Type of agent:** Insect: Beetle, weevil (Coleoptera: Curculionidae).

**Native distribution:** South America.

**Original source:** Argentina.

## Biology

**Generations per year:** Continuously brooded, multiple overlapping generations.

**Overwintering stages:** All stages, if plant material remains present.

**Egg stage:** Eggs are whitish, ovoid, about 0.75 mm long, and deposited singly, embedded in leaf tissue or subtending stipules.

**Larval stage:** Larvae are white or cream-colored with a yellow-orange head. They have no legs or prolegs, only enlarged pedal lobes, each of which bears an apical seta.

407

Larvae of *N. eichhorniae* can be distinguished from *N. bruchi* by the absence of setal-bearing protuberances on the pedal lobes. Neonates are about 2 mm long and cylindrical. The more grub-like fully grown third-stage larva is 8 to 9 mm long. The posterior end of the abdomen is blunt with a terminal pair of spur-like spiracles projecting upward.

**Pupal stage:** Pupae are white and enclosed in cocoons formed among the lateral rootlets. The cocoons are attached to main roots below the water surface. These appear as small balls or nodules about 5 mm in diameter.

**Adult stage:** Adults are long-lived (one year) and occur year-round. They can usually be distinguished from *N. bruchi* by the grayish-brown color and pattern of the elytral scales,

*N. eichhorniae* adult. (Photo credit: A. Cofrancesco, U.S. Army Corps of Engineers)

as well as by genital features. Short, shiny tubercles on either wing cover near the midline are longer and further forward than on *N. bruchi*.

## Effect

**Destructive stages:** Adult feeding leaves characteristic rectangular scars, about 2 to 4 mm wide and of variable length. Larvae tunnel downward in the leaf petioles to the crown, often destroying meristematic tissues.

**Additional plant species attacked:** None.

**Site of attack:** Leaves and lateral buds.

**Impact on the host:** Adult feeding reduces photosynthetic area, sometimes girdling the distal end of the petiole and severing the leaf blade. Severe infestations result in the production of curled leaves nearly covered with feeding scars. This desiccation may cause collapse of the mats. Moderate to severe infestations lead to shorter plants with smaller leaves, fewer offsets and flowers, lower tissue nutrient content, and reduced vigor.

**Nontarget effects:** Adult feeding often occurs on the related pickerelweed (*Pontederia cordata*, Pontederiaceae), and some larval development may occur, but populations do not proliferate within isolated stands. Damage is most frequently observed when pickerelweed and waterhyacinth are intermixed or in close proximity. Negative population-level effects to pickerelweed stands do not occur. Damage sometimes occurs on assorted nontarget plants, especially succulents, present under light sources where flying adults accumulate. Adults normally do not fly but sometimes swarm at lights after induction of indirect flight muscles. The damage probably results from dehydrated adults seeking a source of water. This occurrence seems associated with extensive die-offs of waterhyacinth mats (e.g., after herbicide treatment) or with overly abundant weevil populations. This phenomenon is temporary and sporadic, and the nontarget plants do not suffer permanent harm.

## Releases

**First introduced into the United States:** 1972, Florida.

**Established in:** This weevil is widely established in the southeastern United States. Its status in Hawaii and California is uncertain.

**Habitat:** This insect's habitat is semiaquatic. Climatic limits are unknown, but populations are likely to endure throughout the range of persistent waterhyacinth populations (i.e., wherever water bodies are likely to remain unfrozen throughout the winter).

**Availability:** Adults can be collected year-round.

**Stage to transfer:** Adult.

**Redistribution:** Populations do not recover rapidly during weed resurgence following herbicide use or severe winter dieback. Local redistribution may then become

Florida waterway infested with waterhyacinth. (Photo credit: USDA-ARS)

Same waterway several years after release of *Neochetina* spp. (Photo credit: USDA-ARS)

necessary. Augmentative serial releases may be useful for increasing impact. Weevils can be hand-collected from plants by pulling back older leaves and searching in the leaf axils under the membranous stipules. Adults often aggregate in the partially open youngest leaf during the day, so special attention should be paid to rosettes bearing a loosely furled central leaf. An alternative method involves submerging the plants and collecting the adults as they rise to the surface. Extraction with Berlese funnels is sometimes useful, but the high moisture content of the plant tissue necessitates modification of the funnel design to vent the humidity and prevent overflow of the container. Sweep netting at night can also be effective, but this requires traversing the mat with "water shoes" or in an airboat. Light trapping can be of value if done during periods of flight activity.

## Comments

*Neochetina eichhorniae* has been released in approximately 30 countries where, in some cases, it has produced remarkable waterhyacinth control. Waterhyacinth has been reduced to one-third of its former abundance in the Gulf Coast states. This reduction has sometimes resulted from direct plant mortality, but more often may be attributed to subtle effects such as growth inhibition following diebacks (winter or herbicidal) or suppression of flower and seed production.

Adults undergo reversible generation and degeneration of indirect flight muscles, related to alternating dispersive and reproductive phases. Plant quality may influence this propensity to switch.

## *Niphograpta albiguttalis (= Sameodes albiguttalis)*

T. D. Center

**Common name:** Waterhyacinth moth.

**Type of agent:** Insect: Moth (Lepidoptera: Pyralidae).

**Native distribution:** South America.

**Original source:** Argentina.

## Biology

**Generations per year:** Continuously brooded, multiple overlapping generations.

**Overwintering stages:** All stages, if plant material remains present.

**Egg stage:** The eggs of *N. albiguttalis* are small (about 0.3 mm), spherical, and creamy-white. Complete development of the embryo requires three to four days at 25° C.

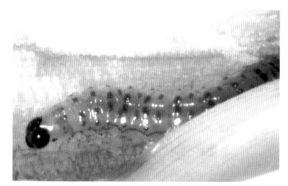

*N. albiguttalis* larva. (Photo credit: U.S. Army Corps of Engineers)

**Larval stage:** The newly emerged, 1.5-mm-long larva is brownish with darker spots and a dark head. The fully grown fifth instar larva (2 cm long) has a dark orange head, a cream-colored body, and is covered with conspicuous dark brown spots. Larval development requires about two weeks.

**Pupal stage:** The fully grown larva pupates within a cocoon in a cavity excavated in a healthy leaf petiole. The cocoon lines the cavity and extends up a short tunnel that leads to the exterior. The pupal stage lasts seven to 10 days.

*N. albiguttalis* adult. (Photo credit: USDA-ARS)

**Adult stage:** After emergence from the pupa, the adult crawls through the silk-lined tunnel and ruptures the leaf epidermis to exit from the petiole. The adults rest on the underside of waterhyacinth leaves. The females are generally darker than the males, but color is variable and darker during cooler months. The forewings range from brown to golden with the hind wings consistently golden. A distinct white spot is present at mid-length toward the leading edge of the forewing and a dark spot is apparent in the center of the hind wing. White hind edges of the abdominal segments give the appearance of white rings around the abdomen. The adults live a week to 10 days. Mating occurs soon after emergence and the female lays most of her eggs the following night. An average female will deposit 450 to 600 eggs. The entire life cycle requires three to four weeks.

## Effect

**Destructive stage:** Larval.

**Additional plant species attacked:** None.

**Site of attack:** Leaves and apical buds.

**Impact on the host:** Larvae tunnel the leaf petioles and plant crown, destroying the apical meristem. Having lost the ability to replace senescent tissue, affected shoots die or lose buoyancy and sink.

**Nontarget effect:** No nontarget effects have been recorded.

## Releases

**First introduced into the United States:** 1977, Florida.

**Established in:** Florida and Texas; Puerto Rico.

**Habitat:** This insect's habitat is semiaquatic. Climatic limits are unknown, but populations are likely to endure throughout the range of persistent waterhyacinth populations (i.e., wherever water bodies are likely to remain unfrozen throughout the winter).

**Availability:** Adults can be collected year-round.

**Stages to transfer:** Adult or egg-laden leaves.

411

**Redistribution:** Populations build rapidly on early phenostage plants during weed resurgence, but usually do not persist on later phenostage plants. Redistribution is probably not necessary in view of the extraordinary dispersive capabilities of this insect. Augmentative releases may be useful for increasing impact, but attention must be given to selecting the appropriate type of plant. Pupae can be collected by hand-inspecting rosettes while looking for the hyaline "window" covering the adult emergence tunnel on leaf petioles. Adults can be collected by placing a mercury-vapor lamp on a white sheet on the foredeck of an airboat while slowly cruising through mats of small plants. The adults fly when disturbed by the boat, then orient toward the lamp and alight on the sheet. They can then be collected by trapping them in petri dishes or other containers. Gravid females (use pairs if males are available) are egged by placement in a container with a waterhyacinth leaf that has been partially stripped of the epidermal layer, exposing the underlying aerenchyma. The females oviposit within the exposed cells, sometimes abundantly. The oviposition substrate should be replaced daily and the egg-laden leaves held until larvae begin to eclose. Field release is best accomplished by tucking these leaves with the neonates within partially furled, newly emerging leaves in the center of rosettes at the site.

## Comments

*Niphograpta albiguttalis* prefers young waterhyacinth plants with bulbous petioles, such as those that typically regrow after herbicide treatment. It establishes quickly, creates a great deal of damage, and then often disappears. This has made it difficult to evaluate in terms of its impact and importance as a biological control agent.

*Niphograpta albiguttalis* populations initially established in southern Florida, then dispersed to extreme northern Florida (500 km) within 18 months. It was found in Mexico during 1993 near Vera Cruz and as far south as Tapachula, near the border with Guatemala on the west coast, but there are no records of its release in Mexico. These populations may have derived from those established in Texas. Likewise, there are no records of *N. albiguttalis* having been released in Puerto Rico, but larvae were found near San Juan in 1995. Its presence had also been noted in Cuba by 1996, suggesting that it "island hopped" from southern Florida to the West Indies.

## References

Center, T. D. 1994. Chapter 23. Biological control of weeds: Waterhyacinth and waterlettuce. Pages 481-521 *in* D. Rosen, F. D. Bennett, and J. L. Capinera, eds. Pest Management in the Subtropics: Biological Control—A Florida Perspective. Intercept Publishing Co., Andover, UK.

Center, T. D., and N. R. Spencer. 1981. The phenology and growth of waterhyacinth (*Eichhornia crassipes* (Mart.) Solms) in a eutrophic north-central Florida lake. Aquat. Bot. 10: 1-32.

Center, T. D., J. K. Balciunas, and D. H. Habeck. 1982. Descriptions of *Sameodes albiguttalis* (Lepidoptera: Pyralidae) life stages with key to Lepidoptera larvae on waterhyacinth. Ann. Entom. Soc. Am. 75: 471-79.

Center, T. D., F. A. Dray, G. Jubinsky, and M. J. Grodowitz. 1999. Biological control of waterhyacinth under conditions of maintenance management: Can herbicides and insects be integrated? Environ. Manage. 23: 241-56.

Center, T. D., F. A. Dray, G. P. Jubinsky, and A. J. Leslie. 1999. Waterhyacinth weevils (*Neochetina eichhorniae* Warner and *N. bruchi* (Hustache)) inhibit waterhyacinth (*Eichhornia crassipes*) colony development. Biol. Control 15: 30-50.

Center, T. D., M. D. Hill, H. Cordo, and M. H. Julien. 2002. Waterhyacinth. Pages 41-64 *in* R. Van Driesche, S. Lyon, B. Blossey, M. Hoddle, and R. Reardon, eds. Biological Control of Invasive Plants in the Eastern United States. USDA Forest Service Pub. FHTET-2002-04. Morgantown, WV.

DeLoach, C. J. 1975. Identification and biological notes on the species of *Neochetina* that attack Pontederiaceae in Argentina (Coleoptera: Curculionidae: Bagoini). Coleop. Bull. 29: 257-65.

DeLoach, C. J., and H. A. Cordo. 1976a. Life cycle and biology of *Neochetina bruchi* and *N. eichhorniae*. Ann. Entomol. Soc. Am. 69: 643-52.

DeLoach, C. J., and H. A. Cordo. 1976b. Ecological studies of *Neochetina bruchi* and *N. eichhorniae* on waterhyacinth in Argentina. J. Aquat. Plant Manage. 14: 53-59.

Gopal, B. 1987. Water Hyacinth. Elsevier, NY.

Grodowitz, M. J., T. D. Center, and J. E. Freeman. 1997. A physiological age-grading system for *Neochetina eichhorniae* Warner (Coleoptera: Curculionidae), a biological control agent of waterhyacinth, *Eichhornia crassipes* (Mart.) Solms. Biol. Control 9: 89-105.

Heard, T. A., and S. L. Winterton. 2000. Interactions between nutrient status and weevil herbivory in the biological control of water hyacinth. J. Appl. Ecol. 37: 117-27.

Julien, M. H., M. W. Griffiths, and J. N. Stanley. 2001. Biological control of waterhyacinth 2. The moths *Niphograpta albiguttalis* and *Xubida infusellus*: biologies, host ranges, and rearing, releasing and monitoring techniques for biological control of *Eichhornia crassipes*. ACIAR Monogr. No. 79. Australian Centre for International Agricultural Research, Canberra, Australia.

Julien, M. H., M. W. Griffiths, and A. D. Wright. 1999. Biological control of water hyacinth. The weevils *Neochetina bruchi* and *N. eichhorniae*: biologies, host ranges, and rearing, releasing and monitoring techniques for biological control of *Eichhornia crassipes*. ACIAR Monogr. No. 60. Australian Centre for International Agricultural Research, Canberra, Australia.

O'Brien, C. W. 1976. A taxonomic revision of the new world subaquatic genus *Neochetina*. Ann. Entomol. Soc. Am. 69: 165-74.

Penfound, W. T., and T. T. Earle. 1948. The biology of the waterhyacinth. Ecol. Monogr. 18: 447-72.

Richards, J. H. 1982. Developmental potential of axillary buds of waterhyacinth, *Eichhornia crassipes* Solms. (Pontederiaceae). Am. J. Bot. 69: 615-22.

Watson, M. A. 1984. Developmental constraints: effect on population growth and patterns of resource allocation in a clonal plant. Am. Nat. 123: 411-26.

Watson, M. A., and C. S. Cook. 1982. The development of spatial pattern in clones of an aquatic plant, *Eichhornia crassipes* Solms. Am. J. Bot. 69: 248-53.

Watson, M. A., and G. S. Cook. 1987. Demographic and developmental differences among clones of waterhyacinth. J. Ecol. 75: 439-57.

Wright, A. D., and T. D. Center. 1984. Predicting population intensity of adult *Neochetina eichhorniae* (Coleoptera: Curculionidae) from incidence of feeding on leaves of waterhyacinth, *Eichhornia crassipes*. Environ. Entomol. 13: 1478-82.

# Waterlettuce

*Pistia stratiotes*

Aroid family—Araceae

*F. A. Dray, Jr.*

**Additional common names:** River lettuce, water-bonnet, water lettuce.

**Native range:** Africa, Asia, and South America.

**Entry into the United States:** Native to North America prior to the Pleistocene, when it was exterminated from the continent, this species was reintroduced (probably accidentally) by the 18th century in the European colonization of Florida.

## Biology

**Life duration/habit:** Waterlettuce is an herbaceous hydrophyte consisting of free-floating rosettes of many leaves. It occurs primarily in stagnant and slow-flowing waters of the tropics and subtropics. It is principally a perennial, but behaves as an annual in temperate regions of its range.

**Reproduction:** Reproduction occurs both by seed and by production of lateral stolons that develop terminal buds which give rise to new rosettes. Up to 15 secondary rosettes may be attached to a single primary plant, and up to four generations of rosettes may be interconnected by stolons. Sexually reproducing populations can be quite prolific, producing crops containing up to 726 seeds/m$^2$. The hydrosoil under such infestations holds more than 4,000 seeds/m$^2$.

**Roots:** Waterlettuce has as abundance of feathery roots comprising a thick cluster of adventitious roots each bearing copious lateral root hairs.

**Stems and leaves:** Wedge-shaped leaves are gray-green and densely pubescent with conspicuous parallel veins. They frequently have thick, spongy tissue at the base, and vary from being slightly broader (at the apex) than long to much longer than broad. They range from 2 to 35 cm long. The leaves of young plants, or those on plants in uncrowded conditions, typically lie flat on the water surface. Leaves of older plants, or on plants in crowded conditions, are generally more upright.

**Flowers:** The inflorescences are inconspicuous pale green spathes near the center of the rosette. Each spathe is constricted near the middle, with a whorl of eight male flowers above and a single female flower below the constriction. A typical rosette contains six inflorescences. Flowers are present year-round, but their greatest abundance occurs April through November.

**Fruits and seeds:** Fruits are many-seeded (up to 40), dehiscent, green berries. Mature seeds have a thick, golden-brown, wrinkled seed coat, and an 84% germination rate.

## Infestations

**Worst infested states:** Subtropical Florida harbors the most abundant waterlettuce populations in the United States. Other principal infestations occur in the warm temperate regions of the Gulf Coast states (excluding Alabama), California, Hawaii, Puerto Rico, and the Virgin Islands. Scattered ephemeral populations occur in Arizona, North Carolina, South Carolina, and Virginia.

Waterlettuce plants. (Photo credit: U.S. Army Corps of Engineers)

**Habitat:** Waterlettuce occurs in lakes, ponds, canals, and slow-flowing streams as dense monocultures or mixed with other floating weeds, primarily waterhyacinth (*Eichhornia crassipes*).

**Impacts:** The principal impact arises when heavy infestations restrict water flow in irrigation and flood control canals, adversely affecting agriculture. Waterlettuce occasionally forms dense, impenetrable floating mats that impede navigation and interfere with recreational uses of affected water bodies. Also, this plant harbors several pathogen-transmitting mosquito species.

Ecological impacts are largely unstudied, but large infestations accelerate siltation rates in waterways and cause thermal stratification of the water column resulting in depleted oxygen levels near the hydrosoil. Both of these impacts are likely to adversely affect fish and benthic macroinvertebrate species. Dense mats also reportedly increase evapotranspiration rates.

## Comments

Historically, waterlettuce was known to form large floating islands nearly blocking upper reaches of the St. Johns River in Florida, but these are uncommon today. Several authors contend that waterlettuce began to be suppressed by waterhyacinth after the latter was introduced into Florida during the late 19th century; competition experiments support this argument.

Waterway infested with waterlettuce. (Photo credit: U.S. Army Corps of Engineers)

415

# *Neohydronomus affinis*

F. A. Dray, Jr.

**Common name:** Waterlettuce weevil.

**Type of agent:** Insect: Beetle, weevil (Coleoptera: Curculionidae).

**Native distribution:** South America.

**Original source:** Brazil (via Australia).

## Biology

**Generations per year:** Three generations in its native range.

**Overwintering stages:** Adult and pupal.

**Egg stage:** Eggs are cream-colored and subspherical (0.33 by 0.40 mm). Females chew a 0.5-mm-diameter hole in the leaf (usually on the upper surface near the leaf margins), deposit a single egg inside this puncture, and close the hole with a black substance. Eggs hatch within four days (at temperatures above 24° C).

**Larval stage:** Larvae progress through two molts in 11 to 14 days. The first molt occurs when larvae are about three days old, the second occurs three to four days later. First instar larvae are very small and pale cream in color. Second instars are slightly larger. Third instars are yellow, 2.4 to 3.0 mm long, and are characterized by a pronounced brown anal shield on the back.

**Pupal stage:** The naked, white pupae are found in the spongy tissue at the bases of the leaves. The pupal stage lasts 10 to 24 days.

**Adult stage:** Adult weevils are small (1.7 to 2.3 mm long) and have a nearly straight snout that is strongly constricted at the base. They range in color from uniform bluish-gray to reddish-brown with a tan, chevron-like band across the forewings. Females are generally larger than males (2.1 versus 1.8 mm, respectively). Females produce one egg per day. Adults in their native range are occasionally parasitized by nematodes.

## Effect

**Destructive stages:** Larval and adult.

**Additional plant species attacked:** None.

**Site of attack:** Newly emerged larvae burrow under the leaf epidermis and work their way toward the spongy portions of the leaf at a rate of about 1.5 to 2.0 cm/day. This feeding produces highly visible brown, serpentine mines in the outer third of the leaf where tissues are thin; mines are less apparent in the central and basal portions of the leaf. Third instars are generally found excavating the spongy portions of the leaf. Adults chew holes (about 1.4 mm in diameter) in the leaf surface and burrow into the spongy tissues of the leaf. Adults and larvae will also feed on meristematic tissues concentrated in the crowns of the plant and on newly emerging shoots.

**Impact on the host:** The characteristic round feeding holes are easily observed when weevil populations are large (several hundred insects per m$^2$), but may be

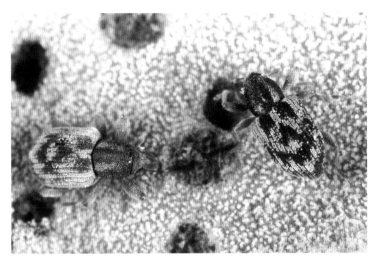

*N. affinis* adults and damage to waterlettuce. (Photo credit: USDA-ARS)

concentrated near leaf margins and more difficult to observe when populations are small. Plants under stress from weevil feeding are typically smaller, have fewer leaves, and grow less rapidly than uninfested plants. Intermediate levels of stress can result in increased shoot production, which spurs further weevil population increases. Heavily stressed plants produce fewer inflorescences. Ultimately, destruction of the buoyancy-providing spongy tissues causes the plants to sink.

**Nontarget effects:** Occasional adult feeding may be observed on duckweed (*Spirodela* and *Lemna*) and frogbit (*Limnobium*), but in laboratory tests, egg laying on these plants is unusual and larvae die without completing development. No other nontarget effects are known.

### Releases

**First introduced into the United States:** 1987, Florida. This insect was also released at a single site in Texas in 1991.

**Established in:** Florida and Louisiana.

**Habitat:** The weevils prefer to attack succulent, well-nourished plants. Climatic limits have not been documented.

**Availability:** The weevil is most easily obtained in southern and central Florida.

**Stage to transfer:** Adult.

**Redistribution:** The most common method of redistribution is to move infested plants.

### Comments

The weevil reduced waterlettuce abundance up to 90% at five sites in Florida and two in Louisiana. Long-term suppression of this weed has not occurred, although at one site in Florida there were annual cycles from 1990 to 1994 in which spring increases in waterlettuce abundance were followed by sharp declines attributable to the weevil.

# *Spodoptera pectinicornis*

F. A. Dray, Jr.

**Common name:** Asian waterlettuce moth.

**Type of agent:** Insect: Moth (Lepidoptera: Noctuidae).

**Native distribution:** Asia.

**Original source:** Thailand.

## Biology

**Generations per year:** In Asia, moth populations are present all year and produce continuous, overlapping generations.

**Overwintering stage:** None.

**Egg stage:** Females lay eggs on both surfaces of waterlettuce leaves in masses of up to 150 eggs. The eggs are covered by a substance produced by the female, perhaps scales from her abdomen. Egg laying lasts two to six days; each female produces up to 990 eggs. The incubation period ranges from three to six days. Eggs are subspherical, about 0.03 mm in diameter, greenish when newly deposited and turning yellow as they develop.

**Larval stage:** First instars are pale yellow to greenish-white. They feed within the leaf on the spongy tissues. Larval development progresses through six instars and requires 17 to 20 days. The last two instars are usually brown intermixed with purplish-brown and green. Fully-grown larvae attain lengths of up to 25 mm. Larvae and pupae both have conspicuous, enlarged spiracles extending laterally from the body.

**Pupal stage:** Pupation occurs in a leaf base, between the leaves, or between the thick ribs on the underside of the leaf. Ventral abdomens of pupae are initially green, but over time they become dark brown. The prepupal period lasts one to two days and the pupal stage lasts 3.5 to 5.5 days.

**Adult stage:** Adults are short-lived with females living three to seven days and males two to four days. The moths have mottled reddish-brown forewings and cream-colored hindwings and a wingspan of 17 to 22 mm. Females are larger than males. Adults are generally active at night and quiescent during the day unless disturbed. They mate while resting on leaves at night.

## Effect

**Destructive stage:** Larval.

**Additional plant species attacked:** None.

**Site of attack:** Caterpillars feed on leaf tissues and on the meristems in the crowns of the rosettes.

**Impact on the host:** Feeding on leaf tissues decreases a plant's buoyancy. Destruction of meristematic tissue prevents leaf replacement and impedes vegetative propagation. Research has shown that 100 caterpillars from one average-sized egg mass can destroy the waterlettuce within a 1 m² area, and that a single caterpillar, can eat two rosettes at a rate of one leaf per day.

*S. pectinicornis* pupa (above), mature larvae (below), and feeding damage. (Photo credit: USDA-ARS)

**Nontarget effects:** No nontarget effects have been reported.

## Releases
**First introduced into the United States:** 1990, Florida.

**Established in:** This insect failed to establish in the United States.

**Habitat:** The moth prefers to attack succulent, well-nourished plants. Climatic limits of this moth have not been documented.

**Availability:** This insect was not available in the United States as of 2003.

**Stage to transfer:** Larval.

**Redistribution:** In Asia, the most common method of redistribution is to move infested plants.

## Comments
In India, peak periods of *S. pectinicornis* occurrence coincide with monsoons and with periods of rapid waterlettuce growth. During these periods, moths infest most sites and the destruction to waterlettuce mats frequently exceeds 75%. During dry periods, fewer sites are infested and smaller proportions of the waterlettuce populations are affected.

Damage to waterlettuce caused by *S. pectinicornis*. (Photo credit: USDA-ARS)

## References

Dray, F. A., Jr., and T. D. Center. 1989. Seed production by *Pistia stratiotes* L. (waterlettuce) in the United States. Aquat. Bot. 33: 155-60.

Dray, F. A., Jr., and T. D. Center. 1992. Biological control of *Pistia stratiotes* L. (waterlettuce) using *Neohydronomus affinis* Hustache (Coleoptera: Curculionidae). Tech. Rep. A-92-1. U.S. Army Engineer Waterways Experiment Station, Vicksburg, MS.

Dray, F. A., Jr., and T. D. Center. 2002. Waterlettuce. Pages 65-78 *in* R. Van Driesche, S. Lyon, B. Blossey, M. Hoddle, and R. Reardon, eds. Biological Control of Invasive Plants in the Eastern United States. USDA Forest Serv. Pub. FHTET-2002-04. Morgantown, WV.

Dray, F. A., Jr., T. D. Center, and G. S. Wheeler. 2001. Lessons from unsuccessful attempts to establish *Spodoptera pectinicornis* (Lepidoptera: Noctuidae), a biological control agent of waterlettuce. Biocontrol Sci. Technol. 11: 301-16.

Dray, F. A., Jr., T. D. Center, D. H. Habeck, C. R. Thompson, A.F. Cofrancesco, and J.K. Balciunas. 1990. Release and establishment in the southeastern U.S. of *Neohydronomus affinis* (Coleoptera: Curculionidae), an herbivore of waterlettuce (*Pistia stratiotes*). Environ. Entomol. 19: 799-803.

Habeck, D. H., and C. R. Thompson. 1994. Host specificity and biology of *Spodoptera pectinicornis* (Lepidoptera: Noctuidae), a biological control agent of waterlettuce. Biol. Control 4: 263-68.

Schmitz, D. C., J. D. Schardt, A. J. Leslie, F. A. Dray, Jr., J. A. Osborne, and B. V. Nelson. 1993. The ecological impact and management history of three invasive alien aquatic plant species in Florida. Pages 173-94 *in* B. N. McKnight, ed. Biological Pollution: The Control and Impact of Invasive Exotic Species. Indiana Acad. Sci., Indianapolis, IN.

Thompson, C. R., and D. H. Habeck. 1989. Host specificity and biology of the weevil *Neohydronomus pulchellus*, biological control agent of waterlettuce (*Pistia stratiotes*). Entomophaga 34: 299-306.

Wheeler, G. S., T. K. Van, and T. D. Center. 1998. Herbivore adaptations to a low-nutrient food: weed biological control specialist *Spodoptera pectinicornis* (Lepidoptera: Noctuidae) fed the floating aquatic plant *Pistia stratiotes*. Environ. Entomol. 27: 993-1000.

Wheeler, G. S., T. K. Van, and T. D. Center. 1998. Fecundity and egg distribution of the herbivore *Spodoptera pectinicornis* as influenced by quality of the floating aquatic plant *Pistia stratiotes*. Entomol. Exp. Appl. 86: 295-304.

# Yellow starthistle

*Centaurea solstitialis*

Sunflower family—Asteraceae

*M. J. Pitcairn, G. L. Piper, and E. M. Coombs*

**Additional common name:** St. Barnaby's thistle.

**Native range:** Southern Europe, Mediterranean basin.

**Entry into the United States:** This plant was introduced into North America from Europe in the 19th century, probably through contaminated grain or alfalfa seed. The first collection of yellow starthistle in the western United States was near Oakland, California, in 1869.

## Biology

**Life duration/habit:** It is usually a winter annual. It is 0.3 to 2.0 m tall at maturity, depending on growing conditions.

**Reproduction:** Seeds. Germination begins with fall rains and continues throughout the winter into early spring.

**Roots:** The plant has a deep taproot.

**Stems and leaves:** A rosette with lobed leaves develops during the fall, winter, and early spring. The plant bolts to produce the flowering stems that are rigidly branched, winged, and covered with a cottony pubescence that imparts a gray-green color to the plant.

**Flowers:** The bright yellow flowers are produced in summer. The flower head bracts bear long, sharp spines. The flowers are highly attractive to honeybees, which are important for seed production.

**Fruits and seeds:** Two types of seeds are produced. Numerous centrally positioned seeds are light-colored, modestly plumose, and tend to drop from the heads rapidly after maturation, whereas peripheral seeds are dark-colored, plumeless, and tend to remain in the heads until winds and rain break down the heads in winter.

Yellow starthistle. (Photo credit: S. Dewey, Utah State University)

421

## Infestations

**Worst infested states:** California, Idaho, Oregon, and Washington.

**Habitat:** The plant is a serious pest of roadsides, rangelands, pastures, vineyards, abandoned croplands, alfalfa and small-grain fields, and natural areas, and is able to grow in various types of soil. While it is found growing at elevations over 2,000 m, it is most abundant at elevations below 1,200 m. It is rarely encountered in areas where annual rainfall is less than 15 cm.

**Impacts:** Yellow starthistle is poisonous to horses—a substance in the foliage, probably a sesquiterpene lactone, causes brain lesions resulting in a nervous disorder called "chewing disease," or nigropallidal encephalomalacia. Spiny heads deter grazing by livestock and are a nuisance to people. The weed is aggressively invasive, especially following disturbance.

Yellow starthistle flower. (Photo credit: R. Hawkes, Oregon Department of Agriculture)

## Comments

Yellow starthistle is the worst rangeland weed in California.

The first agent released was *Urophora jaculata* in 1969, but it failed to establish. Five insect enemies have established in the United States; they were released between 1984 and 1992. Field studies at three locations in California where four of five insects are established have yet to show substantial reduction in adult plant densities. The hairy weevil (*Eustenopus villosus*) and the false peacock fly (*Chaetorellia succinea*) appear to cause most of the seed destruction and are considered the most promising.

The false peacock fly was unintentionally released near Merlin, Oregon, in 1991 with the peacock fly (*Chaetorellia australis*). Limited host testing on this insect had been done prior to 1991 by the United States Department of Agriculture (USDA-ARS) European Biological Control Laboratory, but the fly was not considered a candidate for biocontrol of yellow starthistle because it could not be adequately distinguished from *Chaetorellia carthami*, a pest of safflower in the Middle East. Recent host specificity tests show that *C. succinea* can oviposit and complete development on several safflower varieties in no-choice laboratory tests. However, flies did not attack safflower in open field tests. A survey of 47 commercial safflower fields in California detected only one field with infested seed heads (infestation rate 1 to 5%). A survey of native *Cirsium* thistles growing near infested yellow starthistle failed to find evidence of attack by this fly. Field surveys have reared *C. succinea* from *Centaurea melitensis* and *C. sulphurea*, two exotic *Centaurea* spp. closely related to yellow starthistle. In the laboratory, the fly has been reared on *Centaurea americana*, a native species growing in Arizona and New Mexico. However, open field tests need to be performed to determine risk of attack for this species. Approval by the USDA –Animal Plant Health Inspection Service is required before collection and distribution of *C. succinea* is allowed.

Yellow starthistle infestation. (Photo credit: California Department of Food and Agriculture)

New biocontrol agents for yellow starthistle undergoing host specificity testing as of 2003 include the beetle *Ceratapion basicorne* and a blister mite, *Aceria solstitialis*.

# *Bangasternus orientalis*

M. J. Pitcairn, G. L. Piper, and E. M. Coombs

**Common name:** Yellow starthistle bud weevil.

**Type of agent:** Insect: Beetle, weevil (Coleoptera: Curculionidae).

**Native distribution:** Southern Europe, Mediterranean basin.

**Original source:** Northern Greece.

## Biology

**Generations per year:** One.

**Overwintering stage:** Adult (outside the host plant).

**Egg stage:** Single eggs covered with a dark mucilage are laid on or near scale leaves beneath the immature head buds at the tips of flowering shoots in late spring to early summer. A female may produce up to 470 eggs.

**Larval stage:** Hatched larvae tunnel through the scale leaf, the flowering stalk, and up into the flower head where they feed on the bracts, receptacle tissue, and developing seeds. There are four larval growth stages.

**Pupal stage:** Pupation occurs within the heads in chambers formed from damaged and undamaged seeds. Usually one, but occasionally two pupal chambers can be found in a head.

423

**Adult stage:** Adults exit from the pupal chambers in late summer to overwinter outside the host plant. The adults are 4 to 6 mm long (not including the snout) and brown with yellow to whitish hairs that impart a somewhat mottled appearance. This genus has a shorter snout than *Eustenopus* and *Larinus*.

## Effect

**Destructive stages:** Larval, adults cause minor damage.

**Additional plant species attacked:** None.

*B. orientalis* pupal chamber and exit hole in yellow starthistle. (Photo credit: E. Coombs, Oregon Department of Agriculture)

**Site of attack:** Within the stem below the flower head and inside the flower head.

**Impact on the host:** Larval feeding reduces seed production. Preliminary data indicate that a single larva destroys 50 to 60% of the seeds in an infested head.

**Nontarget effects:** No nontarget effects have been reported.

## Releases

**First introduced into the United States:** 1985, California, Idaho, Oregon, and Washington.

**Established in:** Arizona, California, Idaho, Oregon, and Washington.

**Habitat:** This weevil is widespread throughout most of the western United States wherever yellow starthistle is found. Specific habitat requirements are unknown.

Adult *B. orientalis.* (Photo credit: USDA-ARS)

**Availability:** The weevil is readily available for collection and redistribution.

**Stage to transfer:** Adult.

**Redistribution:** Adults can be swept or handpicked from the host plants in spring through early summer.

## Comments

Numbers of *Bangasternus orientalis* were quite high for a few years following its initial release in 1985. Since then its numbers have declined, partially due to displacement by the other flower head insects, especially *Eustenopus villosus*. Also, many of the eggs are destroyed by an unidentified predator. It now occurs in low numbers, rarely exceeding 25% attack rate despite the high number of flower heads receiving eggs. *Bangasternus orientalis* larvae are parasitized to a small degree in California by a native wasp, *Microdontomerus anthonomi*.

# *Chaetorellia australis*

M. J. Pitcairn, G. L. Piper, and E. M. Coombs

**Common name:** Yellow starthistle peacock fly.

**Type of agent:** Insect: Fly (Diptera: Tephritidae).

**Native distribution:** Southern Europe, Mediterranean basin.

**Original source:** Northern Greece.

## Biology

**Generations per year:** At least two.

**Overwintering stage:** Larval (in the seed heads).

**Egg stage:** The eggs are laid beneath the bracts of mature, closed head buds. The cylindrical eggs have a long, terminal filament. A female may produce up to 240 eggs.

**Larval stage:** The larvae feed inside the seed heads, tunneling through the developing seeds.

**Pupal stage:** Pupation occurs within the damaged seed heads.

*C. australis* adult on yellow starthistle. (Photo credit: USDA-ARS)

**Adult stage:** The adults begin to emerge from their

425

puparia in early April. Adults are golden straw-colored with black spots on the back of the thorax and abdomen and the eyes have a multicolored metallic appearance. The wings have straw-colored bands. Males are 3 to 4 mm long, females measure 4 to 6 mm, including the ovipositor. *C. australis* can be easily distinguished by its single black dot on the thorax in front of the wing, whereas *C. succinea* has two.

## Effect

**Destructive stage:** Larval.

**Additional plant species attacked:** Bachelor's button or cornflower (*C. cyanus*), an early-flowering exotic species related to yellow starthistle.

**Site of attack:** Seed head interior.

**Impact on the host:** Larval feeding reduces seed production 80 to 90%.

**Nontarget effects:** No nontarget effects have been reported.

## Releases

**First introduced into the United States:** 1988, California and Oregon.

**Established in:** California, Idaho, Oregon, and Washington.

**Habitat:** Cool climates (coastal, higher elevations and latitudes) are unfavorable for development.

Yellow starthistle (bluish-green) infestation near Myrtle Creek, Oregon, in 1989, when *E. villosus* was released. (Photo credit: E. Coombs, Oregon Department of Agriculture)

Same vista in 2000, after control by *E. villosus* and *C. australis*. (Photo credit: E. Coombs, Oregon Department of Agriculture)

**Availability:** Readily available in California, Idaho, Oregon, and Washington.

**Stages to transfer:** Adult or overwintering larva in infested seed heads.

**Redistribution:** Collecting adults with sweep nets can cause some mortality when they are damaged by grass seeds and other debris in the nets. Infested seed heads can be collected in late winter and moved to new sites for spring adult emergence, or adults can be reared from the infested seed heads in the laboratory and then collected and transported to the field for release. On rare occasions, parasitoids have been reared from *C. succinea*; these parasitoids will likely attack *C. australis*. Use caution in moving seed heads from one place to another to avoid moving parasitoids and introducing new biotypes of yellow starthistle into new areas.

### Comments

The first generation of *C. australis* often attacks bachelor's button (*Centaurea cyanus*). The next generation attacks yellow starthistle from June through October. In some areas it is primarily associated with *C. cyanus* and may not readily establish where bachelor's button does not co-occur with yellow starthistle. Bachelor's button is a minor exotic weed and ornamental in the Pacific Northwest and northern California.

## *Eustenopus villosus*

M. J. Pitcairn, G. L. Piper, and E. M. Coombs

**Common name:** Yellow starthistle hairy weevil.

**Type of agent:** Insect: Beetle, weevil (Coleoptera: Curculionidae).

**Native distribution:** Southern Eurasia, Mediterranean basin.

**Original source:** Northern Greece.

### Biology

**Generations per year:** One.

**Overwintering stage:** Adult (outside the host plant).

**Egg stage:** Single eggs are laid inside mature, closed head buds in early summer to midsummer. Females chew a hole in the side of the head for oviposition. The hole is usually capped with dark mucilage. The hole or cap is visible. The oviposition hole may cause the flower head to become distorted. The eggs hatch in three days.

**Larval stage:** The larvae feed on the receptacle and developing seeds. Larval development is reported to take 16 days.

**Pupal stage:** Pupation occurs within the seed head in a chamber fashioned from chewed seeds and pappus hairs. The pupal stage lasts from eight to 13 days.

**Adult stage:** The adults are present from late May through August. They are brown with whitish stripes and long hairs on the back. They have long, slender snouts and are 4 to 6 mm long, not including the snout.

427

*E. villosus* larva in seed head. (Photo credit: USDA-ARS)

*E. villosus* adult on yellow starthistle. (Photo credit: N. Poritz, bio-control.com)

### Effect

**Destructive stages:** Adult and larval.

**Additional plant species attacked:** Adults have been observed to feed on other exotic *Centaurea*, including *C. stoebe* ssp. *micranthos*, *C. melitensis*, *C. sulphurea*, and occasionally on *Acroptilon repens*.

**Site of attack:** Seed heads.

**Impact on the host:** Adults feed on small seedhead buds (whicch turn grayish with a slight crook), usually destroying a high percentage. Adults also feed on mature buds which causes limited damage. The larvae feed inside the flower heads and reduce seed production by as much as 90 to 100% in infested heads.

**Nontarget effects:** No nontarget effects have been reported.

### Releases

**First introduced into the United States:** 1990, California, Idaho, Oregon, and Washington.

**Established in:** Arizona, California, Idaho, Nevada, Oregon, and Washington.

**Habitat:** This weevil is widespread throughout most of the western United States wherever yellow starthistle is found.

428

**Availability:** This weevil is readily available.

**Stage to transfer:** Adult.

**Redistribution:** Adults can be swept or handpicked from the host plants in early summer.

## Comments

*Eustenopus villosus* is a double threat to yellow starthistle because it attacks both buds and flower heads. It has readily established and built up high numbers at most release sites. Because of this, the weevil has excellent biocontrol potential. Several collection sites in Oregon and Washington are no longer available because of the decline of the host weed.

# *Larinus curtus*

M. J. Pitcairn, G. L. Piper, and E. M. Coombs

**Common name:** Yellow starthistle flower weevil.

**Type of agent:** Insect: Beetle, weevil (Coleoptera: Curculionidae).

**Native distribution:** Southern Europe, Mediterranean basin.

**Original source:** Northern Greece.

## Biology

**Generations per year:** One.

**Overwintering stage:** Adult (outside the host plant).

**Egg stage:** The eggs are laid at the base of flowers, in heads with open flowers. The incubation period is four days.

**Larval stage:** The larvae feed on developing seeds. There are four larval growth stages; the total period of larval development lasts 17 to 20 days.

**Pupal stage:** Pupation occurs in chambers formed in the seed heads. The duration of the pupal stage is four to five days.

**Adult stage:** The adults feed on flowers and pollen of yellow starthistle. They occur on the plant from June to early August. Adults are dark brown to black with a somewhat mottled appearance due to patches of light-colored hairs on the back that are sometimes covered with pollen. Body length is about 5 to 6 mm,

*L. curtus* larva. (Photo credit: USDA-ARS)

429

not including the snout. It is the most robust of the weevils that attack yellow starthistle. This weevil characteristically inserts its head deep into the flowers leaving the posterior exposed to view.

*L. curtus* adult. (Photo credit: USDA-ARS)

## Effect

**Destructive stage:** Larval.

**Additional plant species attacked:** None.

**Site of attack:** Seed heads (interior).

**Impact on the host:** Larval feeding on developing seeds can reduce seed production up to 100%. Adults feed on flowers and pollen, but this probably has little impact on the plant.

**Nontarget effects:** No nontarget effects have been reported.

## Releases

**First introduced into the United States:** 1992, California, Idaho, Oregon, and Washington.

**Established in:** California, Idaho, Oregon, and Washington.

**Habitat:** The habitat requirements for this weevil are unknown.

**Availability:** Availability of this weevil is limited in southwestern Oregon and California. It can be readily collected in western Idaho, eastern Oregon, and southcentral Washington.

**Stage to transfer:** Adult.

**Redistribution:** Adults can be swept or handpicked from yellow starthistle in June and July. Adults are most easily collected by handpicking them from flowering plants at about 10% bloom.

## Comments

In California and some sites in Washington and southwestern Oregon, *Larinus curtus* has failed to build significant populations where established. Thus, its potential as a biological control agent appears limited in those areas. Populations appear to fare better in the Columbia Gorge and Snake River areas of Oregon and Washington.

# *Puccinia jaceae* var. *solstitialis*

D. M. Woods and W. L. Bruckart, III

**Common name:** Yellow starthistle rust.

**Type of agent:** Rust fungus (Uredinales: Pucciniaceae).

**Native distribution:** Southern Eurasia and the Mediterranean basin.

**Original source:** Turkey.

## Biology

**Generations per year:** Multiple.

**Overwintering stage:** Teliospore.

**Life cycle of the fungus:** The rust has five spore stages that are all completed on a single host. Urediniospores are rust-colored and are produced on leaves and stems during most of the plant's life. Spores are spread by wind. When appropriate levels of moisture and temperature are present, the spores germinate and infect new plants producing new infections and more urediniospores. With ideal weather conditions, this portion of the life cycle can repeat every two weeks. By midsummer, as plants mature, teliospores are formed in place of the urediniospores. These darker, thick-walled, two-celled spores are the overwintering stage for the rust. Teliospores germinate in the spring producing basidiospores which initiate the sexual process. Infections resulting from the sexual process produce aecia and aeciospores. The aecial stage is followed by the uredinial stage of the fungus which is visually indistinguishable from the aecial stage.

**Urediniospore description:** This is the most characteristic spore of the fungus. Spores are round or slightly flattened, about 25 microns in diameter, with a thick wall that is thicker at the point of attachment (hilum), nearly always with two germpores above the "equator" of the spore (halfway between the hilum and the top of the spore), and the spore surface is covered with short spines (echinulations). A mass of spores, as would be seen in a pustule on an infected leaf, is dark to reddish-brown and powdery in appearance.

## Effect

**Destructive stage:** The uredinial spore stage is the repeating infective stage and likely has the greatest impact on plant development.

**Additional plant species attacked:** None.

**Site of attack:** Foliage and green stems.

**Impact on the host:** The rust is anticipated to reduce plant vigor.

*P. jacea* var. *solstitialis* pustules on leaves. (Photo credit: California Department of Food and Agriculture)

431

**Nontarget effects:** No nontarget impacts have been reported.

## Releases

**First introduced into the United States:** 2003, California.

**Established in:** Recovered first generation in California.

**Habitat:** Climates with dew periods that allow moisture to form on foliage are the most favorable for the development of the fungus.

**Availability:** Not available for large-scale release as of 2003, but the fungus may become more readily available in following years.

**Stage to transfer:** Urediniospores.

**Redistribution:** Urediniospores are vacuumed from infected leaves, suspended in water with a wetting agent, and then sprayed on foliage prior to an extended dew period.

## Comments

Plant pathogens, even those used as biological control agents, are subject to a different set of regulations than arthropod biological control agents. In California, United States Environmental Protection Agency (EPA) guidelines are enforced by California EPA, and an experimental use permit is filed for each field release.

# *Urophora sirunaseva*

M. J. Pitcairn, G. L. Piper, and E. M. Coombs

**Common name:** Yellow starthistle gall fly.

**Type of agent:** Insect: Fly  (Diptera: Tephritidae).

**Native distribution:** Southern Europe, Mediterranean basin.

**Original source:** Northern Greece.

## Biology

**Generations per year:** Two.

**Overwintering stage:** Larval (in seed head galls).

**Egg stage:** Egg laying begins within three days after adult emergence. The eggs are laid in the spring and summer among the immature florets of intermediate-stage closed head buds with vertical spines. Up to 270 eggs can be deposited by a female. The eggs are white and spindle-shaped.

**Larval stage:** A one-chambered, woody gall is formed around each larva within a seed head. Multiple galls occur within a single head.

**Pupal stage:** Pupation occurs within the galls in the seed heads.

**Adult stage:** Adults emerge from galls in the spring (overwintering generation) and summer. The fly is black with a yellow spot on the thorax. Its wings are marked with dark crossbands. Males are about 3 to 4 mm long and females are 4 to 6 mm long, including the ovipositor.

*U. sirunaseva* larva (Photo credit: USDA-ARS)

*U. sirunaseva* adult. (Photo credit: USDA-ARS)

## Effect

**Destructive stage:** Larval.

**Additional plant species attacked:** None.

**Site of attack:** Seed heads (interior).

**Impact on the host:** Galls displace seeds and act as a nutrient sink, thereby reducing seed production.

**Nontarget effects:** No nontarget effects have been reported.

## Releases

**First introduced into the United States:** 1984, California.

**Established in:** California, Idaho, Oregon, and Washington.

**Habitat:** This gall fly is widespread throughout most of the western United States wherever yellow starthistle is found. This species does not seem to do well in windy areas such as in the Columbia River Gorge in Oregon and Washington. Specific habitat requirements are unknown.

**Availability:** This fly is readily available for collection and redistribution.

**Stages to transfer:** Adult and larval.

**Redistribution:** Adults can be collected with a sweep net from May through August. It is best to time collections with peak emergence, usually late May and July. Infested seed heads can be collected in late winter and moved to new sites for spring adult emergence; however this method can also move several larval parasitoids known to attack this species and is not recommended for areas where the fly is free of

433

parasitoids. A better alternative is to rear adults from the infested seed heads in the laboratory and field-release them.

## Comments

While up to 12 galls per head have been recorded in the field in California, usually only one to two galls per head are observed. This species occurs in low numbers, rarely exceeding 25% attack rate, thus its potential as a biological control agent appears limited.

## References

Balciunas, J. K., and B. Villegas. 2001. Unintentionally released *Chaetorellia succinea* (Diptera: Tephritidae): Is this natural enemy of yellow starthistle a threat to safflower growers? Environ. Entomol. 30: 953-63.

Bruckart, W. L. 1989. Host range determination of *Puccinia jaceae* from yellow starthistle. Phytopath. 79: 155-60.

Callihan, R. H., F. E. Northam, J. B. Johnson, E. L. Michalson, and T. S. Prather. 1989. Yellow starthistle: biology and management in pasture and rangeland. Univ. Idaho College of Agric. Coop. Ext. Serv. Curr. Info. Ser. No. 634, Moscow, ID.

Campobasso, G., R. Sobhian, L. Knutson, and G. Terragitti. 1998. Host specificity of *Bangasternus orientalis* Capiomont (Coleoptera: Curculionidae), introduced into the United States for biological control of yellow starthistle (*Centaurea solstitialis* L., Asteraceae: Carduae). Environ. Entomol. 27: 1525-30.

Fornasari, L., C. E. Turner, and L. A. Andres. 1991. *Eustenopus villosus* (Coleoptera: Curculionidae) for biological control of yellow starthistle (Asteraceae: Cardueae) in North America. Environ. Entomol. 20: 1187-94.

Jette, C., J. Connett, and J. P. McCaffrey. 1999. Biology and Biological Control Agents of Yellow Starthistle. USDA Forest Serv. Forest Health Technol. Enterprise Team. FHTET-98-17. Morgantown, WV.

Lanini, W. T., C. Thomsen, T. S. Prather, C. E. Turner, M. J. Smith, C.L. Elmore, M. Vayssieres, and W. A. Williams. 1994. Yellow starthistle. Univ. Calif. Div. Agric. Nat. Res., Pest Notes No. 3.

Maddox, D. M., A. Mayfield, and C. E. Turner. 1990. Host specificity of *Chaetorellia australis* (Diptera: Tephritidae) for biological control of yellow starthistle (*Centaurea solstitialis*, Asteraceae). Proc. Entomol. Soc. Wash. 92: 426-30.

Maddox, D. M., R. Sobhian, D. B. Joley, A. Mayfield, and D. Supkoff. 1986. New biological control for yellow starthistle. Calif. Agric. 40 (11 and 12): 4-5.

Rosenthal, S. S., G. Campobasso, L. Fornasari, R. Sobhian, and C. E. Turner. 1991. Biological control of *Centaurea* spp. Pages 292-302 *in* L. F. James, J. O. Evans, M. H. Ralphs, and R. D. Child, eds. Noxious Range Weeds. Westview Press, Boulder, CO.

Turner, C. E. 1994. Host specificity and oviposition of *Urophora sirunaseva* (Hering) (Diptera: Tephritidae), a natural enemy of yellow starthistle. Proc. Entomol. Soc. Wash. 96: 31-36.

Turner, C. E., J. B. Johnson, and J. P. McCaffrey. 1995. 72. Yellow starthistle, *Centaurea solstitialis* L. (Asteraceae). Pages 270-75 *in* J. R. Nechols, L. A. Andres, J. W. Beardsley, R. D. Goeden, and C. G. Jackson, eds. Biological Control in the Western United States: Accomplishments and Benefits of Regional Research Project W-84, 1964-1989. Univ. Calif. Div. Agric. Nat. Res. Pub. 3361. Oakland, CA.

Turner, C. E., G. L. Piper, and E. M. Coombs. 1996. *Chaetorellia australis* (Diptera: Tephritidae) for biological control of yellow starthistle, *Centaurea solstitialis* (Compositae), establishment and seed destruction. Bull. Entomol. Res. 86: 177-82.

Turner, C. E., R. Sobhian, D. B. Joley, E. M. Coombs, and G.L. Piper. 1994. Establishment of *Urophora sirunaseva* (Diptera: Tephritidae) for biological control of yellow starthistle in the western United States. Pan-Pac. Entomol. 70: 206-11.

Villegas, B., F. Hrusa, and J. Balciunas. 2001. *Chaetorellia* seed head flies and other seed head insects on *Cirsium* thistles in close proximity to *Centaurea* spp. Pages 76-77 *in* D. M. Woods, ed. Biological Control Program Annual Summary, 2000. Calif. Dept. Food and Agric., Plant Health and Prevention Services, Sacramento, CA.

Woods, D. M., and V. Popescu. 2000. Seed destruction in Sicilian starthistle by yellow starthistle biological control insects. Pages 64-66 *in* D. M. Woods, ed. Biological Control Program Annual Summary, 1999. Calif. Dept. Food and Agric., Plant Health and Prevention Services, Sacramento, CA.

Woods, D. M., B. Villegas, and V. Popescu. 2001. Attack of Napa thistle by yellow starthistle biological control insects. Page 78 *in* D. M. Woods, ed. Biological Control Program Annual Summary, 2000. Calif. Dept. Food and Agric., Plant Health and Prevention Services, Sacramento, CA.

# New and Ongoing Biological Control Projects in the United States

Classical biological control research is being conducted on a number of new plant targets that are nonnative and invasive in the United States. Following is a synopsis of these new biological control research projects as of 2003. In some cases, research is in full swing; in others, research has been suspended and may be continued in the future. For more information, contact project cooperators and/or research agencies conducting the work.

## *Acknowledgments*

Thanks to Hariet Hinz of CABI Bioscience Switzerland Centre; Luke Skinner, Minnesota Department of Natural Resources; Dick Shaw, CABI Bioscience UK; and Chuck Quimby, Jr., USDA-ARS European Biological Control Laboratory for their reviews of and suggestions for Section III.

# Brazilian Peppertree                                          *Schinus terebinthifolius*

Sumac family—Anacardiaceae

*J. P. Cuda, D. H. Habeck, S. D. Hight, J. C. Medal,*
*and J. H. Pedrosa-Macedo*

Brazilian peppertree (*Schinus terebinthifolius*), also known as Christmasberry or Florida holly, is an invasive, toxic evergreen shrub or small tree that is threatening the biodiversity of Florida, California, and Hawaii. Native to Argentina, Paraguay, and Brazil, Brazilian peppertree was introduced into Florida as a landscape ornamental in the late 19th century. Its popularity as an ornamental plant can be attributed to the numerous bright-red drupes (fruits) produced during the October to December holiday season. Brazilian peppertree now dominates entire ecosystems in southcentral Florida and its invasive potential has been documented in California as well as Hawaii. The plant readily invades disturbed sites as well as natural communities where it forms dense thickets that completely shade out and displace native vegetation.

Brazilian peppertree also is considered an invasive weed in at least 20 countries throughout subtropical regions worldwide. Its invasiveness is attributed to its enormous reproductive potential. Large quantities of drupes are produced per plant, and wildlife disperse the seeds in their droppings. Brazilian peppertree outcompetes native plants because of its tolerance to extreme moisture conditions, its capacity to grow in shady environments, and possible allelopathic effects on neighboring plants. In Florida, the plant readily invades disturbed sites (e.g., fallow farmlands) as well as natural communities such as pinelands, hardwood hammocks, and mangrove forests, and is a major invader of the Everglades National Park. In the early 1990s, it was estimated that more than 400,000 ha in Florida were infested with Brazilian peppertree.

A lack of natural enemies on Brazilian peppertree in Hawaii was the rationale for initiating a classical biological control program in the 1950s. Surveys were conducted in South America and three insects (two moths and one beetle) were screened and eventually released in Hawaii. Two of the insects established, but they apparently have had little effect on Brazilian peppertree population levels in Hawaii.

During the 1980s, a classical biological control program was initiated in Florida following the completion of two domestic surveys of the insect fauna associated with Brazilian peppertree. Exploratory surveys for promising natural enemies in the native range of Brazilian peppertree were initiated in 1987 in South America. Several insects were identified as potential biological control agents.

The defoliating sawfly *Heteroperreyia hubrichi* (Hymenoptera: Pergidae) was initially selected as a candidate for further study because the larvae visibly damaged the plant in its native range, and because the insect was collected in South America only from Brazilian peppertree. Field studies in Brazil indicated the sawfly is capable of developing high population densities and the voracious larvae may completely defoliate Brazilian peppertrees up to 6 m in height. This type of feeding damage could severely injure or kill young plants and prevent older plants from reproducing, thereby reducing the competitive advantage that Brazilian peppertree currently holds over native vegetation.

**439**

The results of Hawaiian screening studies showed that although Brazilian peppertree was the preferred host for the sawfly, a native sumac (*Rhus sandwicensis*) also could be a potential host plant. Because of the perceived risk to the native Hawaiian sumac, the introduction of the sawfly into Hawaii was postponed until additional information about the insect's host range could be obtained from multiple-choice laboratory tests or preferably open field tests conducted in Brazil. These studies have been postponed due to a lack of funding and public support for the project.

A petition to release the sawfly in Florida was submitted to the Technical Advisory Group for Biological Control Agents of Weeds (TAG) in 1996. After reviewing the petition, the TAG considered the insect to be sufficiently host-specific to introduce into Florida. Based on the TAG recommendation, an Environmental Assessment was prepared by APHIS-PPQ and submitted to the U.S. Fish and Wildlife Service for consultation. The U.S. Fish and Wildlife Service expressed concern that the sawfly could pose a risk to Michaux's sumac (*Rhus michauxii*), a federally listed endangered species. Supplemental host range tests showed that *H. hubrichi* will not attack this sumac. However, field release of *H. hubrichi* has been delayed following the discovery that the larvae

> Additional information on the management of Brazilian peppertree can be found in:
>
> Ferriter, A. P., ed. 1997. Brazilian pepper management plan for Florida. Florida Exotic Pest Plant Council, Brazilian Pepper Task Force, SFWMD, West Palm Beach, FL. http://www.fleppc.org/Manage_Plans/schinus.pdf.
>
> Hight, S. D., J. P. Cuda, and J. C. Medal. 2002. Brazilian peppertree. Pages 311-21 *in* R. Van Driesche, S. Lyon, B. Blossey, M. Hoddle, and R. Reardon, eds. Biological Control of Invasive Plants in the Eastern United States. USDA Forest Service Pub. FHTET-2002-04. Morgantown, WV.
>
> Randall, J. M. 2000. *Schinus terebinthifolius* Raddi. Pages 282-87 *in* C. C. Bossard, J. M. Randall, and M. C. Hoshovsky, eds. Invasive Plants of California's Wildlands. Univ. California Press, Berkeley, CA.

contain the amino acids lophyrotomin and pergidin, which are toxic to some vertebrates. The presence of these toxins has created a risk issue that is unprecedented in the field of weed biological control because of the potential for poisoning of susceptible native wildlife and domesticated animals that may consume the sawfly larvae. A pilot release of unmated females, which will produce only male progeny, has been proposed for a field risk assessment study.

Another promising natural enemy of Brazilian peppertree is the thrips *Pseudophilothrips ichini* (Thysanoptera: Phlaeothripidae). The biology and field host range of *P. ichini* were studied in southeastern Brazil, and its host range was investigated in Florida quarantine. A petition to release the thrips was submitted to the TAG in 2002. A permit to release *P. ichini* in Florida is expected, pending TAG approval and an Environmental Assessment in consultation with the U.S. Fish and Wildlife Service.

The host range of the leafroller moth *Episimus utilis* (Lepidoptera: Tortricidae) that was introduced into Hawaii for classical biological control of Brazilian peppertree in the 1950s currently is under investigation in Florida quarantine. The biology of *E. utilis* and development of a rearing procedure for maintaining a laboratory colony for host specificity studies were completed in 2001.

The adventive torymid wasp *Megastigmus transvaalensis* (Hymenoptera: Torymidae) attacks the drupes and is the only insect currently causing some damage to the plant in California, Florida, and Hawaii. This wasp was originally described from South Africa and was probably introduced accidentally into the United States in Brazilian peppertree seeds sold as spices in some food shops. A detailed study on the distribution and effect of the wasp on Brazilian peppertree was completed in Florida in 2001.

Research on biological control of Brazilian peppertree in Florida is being supported by the Florida Department of Environmental Protection, the South Florida Water Management District, and the UF/IFAS Center for Aquatic and Invasive Plants. This is Florida Agricultural Experiment Station Journal Series No. N-02429.

## Cape ivy               *Delairea odorata (=Senecio mikanoides)*

Sunflower family—Asteraceae

*J. K. Balciunas*

Cape ivy (*Delairea odorata*), also known as German ivy, is a native of South Africa and has become one of the most pervasive and damaging nonnative plants to invade coastal areas of the western United States. This vine has the potential to cause serious environmental problems by overgrowing riparian and coastal vegetation, including endangered plant species, and is reputed to be poisonous to aquatic organisms. In California, Cape ivy is spreading in riparian forests, coastal scrubland, coastal bluff communities, and seasonal wetlands. Cape ivy was also introduced into the Big Island of Hawaii around 1909 and has become a serious weed in a variety of upland habitats there.

Research on biological control of Cape ivy began in 1998 when a USDA-ARS scientist joined South African colleagues to survey for natural enemies of this plant. More than 230 species of plant-injuring insects were collected over the next two years. Thus far, two insects have shown the most promise as safe and effective biological control agents in the United States. The host ranges of the gall fly *Parafreutreta regalis* and the stem-boring moth *Digitivalva delaireae* were extensively tested by USDA-ARS scientists in Albany, California, between 2001 and 2004. Experiments indicated that the very selective *P. regalis* fly induced galls that reduced the biomass and structure of Cape ivy vines. The minute *Digitivalva* caterpillars frequently kill leaves, stems, and occasionally entire plants of Cape ivy. Testing will continue on both potential agents.

In addition, in South Africa a flower-feeding phalacrid beetle as well as a pathogen that damages Cape ivy leaves show some promise as potential biological control agents and may be tested.

This project has been conducted cooperatively with the California Exotic Pest Plant Council (now called California Invasive Plant Council) and the California Native Plant Society, as well as numerous colleagues in South Africa. The annual reports for the USDA-ARS Biological Control of Cape Ivy Project, by Joe Balciunas, are available online at http://wric.ucdavis.edu/exotic/exotic.htm.

## Common buckthorn

*Rhamnus cathartica*

Buckthorn family—Rhamnaceae

Common buckthorn (*Rhamnus cathartica*) is an introduced shrub or small tree in North America that causes significant economic and environmental damage. It was probably introduced to North America in the 1800s for use as hedges, shelter belts, and wildlife habitat. Today it can be found throughout the northeastern and midwestern United States in pastures, fencerows, roadsides, and ravines. Common buckthorn invades logged or grazed woodlands, disturbed forest edges or openings, prairie thickets, and open oak woodlands. It has a long growing season and a rapid growth rate. It vigorously sprouts from buds below the soil surface and is a prolific seed producer.

In 2002 and 2003, CABI Bioscience Switzerland Centre identified suitable collection sites and times for several promising biological control agents of exotic buckthorns. Potential agents have been prioritized according to: 1) host acceptance; 2) potential host specificity; 3) phenology and food niche; and 4) potential availability and rearing facility. The leaf margin galler *Trichodermes walkeri*, the shoot-tip miner *Sorhagenia janiszewskae*, the lepidopterous defoliator *Philereme vetulata*, the stem-borer *Oberea pedemontana*, and the root-boring sesiid moth *Synanthedon stomoxiformis* have been identified for further study. New emphasis will be put on field surveys of flower- and fruit/seed-feeding insects as well as on *O. pedemontana*. In 2004, preliminary screening tests will be carried out with *P. vetulata*, *S. janiszewskae*, and *T. walkeri*.

Excerpted, in part, from CABI Bioscience Switzerland Centre Annual Report 2002, available online at http://www.cabi-bioscience.org/html.ch.htm.

This project has been conducted cooperatively with Luke Skinner, Minnesota Department of Natural Resources.

## Common teasel

*Dipsacus fullonum*

Teasel family—Dipsacaceae

Common teasel (*Dipsacus fullonum*) is endemic to Europe, but was introduced to North America possibly as early as the 1700s. Teasel grows in open, sunny habitats such as prairies, savannas, seeps, and sedge meadows, although roadsides, dumps, and heavily disturbed areas are the most common habitats infested. In the last several decades, common teasel has spread rapidly and invaded 40 of the 50 U.S. states.

In 2003, teasel seed heads from different European countries were collected by USDA-ARS scientists for study at the USDA laboratory in France. Tortricid moths from France, Turkey, Spain, and Greece were found in dry flower heads, but did not seem to significantly reduce seed production. A flea beetle was collected from teasel rosettes in Russia and identified as *Galleruca fuliginosa*. In addition, the beetle *Galleruca pomonae* was found feeding on teasel in France. Both beetles will continue to be tested for host specificity in field and greenhouse trials.

Excerpted from USDA-ARS European Biological Control Laboratory Research Progress report 2003 (http://www.ars-ebcl.org/).

This project has been conducted cooperatively with the University of Illinois (Natural History Survey) and the University of Missouri.

# Garlic mustard

*Alliaria petiolata*

Mustard family—Brassicaceae

Garlic mustard (*Alliaria petiolata*) is a biennial cruciferous plant of European origin that was presumably introduced into North America for medical use and as a green vegetable. It has spread throughout northeastern and midwestern North America and is now recorded in 28 states in the United States. Garlic mustard invades natural forest communities and is thought to decrease biodiversity of native herbaceous communities.

In 1998, CABI Bioscience Switzerland Centre began investigations into the biological control of garlic mustard. As of 2003, European scientists were studying four weevils: the stem-miners *Ceutorhynchus alliariae* and *C. roberti*, the root-miner *C. scrobicollis*, and *C. constrictus*, which develops in the seeds of garlic mustard. Host specificity tests have been focusing on the Brassicaceae family, as well as plants indigenous to North America in other families growing intermixed with garlic mustard. Tests with all species are quite advanced and successful development to the adult stage has been restricted to a limited number of very closely related plants. Since it has been shown to be the most damaging agent, *C. scrobicollis* has been prioritized and will be sent to the United States for additional tests in quarantine.

Excerpted from CABI Bioscience Switzerland Centre Annual Report 2002, available online at http://www.cabi-bioscience.org/html.ch.htm.

This project has been conducted cooperatively with Bernd Blossey, Cornell University, with financial support from the Strategic Environmental Research Development Programme and USDA Forest Service.

# Giant reed

*Arundo donax*

Grass family—Poaceae

Giant reed (*Arundo donax*) is a stout, tall, perennial grass native to the Mediterranean region. The plant has invaded the warmer coastal drainages on both coasts of the United States. The plant spreads aggressively by creeping rhizomes; its rapid growth and ability to recover after fires have allowed it to outcompete native vegetation. Extensive monocultures of giant reed pose serious ecological problems such as evapotranspiration/water loss, flood and debris risks, erosion from undercutting, sediment retention, and loss of biodiversity.

In 2003, at least one new species of chloropid fly that destroys new giant reed shoots was found in France. Studies have been initiated by USDA-ARS to determine the effect of these chloropid flies on plant populations. Parallel studies will be conducted in the United States to determine the effect of native North American natural enemies on giant reed populations.

Excerpted from USDA-ARS European Biological Control Laboratory Research Progress report 2003 (http://www.ars-ebcl.org/).

USDA-ARS scientists in France are working closely with USDA-ARS scientists in Albany, California, and Temple, Texas, on this project.

# Hawkweeds

*Hieracium* spp.

Sunflower family—Asteraceae

*L. M. Wilson and M. Schwarzländer*

Hawkweeds (*Hieracium* spp.) are herbaceous, perennial plants that, like dandelion and chicory, contain a milky juice. There are about 11 species of nonindigenous hawkweeds in North America, six of which are weedy. All are yellow-flowered, except one orange-flowered species, orange hawkweed (*H. aurantiacum*), which probably has the widest distribution because it has been sold and planted as an ornamental. Meadow hawkweed (*H. caespitosum [= H. pratense]*) and the closely related king devil hawkweed (*H. floribundum*) are considered the most invasive. A similar species, yellow-devil hawkweed (*H. glomeratum*) is known from northeastern Washington and southeastern British Columbia. Mouse-ear hawkweed (*H. pilosella*) is abundant throughout the eastern United States and Canada, but also occurs in coastal Oregon and Washington. A less widespread hawkweed, tall hawkweed (*H. piloselloides*), occurs in western Montana, northeastern Idaho, southeastern British Columbia, and southwestern Alberta. These invasive hawkweeds are classified in the subgenus *Pilosella*; all are polyploid and reproduce vegetatively and by asexual seed production (apomixis). Three additional species, also nonindigenous but belonging to the subgenus *Hieracium*, are common hawkweed (*H. vulgatum*), European hawkweed (*H. saubaudum*), and wall hawkweed (*H. mumorum*). Taxonomic confusion resulting from asexual reproduction and polymorphism renders classification of this subgenus in North America difficult and arguably tentative.

The complex of weedy nonindigenous hawkweeds in North America is matched by a large and diverse complex of native species. About 25 native hawkweeds occur throughout the United States and Canada, mostly in cool, mesic, forest environments from the eastern seaboard throughout the Midwest, the intermountain West, and the desert mountains of the Southwest. Taxonomically, these species belong to the subgenus *Stenotheca*; they are not stoloniferous and all are sexual diploids.

Development of a biological control program for invasive hawkweeds is being coordinated through the Hawkweed Biological Control Consortium in cooperation with CABI Bioscience Switzerland Centre and the Bozeman Quarantine Facility at Montana State University. Completion of host specificity testing is expected in 2005. Four insect species are being tested. They include the root-feeding hover fly *Cheilosia urbana* and the hawkweed crown-feeding hover fly *C. psilophthalma* (Diptera: Syrphidae), the hawkweed gall midge *Macrolabis pilosellae* (Diptera: Cecidomyiidae), and the hawkweed gall wasp *Aulacidea subterminalis* (Hymenoptera: Cynipidae).

# Houndstongue

*Cynoglossum officinale*

Borage family—Boraginaceae

*J. M. Story*

Houndstongue is an Eurasian biennial weed that was introduced into North America as a contaminant of cereal seed in the late 1800s. The weed has become widely distributed in the temperate areas of the United States and Canada and is increasing in much of the western region. The plant colonizes disturbed sites and heavily grazed areas in riparian zones.

Houndstongue is a serious problem on rangeland and pasture. The weed is invasive and significantly reduces forage. In addition, the barbed seeds (or burrs) of the plant readily adhere to hair, wool, and fur, reducing the value of sheep wool and causing irritation and behavioral problems in cattle. Even more importantly, houndstongue contains large quantities of pyrrolizidine alkaloids that are toxic to cattle and horses. Cattle avoid feeding on green plants, but deaths have been reported in the United States when the animals fed on houndstongue-contaminated hay.

Beginning in 1988, British Columbia, with later support from Montana and Wyoming, funded the identification and screening of potential biological control agents of houndstongue in Europe. To date, five promising insects have been identified and are in varying stages of host specificity testing or release. These include a root weevil, *Mogulones cruciger*; a seed weevil, *M. borraginis*; a stem weevil, *M. trisignatus*; a root beetle, *Longitarsus quadriguttatus*; and a root fly, *Cheilosia pasquorum*. The screening of *M. cruciger* and *L. quadriguttatus* was completed and the insects were released in British Columbia in 1997 and 1998, respectively. The remaining three insect species are in the latter stages of host specificity testing at CABI Bioscience Switzerland Centre as of 2003.

*Mogulones cruciger* is proving to be an excellent biological control agent in British Columbia. The insect has rapidly developed large populations and is causing noticeable declines in houndstongue density. *Longitarsus quadriguttatus* is also established in British Columbia, but its population increase has been slower.

Petitions to release both *M. cruciger* and *L. quadriguttatus* in the United States have been submitted to USDA-APHIS. Due to concerns by the U.S. Fish and Wildlife Service about the potential threat of *M. cruciger* to an endangered plant species in Texas, review of the petitions has been suspended. Additional testing of these two insects on additional U.S. native plants, including the endangered plant species, is planned to demonstrate host specificity.

## Japanese knotweed
*Fallopia japonica*

Buckwheat family—Polygonaceae

For more information, see:

Shaw, R. H., and L.A. Seiger. 2002. Japanese knotweed. Pages 159-66 *in* R. Van Driesche, S. Lyon, B. Blossey, M. Hoddle, and R. Reardon, eds. Biological Control of Invasive Plants in the Eastern United States. USDA Forest Serv. Pub. FHTET-2002-04. Morgantown, WV.

The rhizomatous perennial Japanese knotweed (*Fallopia japonica*) is a relatively recent arrival to the United States from Asia. It grows to a height of 2 to 3 m with bamboo-like stems. The orange to brown, woody, dead stems persist throughout the winter and new shoots, produced from the extensive rhizome system, grow up among these the following spring to form dense thickets. The dead stems and leaf litter decompose very slowly and form a deep organic layer that prevents the seeds of native plants from germinating. Once established, Japanese knotweed rapidly increases in area and soon forms monoculture stands. Knotweed is capable of pushing through asphalt and even concrete.

Funding from the Welsh Development Agency (UK) and the U.S. Forest Service enabled a team of British scientists from CABI Bioscience and Leicester University (UK) to carry out an initial survey for natural enemies in Japan in 2000. A large number of arthropod and fungal agents were identified. A rust fungus, in particular, has proven to be very damaging in the laboratory. The full four-year project for the UK, funded by a consortium of partners, is in place and work began in May 2003. A U.S. consortium is being organized through Cornell University.

Excerpted from CABI Bioscience (www.cabi-bioscience.org/) and the Japanese Knotweed Alliance (http://www.cabi-bioscience.org/html/japanese_knotweed_alliance.htm).

## Medusahead ryegrass     *Taeniatherum caput-medusae* ssp. *asperum*
Grass family—Poaceae

This winter annual grass originated in areas bordering the Mediterranean Sea and was introduced into the United States in the late 1880s. Medusahead is now invasive across millions of hectares of intermountain rangelands in the western United States. It favors sites that have soils with high clay content and semiarid areas.

Two smut diseases that eliminate seed production were identified by USDA-ARS scientists in 2002-2003. A cooperative agreement with a university in Turkey was established by the USDA-ARS laboratory in France to begin field and possible alternate host testing.

Excerpted from USDA-ARS European Biological Control Laboratory Research Progress report 2003 (http://www.ars-ebcl.org/).

USDA-ARS scientists in France are working closely with USDA-ARS scientists in Frederick, Maryland, Albany, California, and Reno, Nevada, on this new project.

# Old World climbing fern

*Lygodium microphyllum*

Climbing Fern family—Lygodiaceae

*R. W. Pemberton, J. A. Goolsby, and A. D. Wright*

Old World climbing fern (*Lygodium microphyllum*) is an aggressive invasive weed of moist habitats in southern Florida. This weed colonizes new areas without the need of habitat disturbance and frequently completely dominates native vegetation. The weed's ability to grow up and over trees and shrubs and to run horizontally allows it to smother whole communities of plants. Thick skirts of old fronds enclose trees and serve as ladders that carry fire into tree canopies, resulting in the death of trees that normally withstand ground fires. The fern, first found to be naturalized in 1965, has become one of the most dangerous weeds in southern Florida and in 2002 infested more than 44,516 ha, almost three times the amount infested three years previously. In 2003, Everglades National Park had an estimated 4,000 ha of the weed, compared to 800 ha in 2002. Because *L. microphyllum* is limited to subtropical and tropical climates, the weed is unlikely to become a problem in other states except perhaps in southern Texas. It could be, however, a significant threat to moist habitats in the Caribbean, Mexico, and other parts of the Neotropics, which it may reach via wind-borne spores.

*Lygodium microphyllum* has an exceptionally large native range, occurring in much of the moist Old World tropics and subtropics from Senegal in west Africa to Southeast Asia, Australia, and the Pacific to Tahiti. This fern and Japanese climbing fern (*L. japonicum*), a related weed invasive in central Florida and the American South, were originally imported and sold as ornamental plants. The only native North American member of the genus is *L. palmatum*, a temperate native species distributed from Appalachia north to New England. *Lygodium microphyllum* can be distinguished from *L. japonicum* by entire versus toothed leaflet margins.

Research on biological control of *L. microphyllum* began in 1997 and has focused on surveys for natural enemies and host specificity testing of candidate biological control agents. Australia and southeast Asia have been most thoroughly examined, but surveys also have been made on *Lygodium* species in west Africa, Argentina, the Dominican Republic, and Japan. More than 20 herbivore species have been found associated with *Lygodium* spp. Among these enemies are complexes of pyralid moths (*Cataclysta*, *Musotima*, and *Neomusotima*) and noctuid moths (*Callopistria*) from the Australia-southeast Asian region that defoliate the plants and appear to be *Lygodium* specialists. Another unidentified pyralid moth from Thailand bores the stems of *Lygodium* spp. A buprestid beetle, *Endelus bakerianus*, mines the fronds of *Lygodium* spp. in southeast Asia. A sawfly, *Neostrombocerus albicomus*, is a defoliator from southeast Asia that appears to have host plant races using different *Lygodium* species. A *Brevipalpis* tenuapalpid mite that damages the leaves is the most widely distributed natural enemy, occurring in Africa, southeast Asia, Australia, and the Neotropics, but it probably uses unrelated ferns. A leaf-galling eriophyid mite in the genus *Floracarus*, occurring in Australia and Southeast Asia, appears to limit growth of the plants. A rust disease (*Puccinia lygodii*) that attacks the plants in the Neotropics was recently discovered in northern Florida on *L. japonicum*.

**447**

Host specificity testing is underway in Brisbane, Australia, and in Gainesville, Florida. Because fern phylogeny is less well defined than in flowering plants, the specificity testing uses representatives of the genera of native and economic ferns that occur in Florida and the Caribbean instead of the usual phylogenetic approach. In addition, four *Lygodium* species native to the Caribbean and the single North American *Lygodium* species are being tested, as are related *Anemia* and *Actinostachys* species. Cold-temperature tests are being used to determine the critical lower thermal limits for survival of the candidate biological control agents so we can exclude those that may be able to live in cold regions of the United States where the rare *L. palmatum* occurs. *Cataclysta camptozonale* has been petitioned for release, and the final testing of *Neomusotima conspurcatalis* was underway in the Gainesville Quarantine Laboratory in 2003. A stem-boring pyralid moth (*Ambia* sp.) from Thailand has been brought into Gainesville for evaluation. The specificity testing of the leaf gall mite *Floraracus* sp. was nearing completion in Brisbane in 2003, and several *Callopistria* moths are being evaluated in that lab. The mite appears to be extremely narrow in its host range and may have geographical races adapted to local forms of *L. microphyllum.*

> Additional information is available:
>
> *Lygodium* Management Plan for Florida, available online at http://www.fleppc.org/Manage_Plans/lymo_mgt.pdf.
> Pemberton, R. W., J. Goolsby, and T. Wright. 2002. Old World climbing fern (*Lygodium microphyllum* (Cav.) R.Br.). Pages 139-47 *in* R. Van Driesche, S. Lyon, B. Blossey, M. Hoddle, and R. Reardon, eds. Biological Control of Invasive Plants in the Eastern United States. USDA Forest Serv. Pub. FHTET-2002-04. Morgantown, WV.

In addition to completing the screening mentioned above, future developmental research will include: 1) establishing cultures of the pyralid stem-borer moth, the defoliating sawfly, and the leaf-mining buprestid beetle in Brisbane quarantine for preliminary screening; 2) finding geographical host races of the *Floracarus* mite; 3) prerelease evaluation of the effect of the mite on *L. microphyllum* biomass production; 4) exploration for new natural enemies in India, Borneo, and the Pacific Islands; 5) completion of DNA analysis of populations of *L. microphyllum* to try to determine the region of origin of the Florida plants to facilitate exploration; 6) completion of DNA analysis of *Lygodium* species to understand the patterns of relationship within the genus to assist risk evaluation; and 7) evaluation of the potential of the *Lygodium* rust to control *L. microphyllum.* In Florida, the development of release plots (treatment and control) in different vegetation types and different infestation levels is underway in anticipation of the releases of biological control agents for *L. microphyllum.*

This project has been conducted cooperatively by USDA-ARS, CSIRO, the University of Florida at Gainesville, and the Thai Department of Agriculture.

# Skunk vine                                                    *Paederia foetida*

Madder family—Rubiaceae

*P. D. Pratt and R. W. Pemberton*

Skunk vine (*Paederia foetida*) is a climbing perennial vine of Asian origin that has become an invasive weed in natural areas in the southeastern United States as well as on the island of Hawaii. The plant is classified as a Category I species (invading and disrupting natural communities), the rating given to the most dangerous invasive weeds by the Florida Exotic Pest Plant Council. The common name of this plant relates to the odor of the leaves, which is due to the presence of sulfur compounds.

Skunk vine is prevalent throughout central and northern Florida with widely separated occurrences in Georgia, Louisiana, North and South Carolina, and Texas. Recent discoveries of the weed in North Carolina and in the more tropical regions of southern Florida demonstrate its continued expansion north and south. In Japan, the northern limit of the plant's native range is the Tohoku region, an area with minimum temperatures of -10 to -20° C. This distribution suggests that skunk vine can tolerate similar temperatures to those found in the U.S. Department of Agriculture Plant Hardiness Zone 6. Using Zone 6 as a northerly limit, the weed can potentially spread to 40° latitude— north of Delaware, Maryland, and the Virginias.

A feasibility study conducted in 2001 found the weed to be suitable for biological control and suggested that herbivores suitable for use as biological control agents should be those whose feeding and development are restricted to the tribe Paederieae because there are no North American (nor Hawaiian) native or economically significant members of this tribe. The first intensive explorations for natural enemies of skunk vine were conducted by the authors during the summer of 2002 in Japan and Nepal. The chrysomelid beetle *Trachyaphthona sordida* was encountered during Japanese surveys and is believed to be a specialist of skunk vine. Because chrysomelid beetles have successfully controlled many weeds including alligatorweed, leafy spurge, tansy ragwort, and purple loosestrife, *T. sordida* is of special interest. It is expected that many insects with tribe- or genus-level host specificity should be associated with *P. foetida* and other *Paederia* species in their native ranges.

Additional information can by found in:

Pemberton, R. W., and P. D. Pratt. 2002. Skunk vine. Pages 343-51 *in* R. Van Driesche, S. Lyon, B. Blossey, M. Hoddle, and R. Reardon, eds. Biological Control of Invasive Plants in the Eastern United States. USDA Forest Serv. Pub. FHTET-2002-04. Morgantown, WV.

The development of a skunk vine biological control program is led by USDA-ARS scientists located at the Invasive Plant Research Laboratory in Fort Lauderdale, Florida. Host specificity testing is conducted in cooperation with the Hawaiian Department of Agriculture-Biological Control. Field surveys in Japan and Nepal are being performed in cooperation with Kyushu University and the Nepalese Department of Agriculture, respectively.

449

# Sulfur cinquefoil                                    *Potentilla recta*

Rose family—Rosaceae

*J. M. Story*

Sulfur cinquefoil (*Potentilla recta*) is a long-lived perennial forb introduced from Eurasia. It is well established in the eastern United States and Canada and is continuing to expand its range in western regions, including the northern Rocky Mountains. It infests disturbed areas, meadows, pastures, and rangeland and can dominate a site within two to three years. Overgrazing, which reduces competition from grass, favors sulfur cinquefoil, which is reported to be very competitive. In some areas of western Montana, for example, sulfur cinquefoil is co-dominant with the aggressive invader spotted knapweed (*Centaurea stoebe* ssp. *micranthos*) at many sites and appears to be replacing knapweed in some areas. The plant is unpalatable to most livestock, possibly because of a high tannin content.

Sulfur cinquefoil is often confused with native cinquefoil species. Long, right-angled hairs on the leaves and stems, many leaves on the stem but few basal leaves on mature plants, and a heavily wrinkled seed coat help distinguish this exotic.

Five insects have been targeted for screening as potential biological control agents for sulfur cinquefoil. These include a root moth, *Tinthia myrmosaeformis*; a flower head weevil, *Anthonomus rubripes*; two gall wasps, *Diastrophus mayri* and *D.* sp. near *mayri*, and an unidentified gall midge.

Screening was completed on *T. myrmosaeformis,* but because the moth attacked domestic strawberry and some native North American species of *Potentilla* in the tests, Montana and British Columbia jointly decided not to request its release in North America. Screening was terminated on *A. rubripes* because its host range was found to be too broad.

The three remaining insects appear to be quite host-specific and may be good candidates for future screening. However, because sulfur cinquefoil has so many close relatives in North America, questions have been raised about its suitability as a target for biological control. As a result, the biological control effort against sulfur cinquefoil in the United States has been suspended.

Six species of native beetles and moths are known to attack sulfur cinquefoil in Montana and Idaho. All are root- or crown-borers; three are known pests of strawberry. A rust fungus, *Phragmidium ivesiae*, recognized by bright neon-orange urediniospores and black teliospores, attacks the plant throughout the Northern Rockies.

Research into the biological control of sulfur cinquefoil at CABI Bioscience Switzerland Centre was supported by Montana and the British Columbia Ministry of Forests.

# Water chestnut

*Trapa natans*

Loosestrife family—Lythraceae

*R. W. Pemberton*

Water chestnut (*Trapa natans*), also known as horned water chestnut or water caltrop, is an annual aquatic weed of the northeastern United States that can dominate ponds, shallow lakes, and river margins. (An unrelated Chinese vegetable, *Eleocharis dulcis*, a sedge in the Cyperaceae, also is called water chestnut.) *Trapa natans* plants, which are rooted in the hydrosoil and have floating rosettes of leaves, displace native vegetation and limit navigation and recreation. The weed occurs from the northeastern United States, west to the Great Lakes, and south to Washington, D. C. *T. natans* is difficult and expensive to control, and if unmanaged can increase dramatically.

*Trapa natans* has often been considered to be a member of the monogeneric Trapaceae, a plant family native to the Eastern hemisphere. If *Trapa* belongs to the Lythraceae, as recent molecular research suggests, then 18 to 20 species in eight genera are confamilial native relatives in North America. Representatives of these taxa would be important test plants in *Trapa* host specificity research on candidate biological control agents. The specific geographic origins of the problematic *T. natans* genotype(s) in the United States are unknown. The weed usually is thought to be from Eurasia but recent work suggests an Asiatic origin.

To date, biological control research on *T. natans* has been limited to surveys and monitoring of *Trapa* spp. to detect and study natural enemy activity and damage. Surveys were made in China, Japan, eastern Russia, and South Korea during 1992 and 1993 on wild *Trapa* spp. and cultivated *T. bicornis* and *T. bispinosa* (grown for their edible nuts). Natural enemy activity associated with *T. japonica* in South Korea was monitored during two seasons. *Trapa natans* was surveyed in France, Germany, Italy, Poland, and Switzerland in 1995. Among the insects found, the leaf beetle *Galerucella birmanica* was the most common and damaging species in Asia, causing complete defoliation of whole populations of plants. Nymphuline pyralid moths also were common and at times damaging. Because the beetle and the moths feed and develop on unrelated plants, they have no potential as biological control agents for *T. natans* in North America. Because of the possibility of sibling *Galerucella* species with different host plants, *G. birmanica* may warrant additional study. Two *Nanophyes* weevils that feed in the floating leaf petioles were found in Asia. They are thought to be specific to *Trapa,* but were not observed to be damaging. In Europe, a similar insect fauna was found, but no species were very damaging to the plant. An Italian weevil, *Bagous rufimanus*, feeds within the fruit stalk, but feeding did not appear to harm the fruit. These weevils might be more damaging at higher population levels.

Further information about this project and species can be found in:

Pemberton, R. W. 1999. Natural enemies of *Trapa* spp. in northeast Asia and Europe. Biol. Control 14: 168-80.

Pemberton, R. W. 2002. Water chestnut. Pages 33-40 *in* R. Van Driesche, S. Lyon, B. Blossey, M. Hoddle, and R. Reardon, eds. Biological Control of Invasive Plants in the Eastern United States. USDA Forest Serv. Pub. FHTET-2002-04. Morgantown, WV.

Biological control research on *Trapa* is inactive as of 2003, although the weed remains a problem. Suggestions for future research are: 1) survey for new natural enemies of *Trapa* in populations growing in the temperate Himalayan region and in the Volga River Delta of Russia; and 2) molecular examination of *G. birmanica* to detect possible sibling species that may have specificity to *Trapa*. Research to date was supported by the U.S. Army Corps of Engineers and USDA-ARS.

## Whitetop / Hoary cress
Mustard family—Brassicaceae

*Lepidium draba* and *L. appelianum*

Whitetops or hoary cresses [*Lepidium draba* (= *Cardaria draba*) and *L. appelianum*] were introduced into North America from Eurasia in the late 19th century. They are now widespread throughout diverse habitats in the United States. *Lepidium draba* is classified as a noxious weed in at least 24 states. Both species are deep rooted, creeping, perennial mustards that spread by seed and vegetative root growth. They are aggressive invaders of crops, rangelands, and riparian lands and are particularly prevalent in disturbed and irrigated areas.

In 2001, the Hoary Cress Consortium was established to investigate classical biological control of hoary cresses, and the CABI Bioscience Switzerland Centre entrusted with the search for suitable agents. During European field surveys in 2001 and 2002, about eighty insect species, one mite, and at least two fungal pathogens were sampled or reared from *Lepidium draba* and *L. latifolium*. Four weevils and a flea beetle have been selected for further study: the shoot-mining weevil *Ceutorhynchus merkli*; the gall-former *C. cardariae*; the seed-feeder *C. turbatus*; *Baris semistriata*, the larvae of which mine in the root crown; and an as yet unidentified flea beetle *Psylliodes* sp. Host specificity tests and experiments on insect-host plant relationships are ongoing. Besides the efforts at CABI, two more potential agents are being studied: the root gall-forming *Ceutorhynchus* sp. nr. *assimilis* (=*C. pleurostigma*) by the USDA-ARS laboratory in France, and the gall-forming eriophyid mite *Aceria draba* in quarantine at Montana State University. The ARS laboratory is conducting genetic characterizations and cross-breeding experiments on different genotypes of *Ceutorhynchus* sp. nr. *assimilis*.

Excerpted from CABI Bioscience Switzerland Centre Annual Report 2002 (available online at http://www.cabi-bioscience.org/html.ch.htm) and the USDA-ARS European Biological Control Laboratory Research Progress report 2003 (http://www.ars-ebcl.org/).

This project has been conducted cooperatively with Mark Schwarzländer, University of Idaho, with financial support from the Idaho Department of Agriculture, Idaho Fish and Game, the Big Horn Drainage Exotic Plant Steering Committee Wyoming, USDI Bureau of Indian Affairs, and USDI Bureau of Land Management through the Panhandle Lakes RC&D Hoary Cress Consortium.

# Authors, Editors, and Contributors

**Joe K. Balciunas**, USDA-ARS Exotic and Invasive Weed Research Unit, 800 Buchanan St., Albany, CA 94710.

**Anton B.A.M. Baudoin**, 417 Price Hall, Department of Plant Pathology, Physiology and Weed Science, Virginia Polytechnic Institute and State University, Blacksburg, VA 24061-0331.

**Dana K. Berner**, USDA-ARS-FDWSRU, 1301 Ditto Ave., Ft. Detrick, MD 21702.

**Bernd Blossey**, Department of Natural Resources, Fernow Hall, Cornell University, Ithaca, NY 14853.

**William L. Bruckart, III**, USDA-ARS-FDWSRU, 1301 Ditto Ave., Ft. Detrick, MD 21702.

**Gary R. Buckingham**, USDA-ARS (ret.), Biocontrol Laboratory, University of Florida, P.O. Box 147100, Gainesville, FL 32614-7100.

**Anthony J. Caesar**, USDA-ARS Northern Plains Agricultural Research Laboratory, P.O. Box 463, Sidney, MT 59270.

**Ray I. Carruthers**, USDA-ARS Exotic and Invasive Weed Research Unit, 800 Buchanan St., Albany, CA 94710.

**Ted D. Center**, USDA-ARS Invasive Plant Research Laboratory, 3205 College Ave., Ft. Lauderdale, FL 33314.

**Janet K. Clark**, Center for Invasive Plant Management, Montana State University, Department of Land Resources & Environmental Sciences, P.O. Box 173120, Bozeman, MT 59717-3120.

**Alfred F. Cofrancesco, Jr.**, U.S. Army Corps of Engineers, Waterways Experiment Station, 3909 Halls Ferry Rd., Vicksburg, MS 39180.

**Nancy Coile**, Florida Department of Agriculture and Consumer Services (ret.), Division of Plant Industry, Gainesville, FL 32614-7100.

**Eric M. Coombs**, Oregon Department of Agriculture, 635 Capitol St. NE, Salem, OR 97301.

**Jack R. Coulson**, USDA-ARS (ret.), Documentation Center, National Agricultural Library, 4th Floor, Beltsville, MD 20705.

**James P. Cuda**, Department of Entomology and Nematology, University of Florida, P.O. Box 110620, Gainesville, FL 32611-0620.

**Ernest S. Delfosse**, USDA-ARS, 5601 Sunnyside Ave., Bldg. 4, Rm 4-2230, Beltsville, MD 20705-2350.

**C. Jack DeLoach**, USDA-ARS Grassland, Soil & Water Research Laboratory, 808 E. Blackland Rd., Temple, TX 76502.

**Steven A. Dewey**, Department of Plants, Soils and Biometeorology, Utah State University, Logan, UT 84322.

**Joseph M. DiTomaso**, Weed Research and Information Center, University of California, 210 Robbins Hall, Davis, CA 95616.

**F. Allen Dray, Jr.**, USDA-ARS Invasive Plant Research Laboratory, 3205 College Ave., Ft. Lauderdale, FL 33314.

**Luca Fornasari**, USDA-ARS-EBCL, Campus Int'l de Baillarguet, CS 90013 Montferrier-sur-lez, 34988 Saint-Gely-du-Fesc, France.

**Wendy Forno**, CSIRO Entomology, 120 Meiers Rd., Indooroopilly, Queensland, Australia 4068.

**Timothy G. Forrest**, Department of Biology, 1 University Heights, University of North Carolina –Asheville, Asheville, NC 28804.

**Daniel Gandolfo**, USDA-ARS, South American Biological Control Laboratory, Hurlingham, Argentina 34034-0001.

**John A. Goolsby**, USDA-ARS, Australian Biological Control Laboratory, 120 Meiers Rd., Indooroopilly, Queensland, Australia 4068.

**Fritzi S. Grevstad**, Spartina Biocontrol Program, University of Washington, Olympic Natural Resources Center, 2907 Pioneer Rd., Long Beach, WA 98631.

**Michael J. Grodowitz**, U.S. Army Engineer Research and Development Center, Waterways Experiment Station, 3909 Halls Ferry Road, Vicksburg, MS 39180.

**Dale H. Habeck**, Department of Entomology and Nematology (ret.), University of Florida, Gainesville, FL 32611-0620.

**Richard W. Hansen**, USDA-APHIS-PPQ-CPHST, National Weed Management Laboratory, 2150 Center Ave., Bldg. B, #3E10, Fort Collins, CO 80526-8117.

**Stephen D. Hight**, USDA-ARS, Center for Biological Control, Perry Paige Bldg., South Florida A&M University, Tallahassee, FL 32307.

**Hariet Hinz**, CABI Bioscience Switzerland Centre, 1 chemin des Grillons, CH-2800, Delémont, Switzerland.

**Tracy Horner**, USDA-APHIS, 4700 River Rd., Unit 149, Riverdale, MD 20737.

**Lavon Jeffers**, U.S. Army Corps of Engineers, Waterways Experiment Station, 3909 Halls Ferry Rd., Vicksburg, MS 39180.

**Jerry Johnson**, BioCollect, 5481 Crittenden St., Oakland, CA 94601.

**Don Joley**, California Department of Food and Agriculture, Integrated Pest Control Branch, Biological Control Program, 3288 Meadowview Rd., Sacramento, CA 95832.

**Jeffrey L. Littlefield**, Montana State University, Department of Entomology, P.O. Box 173020, Bozeman, MT 59717.

**George P. Markin**, U.S. Forest Service, Forestry Sciences Lab, Montana State University, Bozeman, MT 59717-2780.

**Aubrey Mayfield**, BioCollect, 5481 Crittenden St., Oakland, CA 94601.

**Peter B. McEvoy**, Department of Botany and Plant Pathology, Oregon State University, Corvallis, OR 97331.

**Julio C. Medal**, Department of Entomology and Nematology, University of Florida/IFAS, Gainesville, FL 32611-0620.

**Jeffrey Mullahey**, West Florida Research and Education Center, University of Florida/IFAS Milton Campus, Milton, FL 32583.

**James R. Nechols**, Department of Entomology, Waters Hall, Kansas State University, Manhattan, KS 66506.

**Thomas Nordblom**, CRC for Weed Management Systems, School of Agriculture, Charles Sturt University, Wagga Wagga, NSW 2678, Australia (current address: NSW Agriculture, Economic Resource Policy/ESU, Wagga Wagga, New South Wales, 2650 Australia).

**Robert M. Nowierski**, USDA-CSREES-PAS, Waterfront Center, Rm. 3424, 800 9th St. SW, Washington, DC 20024.

**Jose Henrique Pedrosa-Macedo**, Federal University of Parana, Curitiba, Parana 80210-170, Brazil.

**Robert W. Pemberton**, USDA-ARS Invasive Plant Research Laboratory, 3205 College Ave., Ft. Lauderdale, FL 33314.

**Gary L. Piper**, Department of Entomology, Washington State University, Pullman, WA 99164-6382.

**Mike J. Pitcairn**, California Department of Food and Agriculture, Integrated Pest Control Branch, Biological Control Program, 3288 Meadowview Rd., Sacramento, CA 95832.

**Noah H. Poritz**, Biological Control of Weeds, Inc., 1418 Maple Dr., Bozeman, MT 59715.

**Paul D. Pratt**, USDA-ARS Invasive Plant Research Laboratory, 3205 College Ave., Ft. Lauderdale, FL 33314.

**Matthew F. Purcell**, CSIRO Entomology, 120 Meiers Rd., Indooroopilly, Queensland, Australia 4068.

**Paul C. Quimby, Jr.**, USDA-ARS-International Programs, 5601 Sunnyside Ave., Rm. 4-1138, Beltsville, MD 20705-5141.

**Hans Radtke**, Department of Agriculture and Resource Economics, Oregon State University, Corvallis, OR 97331.

**Min B. Rayamajhi**, USDA-ARS Invasive Plant Research Laboratory, 3205 College Ave., Ft. Lauderdale, FL 33314.

**Benjamin Rice**, Department of Forest Science, 321 Richardson Hall, Oregon State University, Corvallis, OR 97331.

**Cindy T. Roché**, 109 Meadow View Dr., Medford, OR 97504.

**Shon S. Schooler,** Department of Botany and Plant Pathology, Oregon State University, Corvallis, OR 97331.

**Mark Schwarzländer**, Biological Weed Control Program, Department of Plant, Soil and Entomological Sciences, University of Idaho, Moscow, ID 83844.

**Richard H. Shaw**, Biocontrol of Weeds and Diseases Group, CABI Bioscience, Silwood Park, Ascot, Berkshire SL5 7TA, UK.

**Judy F. Shearer**, U.S. Army Corps of Engineers, Waterways Experiment Station, 3909 Halls Ferry Rd., Vicksburg, MS 39180.

**Luke C. Skinner**, Exotic Species Program, Minnesota Department of Natural Resources, 500 Lafayette Road, Box 25, St. Paul, MN 55155-4025.

**Neal R. Spencer**, USDA-ARS-PPRU, Plant, Soil and Nutrition Lab., Tower Rd., Ithaca, NY 14853.

**Jim M. Story**, Western Agricultural Research Center, Montana State University, 580 NE Quast Lane, Corvallis, MT 59828.

**Phil W. Tipping**, USDA-ARS Invasive Plant Research Laboratory, 3205 College Ave., Ft. Lauderdale, FL 33314.

**Thai K. Van**, USDA-ARS Invasive Plant Research Laboratory, 3205 College Ave., Ft. Lauderdale, FL 33314.

**Baldo Villegas**, California Department of Food and Agriculture, Integrated Pest Control Branch, Biological Control Program, 3288 Meadowview Rd., Sacramento, CA 95832.

**Timothy L. Widmer**, USDA-ARS European Biological Control Laboratory, Montferrier-sur-lez, 34988 Saint-Gely-du-Fesc, France.

**Linda M. Wilson**, Department of Plant, Soil and Entomological Sciences, University of Idaho, Moscow, ID 83844.

**Susan Wineriter**, USDA-ARS Biocontrol Laboratory, University of Florida, P.O. Box 147100, Gainesville, FL 32614-7100.

**Dale M. Woods**, California Department of Food and Agriculture, Integrated Pest Control Branch, Biological Control Program, 3288 Meadowview Rd., Sacramento, CA 95832.

**A. D. (Tony) Wright**, CSIRO Entomology, Long Pocket Laboratories, 120 Meiers Rd., Indooroopilly, Queensland, Australia 4068.

# Index

furze. *See* gorse
*Fusarium,* 55
FWS. *See* U. S. Department of Interior

# G

*Galerucella,* 23
*Galerucella calmariensis,* 20, 109, 282–84
*Galerucella pusilla,* 20, 109, 285–87
gall flies. *See Urophora affinis, Urophora cardui,*
    *Urophora sirunaseva, Urophora stylata*
gall midges. *See Cystiphora schmidti, Spurgia*
    *esulae, Zeuxidiplosis giardi*
gall mites. *See Aceria malherbae, Eriophyes*
    *chondrillae*
Gandolfo, D., 396, 399
garden cornflower. *See* cornflower
garlic mustard, 443
GenBank, 29
*Genista monspessulana. See* French broom
Geographic Information System (GIS), 49
giant reed, 443
giant salvinia, 75, 174–77
goat head. *See* puncturevine
golden loosestrife beetle. *See Galerucella pusilla*
Goolsby, J. A., 27, 447
gorse, 178–83
gorse seed weevil. *See Exapion ulicis*
gorse spider mite. *See Tetranychus lintearius*
*Gratiana boliviana,* 399–401
green clearwing fly. *See Terellia virens*
Grevstad, F. S., 317, 319
grey-winged root moth. *See Pterolonche inspersa*
Grodowitz, M. J., 186, 188, 190, 192
ground bur nut. *See* puncturevine
*Gymnetron antirrhini,* 389–91
*Gymnetron linariae,* 391–92

# H

Habeck, D. H., 439
*Hadroplontus litura. See Ceutorhynchus litura*
Hansen, R. W., 59, 233, 235, 237, 239, 241, 244,
    246, 248, 250, 252, 254, 257, 259
Hawaii, 17
hawkweeds, 444
hazard, 53
hedge bindweed, 151–52
Hemiptera, 19
hemp sesbania, 56
herbal remedies, 17
herbivores, 29
    host range, 36
*Hieracium* spp. *See* hawkweeds
Hight, S. D., 439
hoary cress, 452

holistic analyses, 20
holy thistle. *See* milk thistle
honeybees, 122
Horner, T., 42
host range, 36, 95, 111
    tests, 41
host specificity testing, 3, 32–37, 132
    aquatic biological control agents, 35–36
    feeding tests, 34
    foreign exploration and, 27, 30
    methods of testing, 33, 107, 111
    musk thistle, 108
    multiple-choice tests, 34
    no-choice tests, 34, 111
    oviposition tests, 34
    pathogens, 53-54
    purple loosestrife, 109
    rejection of agent, 32–33
    risk assessment, 5
    TAG, 40-44
    testing locations, 34–35
host susceptibility to disease, 50
houndstongue, 445
hurtsickle. *See* cornflower
*Hydrellia balciunasi,* 75, 190–92
*Hydrellia* fly larvae, 72
*Hydrellia pakistanae,* 75, 81, 82, 192–94
*Hydrellia* spp., 75
hydrilla, 72, 81, 184–95
*Hydrilla vericillata. See* hydrilla
*Hylemyia seneciella. See Botanophila seneciella*
*Hyles euphorbiae,* 87, 254–56
*Hylobius transversovittatus,* 288–90
*Hypericum perforatum. See* St. Johnswort

# I

importing agents, 42–44, 59–62
Indian hydrilla leaf-mining fly. *See Hydrellia*
    *pakistanae*
Indian hydrilla tuber weevil. *See Bagous affinis*
infection, 51–52
    systemic *vs.* local, 51
information collection and analysis. *See*
    database analyses; documentation
injury, definition of, 50
insects
    commercially provided, 127–29
    flowchart, 60
    handling, 59–70
    insect extraction devices, 29
    insect feeding, 50
    life stages, 59, 68
    plant pathogen interaction and, 51
    shipments, 59–60, 68–69, 72